Lech Polkowski

Approximate Reasoning by Parts

T0137868

Intelligent Systems Reference Library, Volume 20

Editors-in-Chief

Prof. Janusz Kacprzyk
Systems Research Institute
Polish Academy of Sciences
ul. Newelska 6
01-447 Warsaw
Poland
E-mail: kacprzyk@ibspan.waw.pl

Prof. Lakhmi C. Jain
University of South Australia
Adelaide
Mawson Lakes Campus
South Australia 5095
Australia
E-mail: Lakhmi.jain@unisa.edu.au

Further volumes of this series can be found on our homepage: springer.com

Lech Polkowski

Approximate Reasoning by Parts

An Introduction to Rough Mereology

 Springer

Prof. dr. Lech Polkowski
Polish-Japanese Institute of Information Technology
and
University of Warmia and Mazury
Koszykowa 86 02–008 Warszawa, Poland
E-mail: polkow@pjwstk.edu.pl

ISBN 978-3-642-26985-1   ISBN 978-3-642-22279-5 (eBook)

DOI 10.1007/978-3-642-22279-5

Intelligent Systems Reference Library   ISSN 1868-4394

Typeset & Cover Design: Scientific Publishing Services Pvt. Ltd., Chennai, India.

Printed on acid-free paper

9 8 7 6 5 4 3 2 1

springer.com

To Maria and Marcin

Preface

And in this way from the old soil–as men say–a new harvest sprouts
(Chaucer, a paraphrase (C.S. Lewis in *The Discarded Image*))

μέρος λέγεται ενα μὲν τρόπον εἰς ο διαιρεθεὶν αν τὸ πὸσον οποσοῦν (ἀεί γὰρ το
ἀφαιρούμενον τοῦ ποσοῦ η πὸσον μέρος λέγεται ἐκείνου οἶον τῶν τριῶν τὰ δυο
μέροσ λέγεται πως), ἀλλον δὲ τρόπον τὰ καταμετροῦντα τῶν τοιοῦτον μὸνον:
διὸ τὰ δύο τῶν τριῶν ἐστι μὲν ὡς λέγεται μέρος ἐστι δ'ως ωὺ. ἐτι εἰς ἁ τὸ εἶδος
διαιρεθείν ἀν ἁνευ τοῦ ποςοῦ και ταῦτα μόρια λέγεται τούτου : διὸ τὰ εἴδη
τοῦ γένους φασιν εἶναι μόρια. ἐτι ἑίς ἁ διαιρεῖται ἡ ἐξ ών σύγκειται τὸ ὁλον,
ἡ τό εἴδος ἡ τὸ ἐχον τὸ εἴδος, οἶον τῆς σφαίρας τῆς χαλκῆς ἡ τοῦ κύβου τοῦ
χαλκοῦ και ὁ χαλκὸς μέρος (τοῦτο δ' ἐστιν ἡ ύλη ἐν ἡ τὸ ἐίδος) και ἡ γονία
μέρος. ἐτι τὰ ἐν τῶ λόγω τῶ δηλοῦντι ἑκαστον και ταῦτα μόρια τοῦ ὁλου: διὸ
τὸ γένος τοῦ εἴδους και μέρος λέγεται, ἀλλος δὲ τὸ εἴδος τοῦ γένους μέρος.

(Aristotle. Metaphysics Book Delta, 1203 b)

Aristotle in the passage quoted, gives, inter alia, a definition of part as an
entity which may be taken from some other entity, also, as a division of an
entity from which some entities may be taken. This intuitive meaning of parts
was taken as a foundation for the theory of *Mereology* by Stanisław Leśniewski
(1916). At the same time, Alfred North Whitehead discussed the dual notion
of *extending over*.

Over the years, Mereology version due to Leśniewski was set in the frame
of a logical calculus of Prototetics and the theory of Ontology whereas the
ideas of Whitehead, stimulated by the work of Th. de Laguna, were later
rendered in the form of Leonard and Goodman's *Calculus of Individuals* and
Clarke's *Connection Calculus*.

Rough Mereology extends Mereology by discussing the notion of a part
to a degree. Mereology is preserved in it by the requirement that a part to
degree of 1 be an ingredient, hence, the theory of part to degree of 1 is the
theory of Mereology.

The notion of a part to a degree carries an analogy to the notion of an element to a degree, which, as proposed by Professor Lotfi A. Zadeh, is the primitive notion of the fuzzy set theory. Rough Mereology replaces this notion with the notion of a part to a degree as Mereology replaces the set–theoretical notion of an element with the notion of a part. One can, hence, say that Rough Mereology in a sense 'fuzzifies' Mereology and is a counterpart of Fuzzy Set Theory within Mereology. The term 'Rough Mereology' comes from the author's work on rough set theory as it had begun from our attempts to characterize in abstract terms partial containment between exact concepts. By considering objects as parts one to another, Mereology is particularly suited to discussion of spatial relations among spatial objects like solids, figures etc., witness the Tarski axiomatization of geometry of solids in terms of Mereology. Rough Mereology extends the scope of this discussion by considering numerical values of containment among spatial objects and allowing for a more precise description of spatial relationships. Hence, Rough Mereology can be also regarded as an extension of Fuzzy Set Theory useful in spatial reasoning.

It is the aim of this book, to give a presentation of main ideas and applications of Rough Mereology, in a simplified language of set theory.

The book consists of nine chapters; in accordance with principles adopted when writing our earlier book *Rough Sets. Mathematical Foundations*, we include preliminary chapters devoted to fundamentals. Chapter 1 brings some basic facts on Set Theory and Aristotelian Theory of Concepts. We put particular emphasis on relations, with relations of tolerance and equivalence highlighted as playing important role in analysis by fuzzy and rough set theories. In Chapter 2 we give an introduction to Topology of Concepts, and we emphasize topics important for fuzzy and rough set theories like topologies on sets and Čech topologies. These chapters provide the reader with the basic facts of set theory and topology, which are useful in later chapters when we discuss some set–theoretical and topological structures induced in the mereological universe.

As Rough Mereology has been conceived as a tool in Approximate Reasoning, we include Chapter 3 on patterns of Deductive Reasoning in which we present basic deductive calculi: calculus of propositions as a model for deductive reasoning, many–valued calculi so important for fuzzy sets–based reasoning, modal calculi immanent in rough sets–based analysis of ambiguity, and predicate calculus. In Chapter 4, we address reductive/inductive reasoning by presenting rough and fuzzy set theories as environments in which to induce decision rules. We discuss rough set methods for induction of decision rules, topology of rough sets with applications to fractal calculus, and rough sets–based reasoning schemes describing modalities induced by indiscernibility relations. In the second part of this chapter, we study in detail t–norms and residual implications which are cornerstone of many–valued fuzzy logics and supply essential tools for Rough Mereology; we conclude with an outline of basic ideas on fuzzy decision rules.

Mereology enters in Chapter 5 in which we present the Leśniewski formulation of Mereology along with the theory of classes so important for our theory of granulation of knowledge; we recapitulate the Tarski–Woodger version of timed mereology and the Woodger theory of cells as important applications of Mereology which are far from being exploited. Passing to the initiated by Whitehead Mereology of Connections, we recall the Clarke Connection Calculus, a modern rendering of Whitehead–de Laguna ideas, and its adaptations suitable for applications like RCC. We stress also topics already known from introductory chapters: topological and algebraic structures hidden in mereological spaces.

Rough Mereology is discussed in Chapter 6. We discuss its primitive notion of a *rough inclusion* which is a ternary predicate/relation asserting that one object is a part of another object to a specified degree and we present known to us methods of inducing rough inclusions. These methods depend on the context of usage and vary from general, e.g., employing a metric, to specific, e.g., employing properties of indiscernibility in decision systems. As with Mereology, we study mereotopology and we induce from rough inclusions predicates of mereogeometry essential in applications in spatial reasoning. Finally, we construct within Rough Mereology predicates of Connection Calculus.

Applications of Rough Mereology in approximate reasoning are presented in Chapter 7. We discuss methods for granulation of knowledge and the theory of rough mereological granules based on mereological theory of classes; this approach allows for Čech granular topologies. We introduce intensional logics in which extensions are evaluated at worlds which are granules; as an application, we propose a rough mereological rendition of the perception calculus introduced by L. A. Zadeh. We introduce reasoning schemes for distributed and many–agent systems. Finally, we discuss the idea of rough mereological perceptrons and their networks.

Applications of rough mereology are also presented in Chapters 8 and 9. In Chapter 8, we give an outline of applications of rough mereology in spatial reasoning taking as an exemplary field the behavioral robotics. We use rough mereogeometry introduced in Chapter 6 in order to define rigorously the notion of a formation of robots and to propose a new method of path planning – rough mereological potential fields. The applicability of these methods is illustrated with experiments with formations of Roomba (intellectual property of iRobot.Inc.) robots.

Chapter 9 is devoted to another essential application of rough mereology, viz., in classifier synthesis. Our approach to granulation of knowledge allows for a data filtration by means of a factorization of data through a granular structure which reduces noise in data and allows for better classifiers of higher accuracy and with strikingly less number of rules. We present the theory underlying this approach as well as a selection of exemplary granular classifiers illuminating benefits of this approach.

We do hope that this book will be used as a source of information and an inspiration to further study of rough mereology and that it will serve as a basis for a course in approximate reasoning. Introductory chapters may serve as a course in basic set theory and topology essential for computer scientists. We do hope it turns out as useful to the research community as our book on rough sets has turned out to be.

We owe a debt of gratitude to many who have helped us by means of encouragement, support, and cooperation.

We would like to express our gratitude to Professor Lotfi A. Zadeh for kind words of encouragement and support.

To Professor Janusz Kacprzyk and Professor Lakhmi C. Jain, Series Editors, we are indebted for kind acceptance of this text into this prestigious series of monographs.

We remember with feeling of deep gratitude the late Professors Helena Rasiowa and Zdzisław Pawlak; without their direct and indirect influence rough mereology would most probably not emerge.

We would like to remember our late Colleagues Professors Cecilia Rauszer and Jan Zytkow whose work is reflected in this book.

Our thanks go to Professor Solomon Marcus, Professor Gheorghe Paun, and Professor Andrzej Skowron for their helpful advice, inspiration, and joint work; Dr Paweł Ośmiałowski and Dr Piotr Artiemjew contributed to the development of applications of rough mereology in behavioral robotics and knowledge discovery, respectively, and we would like to thank them for this cooperation.

We would like to thank Dr Thomas Ditzinger, Dr Dieter Merkle, Mr Holger Schäpe, Ms Jessica Wengrzik and the staff of Engineering Editorial of Springer Verlag for their kind help in all stages of preparation of this book.

The book is dedicated to two people whose help and support have been invaluable.

Lech Polkowski
Warszawa, Poland

Contents

Chapter 1
On Concepts. Aristotelian and Set–Theoretic Approaches

In this chapter the reader is introduced to the idea of a *concept*. Historically, two main approaches can be discerned toward formalization of this idea. First, a holistic approach by Aristotle, in which a concept is construed as a term (category) by which any object can be either styled or not. This fact of calling an object a by a name term A is denoted by the formula a *is* A. With those formulas, Aristotle built a complete theory of concepts, the Calculus of Syllogisms, briefly exposed in section 1.1, below.

The alternative approach emerged with Georg Cantor's set theory in mid–19th century and its essence was a representation of a concept by its *elements*; the relation of being an element was expressed as with Aristotle by the predicate *is* written down as ϵ (the 'esti' symbol): the formula $a\epsilon A$ reads *an object a is an element of the concept (set) A*. We offer in section 1.2 and following, an outline of modern set theory, aimed at establishing a terminology and basic facts of set theory, which seem to be indispensable to any researcher in computer science.

1.1   An Aristotelian View on Concepts

It is a property of human mind that it classifies its perceptions and, in a sense, granulates them into *concepts* denoted with names. To this peculiar property, each theory of reality is by necessity bound to relate, see Bocheński [4] for a discussion. Assuming that reality consists of *objects (entities)*, those objects fall unto concepts and the basic logical construct is the copula 'is': an expression a *is* A means that an object denoted a conforms to the concept denoted A, like in the classical exemplary phrase 'Socrates is man', where 'Socrates' is an object and 'man' a concept denoting the generality of mankind. In terms of concepts and statements of belonging into concepts, relations among objects, concepts and objects, and among concepts are described, and true statements about these relations constitute *knowledge*, see Bocheński, op. cit.

The first logical system ever conceived was constructed on these lines, due to Aristotle [1], see Łukasiewicz [15], [14], Słupecki [24]. Aristotle's *Syllogistic* has as primitive terms expressions of the form Xa *is* Yb which ramify into the following four specialized cases

L. Polkowski: Approximate Reasoning by Parts, ISRL 20, pp. 1–43.
springerlink.com   © Springer-Verlag Berlin Heidelberg 2011

1. *All a is b*;

2. *Some a is b*;

3. *Some a is not b*;

4. *No a is b*,

where 'All' and 'Some' have the intuitive meaning of nowadays' 'every' and 'exists', respectively.

Aristotle's syllogisms, i.e., meaningful expressions of Syllogistic, are of the form of an implication

$$\text{If } p \text{ and } q \text{ then } r \tag{1.1}$$

where p, q, r are of one of four possible forms, written down in traditionally slightly modified shape.

In case 1: Aab stands for *all a is b* (universal affirmative qualification).

In case 2: Iab stands for *some a is b* (existential affirmative qualification).

In case 3: Oab stands for *some a is not b* (existential negative qualification).

In case 4: Eab stands for *no a is b* (universal negative qualification).

In syllogisms of the form (1.1), either of four types of statements may appear for any of p, q, r. This uniformity had been one of reasons which prompted Aristotle to exclude singular, individual terms from his Syllogistic (in spite of some examples in which individual terms do appear), see Łukasiewicz [15]. Also, we have to presume that all terms we consider are non–empty, cf., loc. cit. Looked at from contemporary point of view, syllogistic may be regarded as an attempt at deriving in a systematic way all rules of arguing with non–empty non–singleton sets (i.e., collective properties) based on containment (i.e., being a subset) and on non–empty intersection.

Aristotle himself had been giving proofs of his syllogisms by means of naïve (i.e., not formalized) propositional calculus and its self–evident rules, by falsifying not accepted syllogisms by means of substitutions of specific terms in place of general variables and by using some auxiliary rules (e.g., laws of conversion, see below). A historic merit of Aristotle's system is also first usage of variables.

The Aristotle Syllogistic had been given a formal axiomatic form in Łukasiewicz [14]; see also Łukasiewicz [15]. In this axiomatization, Łukasiewicz had exploited propositional logic, including its axioms T1–T3, see Ch. 3, to form a part of the axiomatic system of Syllogistic. In addition to axioms of propositional logic, the following statements form this axiomatic system

S1 Aaa

S2 Iaa

S3 *If* Amb *and* Aam, *then* Aab

S4 *If* Amb *and* Ima, *then* Iab

Axioms S1, S2 are clearly not genuine syllogisms but their inclusion is motivated by historic reasons as well as by technical reasons (see a discussion in Łukasiewicz [15]). Axioms S3, S4 are genuine syllogisms, which may be regarded as self–evident and not requiring any proof. Traditionally, syllogisms are known by their names in which functors appearing in them are listed. Thus S3 has been called **Barbara** and S4 has been called **Datisi**.

Rules of derivation are the following

Detachment (Modus Ponens) Once α, *If* α *then* β *are derived (or, accepted), then* β *is derived (resp., accepted).*

Substitution In theses of syllogistic, we may substitute for equiform propositional variables meaningful expressions containing constants of propositional logic as well as of syllogistic.

Replacement by equivalents We may replace expressions on right–hand sides of the following definitions with their left–hand sides equivalents

D1 $Oab = NAab$

D2 $Eab = NIab$

The proof of completeness (see Ch. 3 for the completeness property) of Syllogistic can be found in Łukasiewicz [15].

Łukasiewicz (op.cit.) derives formally all the moods:

Barbara, Celarent, Darii, Ferioque prioris,

Cesare, Camestres, Festino, Baroco secundae,

Tertia Darapti, Disamis, Datisi, Felapton, Bocardo, Ferison, habet

Quarta insuper addit Bramantip, Camenes, Dimaris, Fesapo, Fresision

These are the names of 19 moods, as the medieval verse has them, apart from five that are sub–alternates (see below) of above

Barbari, Celaront, Cesaro, Camestrop, Camenop

Altogether there are 24 valid moods, collected in four Figures (prima, secunda, tertia, quarta). The terminology related to syllogisms is as follows. In the statement Xab, a is the *subject* and b is the *predicate*, where $X = A, I, O, E$. For a syllogistic figure of the form (1.1), statements p, q are *premises* of which p is the *major*, q is the *minor*, and r is the *conclusion*. The predicate in the conclusion is the *major* term, the subject in the conclusion is the *minor* term, and the remaining term is the *middle* term.

All valid syllogisms are traditionally divided in four *Figures* depending on the position of the middle term m in premises. We list Figures in their symbolic representation.

Figure 1: m is the subject in the major premise and the predicate in the minor premise; the symbolic representation is $\frac{mb, am}{ab}$.

Figure 2: m is the predicate in both premises; the symbolic representation: $\frac{bm, am}{ab}$.

Figure 3: m is the subject in both premises; symbolically, $\frac{mb, ma}{ab}$.

Figure 4: m is the predicate in the major premise and the subject in the minor premise; symbolically, $\frac{bm, ma}{ab}$.

An easy count shows that each Figure may consist of at most 4^3 distinct syllogisms, one for each of 4^3 ways to select three out of the four functors A, I, O, E. Thus the total number of syllogisms does not exceed $4 \times 4^3 = 256$. It turns out that only 24 of them are valid, as the remaining ones may be falsified by examples.

It is worthwhile to look at these moods; they reveal quite ingenious mechanism of logical inference due to medieval followers of Aristotle, a mechanism comparable to our present-day designs of inference engines.

The *Square of Opposition* pairs expressions Aab, Iab, Eab, Oab as follows

Contraries: Aab, Eab: may be simultaneously false, but never simultaneously true;

Sub–contraries: Iab, Oab: may be simultaneously true, but never simultaneously false;

Sub–alternates: Aab, Iab; Eab, Oab. If Aab is true, then Iab is true; if Eab is true, then Oab is true;

Contradictories: Aab, Oab; Eab, Iab. In each pair, exactly one is true and exactly one false

Hence, we have: Barbara $\frac{Amb,Aam}{Aab}$, Celarent $\frac{Emb,Aam}{Eab}$, Darii $\frac{Amb,Iam}{Iab}$, Ferio $\frac{Emb,Iam}{Oab}$, Barbari $\frac{Amb,Aam}{Iab}$, Celaront $\frac{Emb,Aam}{Oab}$ in First Figure.

Cesare $\frac{Ebm,Aam}{Eab}$, Camestres $\frac{Abm,Eam}{Eab}$, Festino $\frac{Ebm,Iam}{Oab}$, Baroco $\frac{Abm,Oam}{Oab}$, Cesaro $\frac{Ebm,Aam}{Oab}$, Camestrop $\frac{Abm,Eam}{Oab}$ belong in Second Figure.

Darapti $\frac{Amb,Ama}{Iab}$, Disamis $\frac{Imb,Ama}{Iab}$, Datisi $\frac{Amb,Ima}{Iab}$, Felapton $\frac{Emb,Ama}{Oab}$, Bocardo $\frac{Omb,Ama}{Oab}$, Ferison $\frac{Emb,Ima}{Oab}$ in Third Figure.

Finally, Bramantip $\frac{Abm,Ama}{Iab}$, Camenes $\frac{Abm,Ema}{Eab}$, Dimaris $\frac{Ibm,Ama}{Iab}$, Fesapo $\frac{Ebm,Ama}{Oab}$, Fresison $\frac{Ebm,Ima}{Oab}$, Camenop $\frac{Abm,Ema}{Oab}$ in Fourth Figure.

What do these names mean? Let us look at them at a close range. At the first glance, we notice that the sequence of vowels in the name corresponds to the sequence of operators in the mood, like *aaa* in *Barbara*. But what about consonants?

For the first letters: The initial capital letter in the name of any mood of Figures 2-4 points to a unique mood in First Figure to which the former can be "reduced". What "reduction" means?

Well, it goes back to Aristotle (loc.cit.): in premises beginning with E or I, terms may be permuted without affecting the truth (the "simpliciter", or *s*, conversion). As for premises starting with A, the conversion goes in two steps: first, change A to I, by force of sub–alternation, and then reduce simpliciter. This two–stage reduction is "per accidens".

Now, given a mood, its name suggests a way of reduction: the initial capital letter points to the mood in First Figure. The letter 's' after a premise capital letter suggests simpliciter conversion of the premise; the letter 'p' suggests the "per accidens" conversion. When 's' or 'p' occur as last letters in the name, this suggests the conversion of the concluding premise. The letter 'm' suggests that one has to change the order of premises (mutare), and the letter 'c' not in the first position suggests a complex way of reducing by means of contradiction (reductio ad absurdum): only Baroco and Bocardo are experiencing this.

So here is an automated (mechanical) system of deduction allowing for reducing moods to moods of First Figure in case they are valid. This system was proved to be a complete logical description of relations among concepts based on intersection and containment, cf., Słupecki [24].

Syllogistic offers, in a sense, a 'holistic' view on concepts. It is a complete theory of concepts, treated with respect to containment and intersection. But, no possibility of having an insight into inner structure of concepts is offered here; and, it is difficult to bear, for a human, not to be able to see inner workings of any mechanism. Thus, another approach prevailed in the long run, and sets emerged as built from elements. This approach has caused serious logical problems, not fully resolved and probably not to be resolved in future, but sets have provided language for mathematics.

1.2   From Local to Global: Set Theory

The problem of theory of classes was taken up by modern logic and mathematics and the result was the emergence of modern Set Theory. Its principal feature seems to be the fact that global properties of sets are obtained from local properties of set elements by augmenting the sentential calculus with new expressions formed by means of *quantifiers*, i.e., by means of *the predicate calculus*, see Ch. 3. We will be now interested in this new language exploiting calculus of predicates which will allow us to discuss individuals and general terms in their mutual relations. This is the language of *Set Theory*. This language will be based on the predicate calculus and on some new notions, notably that of a *set*.

We will apply new operators: $\forall x..$ (read: *for every x ......*) and $\exists x...$ (read: *there exists x such that ...*). The intuitive meaning of these expressions will be: $\forall x \alpha(x)$ will mean that $\alpha(c)$ is true for each substitution x/c where c belongs in a specified domain (*the scope of the quantifier $\forall$*) and $\exists x \alpha(x)$ will mean that for some substitution x/c the formula $\alpha(c)$ is true. We will often use *restricted quantification* of the form $\forall x \in U...$, $\exists x \in U...$ meaning that substitutions described above are restricted to elements $c \in U$, U a specified domain (set).

In development of Set Theory first the so–called Naïve Set Theory prevailed, see Cantor [6]; a set was conceived intuitively as a collection of objects bound by a property. In a later approach, it was based on the conviction that sets and properties are equivalent notions, i.e., to each property there corresponds a set of objects having this property and vice versa, see Frege [9]. This approach led to some contradictions (*antinomies*) and had to be replaced by a formal approach based on axiom schemes and derivations rules. Nevertheless, Naïve Set Theory retains its usefulness as in practical considerations and applications one faces a very slight chance to come at an antinomy.

We will first present a naïve exposition of set theory concentrating above all on set operations and various related notions and then we will give a more advanced treatment in which some formal issues will be discussed.

1.2.1 Naïve Set Theory

Language of elementary set theory uses two primitive expressions: *set* and *to be an element* and it makes usage of constants and variables which may be divided into the following types (see Ch.3 for a formal rendering of propositional as well as predicate calculi).

1. *Variable symbols for sets*: $X, Y, Z, ..., x, y, z...$;
2. *Predicate constant* $=$ (*identity*);
3. *Predicate constant* $\in$ (*esti symbol: to be an element*);
4. *Symbols for propositional connectives*: $\Rightarrow, \neg, \vee, \wedge, \Leftrightarrow$;
5. *Symbols for quantifiers*: $\forall, \exists$;
6. *Auxiliary symbols (parentheses, commas, dots etc.)*

We recall that the symbol $\forall x$ reads *for all x ...* and the symbol $\exists x$ reads *there exists x such that ...* . Meaningful expressions of set theory are derived from *elementary meaningful expressions*: $x \in Y$ (*the set x is an element of the set Y*), $X = Y$ (*the set X is identical to the set Y*) by the following inductive rules.

If α, β are meaningful expressions then

1. $\alpha \Rightarrow \beta$, $\neg\alpha$, $\alpha \vee \beta$, $\alpha \wedge \beta$ *are meaningful expressions*;

2. $\forall x \alpha$, $\exists x \alpha$ *are meaningful expressions*.

1.2.2 Algebra of Sets

First of all, we state when two sets are identical to each other. This will happen when those sets have the same elements (extensionality property). Formally

$$X = Y \Leftrightarrow \forall x (x \in X \Leftrightarrow x \in Y) \qquad (1.2)$$

It follows from this equivalence that if a set is specified by a condition imposed on its elements, then this condition determines the set uniquely. We will apply this in practice constantly. An important application is the *Algebra of sets* concerned with some operations on sets leading to possibly new sets. Having some sets, we may construct from them new sets by means of logical operations. Assume we are given two sets X, Y. Then we may form the following sets.

1. $X \cup Y$ (the *union* of sets X and Y) defined as follows by virtue of set identity condition

$$x \in X \cup Y \Leftrightarrow \forall x (x \in X \vee x \in Y) \qquad (1.3)$$

$X \cup Y$ consists of those sets which are elements either in X or in Y.

2. $X \cap Y$ (the *intersection* of sets X and Y) defined as follows

$$x \in X \cap Y \Leftrightarrow \forall x(x \in X \wedge x \in Y) \qquad (1.4)$$

$X \cap Y$ consists of those sets which are elements in X and in Y. Sometimes the intersection $X \cap Y$ is referred to as the *common part* of X and Y.

3. $X \setminus Y$ (the *difference* of X and Y) is defined via

$$x \in X \setminus Y \Leftrightarrow \forall x(x \in X \wedge \neg(x \in Y)) \qquad (1.5)$$

$X \setminus Y$ consists of those sets which are elements in X but they are not elements in Y. The difference $X \setminus Y$ is sometimes referred to as the *complement to Y in X*.

4. The union $X \setminus Y \cup Y \setminus X$ is referred to as the *symmetric difference* denoted $X \triangle Y$.

The operations of union, intersection and difference allow for a rich structure which may be created on the basis of given sets. We may apply known to us laws of propositional and predicate calculi in order to establish general laws governing operations on sets. Let us list those general laws of *algebra of sets*. We will rely on theses of propositional calculus applied to statements of the form $x \in X$ for a given x.

We have

$$X \cup Y = Y \cup X \qquad (1.6)$$

which follows from the thesis (i) $p \vee q \Leftrightarrow q \vee p$ of propositional calculus as $x \in X \cup Y$ is equivalent to $x \in X \vee x \in Y$ and $x \in Y \vee X$ is equivalent to $x \in Y \vee x \in X$ so we get (1.6) by substituting in (i) $p/x \in X$, $q/x \in Y$. In what follows we will only point to the respective thesis of propositional calculus leaving the verification to the reader.

$$X \cup (Y \cup Z) = (X \cup Y) \cup Z \qquad (1.7)$$

follows from the thesis (ii) $p \vee (q \vee r) \Leftrightarrow (p \vee q) \vee r$.

$$X \cup X = X \qquad (1.8)$$

follows from the thesis (iii) $p \vee p \Leftrightarrow p$. (1.6) is the *commutativity law* for the union, (1.7) is the *associativity law* for the union and (1.8) is the *idempotency law* for the union.

Similarly, we have the corresponding laws for the intersection based on dual theses of propositional calculus in which we replace $\vee$ with $\wedge$.

$$X \cap Y = Y \cap X \qquad (1.9)$$

which follows from the thesis (iv) $p \wedge q \Leftrightarrow q \wedge p$ of propositional calculus.

$$X \cap (Y \cap Z) = (X \cap Y) \cap Z \qquad (1.10)$$

follows from the thesis (v) $p \wedge (q \wedge r) \Leftrightarrow (p \wedge q) \wedge r$ of propositional calculus.

$$X \cap X = X \tag{1.11}$$

follows from the thesis (vi) $p \wedge p \Leftrightarrow p$.

Properties (1.9), (1.10), (1.11) are called, respectively, the *commutativity,*
associativity, idempotency laws for the intersection of sets.

Relations between the two operations are expressed by the *distributivity*
laws which are

$$X \cap (Y \cup Z) = (X \cap Y) \cup (X \cap Z) \tag{1.12}$$

which follows by the thesis (vii) $p \wedge (q \vee r) \Leftrightarrow (p \wedge q) \vee (p \wedge r)$ of propositional
calculus and

$$X \cup (Y \cap Z) = (X \cup Y) \cap (X \cup Z) \tag{1.13}$$

which follows by the thesis (viii) $p \vee (q \wedge r) \Leftrightarrow (p \vee q) \wedge (p \vee r)$ of propositional
calculus dual to (vii).

These operations may be performed on arbitrary given sets; up to now we
did not mention any specific set. So now we introduce an example of a set.

The set denoted by the symbol $\emptyset$ does satisfy the following condition

$$x \in \emptyset \Leftrightarrow \neg(x = x) \tag{1.14}$$

Thus, the empty set has no element, it is void. Clearly, it plays the role of
the *zero element* by the union and intersection of sets

$$X \cup \emptyset = X \tag{1.15}$$

which follows by the thesis (ix) $p \vee (q \wedge \neg q) \Leftrightarrow p$, and

$$X \cap \emptyset = \emptyset \tag{1.16}$$

which follows by the thesis (x) $p \wedge (q \wedge \neg q) \Leftrightarrow q \wedge \neg q$.

We say that sets X, Y are *disjoint* when $X \cap Y = \emptyset$.

We may yet bring the difference operator into play and observe its properties.

$$X \cap (Y \setminus Z) = X \cap Y \setminus Z \tag{1.17}$$

follows from the thesis (xi) $p \wedge (q \wedge r) \Leftrightarrow (p \wedge q) \wedge r$,

$$X \setminus (Y \cup Z) = (X \setminus Y) \cap (X \setminus Z) \tag{1.18}$$

follows from the thesis (xii) $\neg(q \wedge r) \Leftrightarrow \neg q \vee \neg r$,

$$X \setminus (Y \cap Z) = (X \setminus Y) \cup (X \setminus Z) \tag{1.19}$$

follows from the thesis (xiii) $\neg(q \vee r) \Leftrightarrow \neg q \wedge \neg r$.

Property (1.17) is the *distributivity law* of intersection with respect to difference and (1.18), (1.19) are *De Morgan laws.*

Properties of the symmetric difference follow easily in a similar manner. We list some of them.

$$X \triangle Y = Y \triangle X \tag{1.20}$$

follows by the commutativity of the union,

$$X \triangle (Y \triangle Z) = (X \triangle Y) \triangle Z \tag{1.21}$$

follows by checking on the basis of already established laws that

$$X \triangle (Y \triangle Z) = [X \setminus (Y \setminus Z)] \cap [X \setminus (Z \setminus Y)] \cup [(Y \setminus Z) \setminus X] \cup [(Z \setminus Y) \setminus X] =$$

$$[X \setminus (Y \cup Z)] \cup [Y \setminus (X \cup Z)] \cup [Z \setminus (X \cup Y)] \cup (X \cap Y \cap Z).$$

As by (1.20) we have $(X \triangle Y) \triangle Z = Z \triangle (X \triangle Y)$ it suffices to replace in the last derived formula symbols of set variables to get that $(X \triangle Y) \triangle Z = [X \setminus (Y \cup Z)] \cup [Y \setminus (X \cup Z)] \cup [Z \setminus (X \cup Y)] \cup (X \cap Y \cap Z)$ and thus (1.21) follows.

Property (1.21) is the *associativity law* for the symmetric difference. A particular case is

$$X \triangle (X \triangle Y) = Y \tag{1.22}$$

Similarly,

$$X \cap (Y \triangle Z) = [(X \cap Y) \triangle (X \cap Z)] \tag{1.23}$$

follows by checking that each of sides is identical to the set $[(X \cap Y) \setminus ((X \cap Z)] \cup [(X \cap Z) \setminus (X \cap Y)]$.

Property (1.23) is the *distributivity law* for intersection with respect to the symmetric difference.

We may also note two simple properties that follow immediately from the definition of the symmetric difference, viz.,

$$\begin{cases} X \triangle X = \emptyset \\ X \triangle \emptyset = X \end{cases} \tag{1.24}$$

Restoring the union and the difference from the intersection and the symmetric difference may be done on the basis of the following identities

$$\begin{cases} X \cup Y = X \triangle (Y \triangle (X \cap Y)) \\ X \setminus Y = X \triangle (X \cap Y) \end{cases} \tag{1.25}$$

Yet another method of expressing ways in which sets are related one to another is to compare them with respect to *inclusion.*

We say that a set X is a *subset* of a set Y which is denoted $X \subseteq Y$ when the following condition is fulfilled

$$\forall x(x \in X \Rightarrow x \in Y) \tag{1.26}$$

Thus being a subset of a set means consisting of possibly not all elements of this set.

Being a subset does satisfy also certain laws of which we may mention the following easy to establish ones

$$X \subseteq Y \wedge Y \subseteq Z \Rightarrow X \subseteq Z \tag{1.27}$$

$$X \subseteq X \tag{1.28}$$

$$\emptyset \subseteq X \tag{1.29}$$

$$X \subseteq Y \wedge Y \subseteq X \Rightarrow X = Y \tag{1.30}$$

where (1.27) does express *transitivity* of inclusion, (1.30) states the *weak symmetry* of it; (1.28) says that any set is included in itself and (1.29) states that the empty set is included in any set (by the fact that implication from false to true has the truth value 1).

In many considerations involving sets, we restrict ourselves to sets which are subsets of a given fixed set U called in such a case the *universe*. In this case we may have some shortcut notation notably the difference $U \setminus X$ for $X \subseteq U$ is denoted X^c and it is called the *complement* of X.

The reader will undoubtedly establish many more properties of operations on sets in case of need; we will meet some in the sequel. We introduce a new notion of a *field of sets*.

By a *field of sets*, we will understand a non–empty set $\mathcal{B}$ of subsets of a certain universe U, which is such that results of operations $\triangle, \cap$ performed on its elements are again its elements and U is its element. More formally, $\mathcal{B}$ does satisfy the following requirements

1. $\forall x(x \in \mathcal{B} \Rightarrow x \subseteq U)$;

2. $U \in \mathcal{B}$;

3. $\forall x, y(x, y \in \mathcal{B} \Rightarrow x \triangle y \in \mathcal{B})$;

4. $\forall x, y(x, y \in \mathcal{B} \Rightarrow x \cap y \in \mathcal{B})$.

Let us observe that as $\mathcal{B}$ contains at least one element, e.g., U, it does contain $U \triangle U = \emptyset$. Also the identity

$$x \cup y = (x \triangle y) \triangle x \cap y$$

shows that

5. $\forall x, y(x, y \in \mathcal{B} \Rightarrow x \cup y \in \mathcal{B})$.

It states that any field of sets is closed with respect to the union of its elements. Similarly, by the identity

$$x^c = U \triangle x$$

we have

6. $\forall x (x \in \mathcal{B} \Rightarrow x^c \in \mathcal{B})$.

This means that any field of sets is closed with respect to complements of its elements. We get a characteristics of field of sets. Clearly, any set of sets closed on union, intersection, and complement is closed on symmetric difference.

Proposition 1.1. *A set $\mathcal{B}$ of subsets of a set U is a field of sets if and only if it is non–empty and it is closed with respect to operations of the union, the intersection and the complement.*

A simple example of a field of sets is the *power set of U* denoted 2^U consisting of all subsets of U, i.e.,

$$x \in 2^U \Leftrightarrow x \subseteq U$$

1.2.3 A Formal Approach

As already observed, we may perform some operations on sets once we are given them, but it has turned out in the historic process of development of set theory that naïve, intuitive approach led to some serious problems and contradictions. This motivated attempts at laying strict foundations for set theory. Studies in this respect led to some axiomatic systems of set theory, e.g., the *Zermelo–Fraenkel* system, introduced in Zermelo [30], [31], and modified, among others (Mirimanoff, Skolem, Gödel, Bernays, von Neumann), in Fraenkel [8]. We will not need the full power of this system, so we first restrict ourselves to a subsystem which guarantees us the existence of some basic means for new set construction.

We review below some axiomatic statements about sets which guarantee the existence of some sets as well as the possibility of performing some operations on them.

A1 (*There exists the empty set*): $\exists X (\forall x (x \in X \Rightarrow x \neq x))$

This axiom guarantees that there exists at least one set.

A2 (*The extensionality axiom*) $\forall x (x \in X \Leftrightarrow x \in Y) \Rightarrow X = Y$

This axiom does express our intuition that sets consisting of the same elements should be identical to each other. A fortiori, it does guarantee the uniqueness of other sets implied by the axioms.

A3 (*The unordered pair axiom*) *Given any two objects* a, b, *the following holds*:

$$\exists X (\forall x (x \in X \Leftrightarrow x = a \lor x = b))$$

This axiom guarantees that for any two objects a, b there exists a set whose only elements are a and b. We denote this set by the symbol $\{a, b\}$ and we call it the *unordered pair* of elements a, b. In case $a = b$, we call this set a *singleton* and we denote it by the symbol $\{a\}$.

From this axiom, the existence of the set $\{\{a\}, \{a, b\}\}$ follows immediately. We denote this set by the symbol $< a, b >$ and call it the *ordered pair* of elements a, b.

The reader will check the following basic property of the ordered pair

Proposition 1.2. *Given elements* a, b, c, d *we have* $< a, b > = < c, d >$ *if and only if* $a = c, b = d$.

For the proof, we hint that $< a, b > = < c, d >$ i.e. $\{\{a\}, \{a, b\}\} = \{\{c\}, \{c, d\}\}$ implies $\{a\} = \{c\}$ or $\{a, b\} = \{c\}$ so $a = c$. Similarly, $\{c, d\} = \{b\}$ or $\{c, d\} = \{a, b\}$ hence $b = c = d$ or $b = d$. So either $a = b = c = d$ or $a = b, c = d$ and the thesis follows.

The procedure of generating ordered pairs may be repeated: given objects $a_1, ..., a_k$, we may define inductively the *ordered k–tuple*

$$< a_1, ..., a_k >$$

by letting

$$< a_1, ..., a_k > = << a_1, ..., a_{k-1} >, a_k >$$

Clearly, inducting on k, we get the following extension of the last proposition: $< a_1, ..., a_k > = < b_1, ..., b_k >$ *if and only if* $a_i = b_i$ for $i = 1, 2, ..., k$.

A4 (*The union of sets axiom*)

$$\forall X (\exists Z \forall x (x \in Z \Leftrightarrow \exists Y (x \in y \land Y \in Z)))$$

This axiom guarantees that for any set X there exists a set (uniquely defined by extensionality) Z whose elements are those sets which are elements in some element of X. The set Z is denoted $\bigcup X$ and it is called the *union* of X. This setting may imply that X is a set whose elements are sets; for this reason, to avoid a phrase like *a set of sets*, X is said to be a *family of sets*, so Z is the union of this family.

We may consider the case when X is looked on as a family of subsets of the universe $\bigcup X$, hence, the set

$$\bigcap X = \bigcup X \setminus \bigcup \{\bigcup X \setminus A : A \in X\} \tag{1.31}$$

is well–defined; we call it the *intersection* of the family X of sets. Clearly,

$$x \in \bigcap X \Leftrightarrow \forall A (A \in X \Rightarrow x \in A) \tag{1.32}$$

Particular applications for this axiom are found in case of the two sets X, Y: clearly, in this case we have $\bigcup \{X, Y\} = X \cup Y$. Another application is in the formation of finite unordered tuples: given objects $a_1, a_2, ..., a_k$, we may form the union of singletons $\{a_i\}$ for $i = 1, 2, ..., k$, by induction as follows: $\{a_1, ..., a_k\} = \{a_1, ..., a_{k-1}\} \cup \{a_k\}$. The set $\{a_1, ..., a_k\}$ is called the *unordered k–tuple* of elements $a_1, a_2, ..., a_k$.

To get more specialized sets, we need new axioms. In particular, given a set X and a predicate $\Psi(x)$ expressing a certain property of objects, we may want to consider the new set whose elements are those elements of X which after being substituted for x satisfy $\Psi(x)$. We need an axiom which would guarantee the existence of such a set.

A5 (*The axiom schema of separation*) *Given a set X and a predicate $\Psi(x)$, the following holds*

$$\exists Y \forall x (x \in Y \Leftrightarrow \Psi(x) \wedge x \in X)$$

The (unique) set Y is denoted often by the symbol $\{x \in X : \Psi(x)\}$. Let us observe for example, that given in addition a set Z, the difference $X \setminus Z$ may be obtained via A5 as the set $\{x \in X : \neg(x \in Z)\}$. Similarly, the intersection $X \cap Z$, is the set $\{x \in X : x \in Z\}$.

We may also want for any set X to form a new set whose elements are all subsets of X, i.e., the familiar to us field of sets 2^X. For this also we need a new axiom which however may be regarded from foundational point of view as too strong.

A6 (*The axiom of the power set*) *Given a set X, the following holds*

$$\forall X (\exists Y \forall x (x \in Y \Leftrightarrow x \subseteq X))$$

The set Y is denoted with the symbol 2^X (*the exponential set of X, the power set of X*). It is unique by extensionality.

Finally, we may want to form a new set from a given one by a transformation via a predicate of functional character (in a sense, we want to *map* a given set onto a new one). For this, we also need a new axiom.

A7 (*The axiom schema of replacement*) *Consider a binary predicate $\Psi(x, y)$ with the following property*

(i) $\forall x, y, z (\Psi(x, y) \wedge \Psi(x, z) \Rightarrow y = z)$

Then, if $\Psi(x, y)$ does satisfy (i), *then*

$$\forall X \exists Y (y \in Y \Leftrightarrow \exists x (x \in X \wedge \Psi(x, y)))$$

The set Y may be called the *image* of the set X by Ψ and we may denote it by the symbol $\Psi(X)$.

The axiom A7 is stronger than A5: given $\Psi(x)$ as in the formulation of A7, we may define a binary predicate Φ as follows: $\Phi(x,y)$ is $x = y$ when $\Psi(x)$ is satisfied and $\Phi(x,y)$ is $x = a$ where a is a fixed element not in the set X when $\Psi(x)$ is not satisfied. Then, the set $\Phi(X)$ is identical to the set $\{x \in X : \Psi(x)\}$.

1.3 Relations and Functions

The notion of an ordered pair allows for a formalization of the important notions of a *relation* as well as of a *function*.

Informally, a (binary) relation is a constraint on ordered pairs $< x,y >$ where $x \in X, y \in Y$; thus, a formalization of a relation is done via a choice of a set of pairs which satisfy this constraint.

First, given sets X, Y, we form the set $X \times Y$ of all ordered pairs of the form $< x,y >$ where $x \in X, y \in Y$. We thus let

$$X \times Y = \{< x,y >: x \in X, y \in Y\} \tag{1.33}$$

The existence of this set may be justified on the basis of axioms: we first may form the set $2^{X \cup Y}$ of all subsets of the set $X \cup Y$ and then the set $2^{2^{X \cup Y}}$, both by axiom A6. Letting $\Psi(A)$ to be satisfied by A if and only if

$$\exists x \exists y (x \in X \wedge y \in Y \wedge A = \{\{x\}, \{x,y\}\})$$

we obtain $X \times Y$ as the set $\{A \in 2^{2^{X \cup Y}} : \Psi(A)\}$ by axiom A7.

The set $X \times Y$ is called the *Cartesian product* of the set X and the set Y. A *binary relation* R on $X \times Y$ is a subset R of the set $X \times Y$. For x, y such that $< x,y >\in R$ we say that x, y *are in the relation* R and we also write xRy to denote this fact. We let

$$domR = \{x \in X : \exists y(< x,y >\in R)\} \tag{1.34}$$

and we call $domX$ the *domain of* R; similarly, we let

$$codomR = \{y \in Y : \exists x(< x,y >\in R)\} \tag{1.35}$$

and we call $codomR$ the codomain of R.

As relations are sets, usual set operations may be performed on relations: for relations R, S on the Cartesian product $X \times Y$, the union $R \cup S$, the intersection $R \cap S$, and the difference $R \setminus S$ are defined in the usual way. There is however more to relations: due to the structure of their elements as ordered pairs, we may have more operations on relations than on ordinary sets and these operations constitute what is called the *Algebra of relations*.

1.3.1 Algebra of Relations

First, for any relation $R \subseteq X \times Y$, we may introduce the *inverse* to R which is the relation $R^{-1} \subseteq Y \times X$ defined as

$$yR^{-1}x \Leftrightarrow xRy \qquad (1.36)$$

Properties of this operation are collected in

Proposition 1.3. *The following are properties of inverse relations*

1. $(R \cup S)^{-1} = R^{-1} \cup S^{-1}$;

2. $(R \cap S)^{-1} = R^{-1} \cap S^{-1}$;

3. $(R \setminus S)^{-1} = R^{-1} \setminus S^{-1}$.

Indeed, for instance in case of Property 1,

$$<y, x> \in (R \cup S)^{-1} \Leftrightarrow <x, y> \in R \cup S \Leftrightarrow$$

which in turn is equivalent to

$$<x, y> \in R \vee <x, y> \in S \Leftrightarrow <y, x> \in R^{-1} \vee <y, x> \in S^{-1}$$

i.e.,

$$<y, x> \in R^{-1} \cup S^{-1}$$

The proof in remaining cases goes along similar lines.

Relations may also be *composed*; for relations $R \subseteq X \times Y$, $S \subseteq Y \times Z$, the *composition* (or, *superposition*) $R \circ S \subseteq X \times Z$ is defined as follows

$$<x, z> \in R \circ S \Leftrightarrow \exists y (y \in Y \wedge <x, y> \in R \wedge <y, z> \in S) \qquad (1.37)$$

The operation of relation composition does satisfy easy to be checked properties.

Proposition 1.4. *Basic properties of the operation of relation composition are*

1. $(R \cup S) \circ T = (R \circ T) \cup (S \circ T)$;

2. $(R \cap S) \circ T = (R \circ T) \cap (S \circ T)$;

3. $(R \circ S)^{-1} = S^{-1} \circ R^{-1}$;

4. $(R \circ S) \circ T = R \circ (S \circ T)$.

These properties are proved similarly to those of the preceding proposition.

For a set X, the relation $id_X = \{<x,x>: x \in X\}$ is called the *identity relation* on X. Given a relation R on $X \times Y$, we have

Proposition 1.5. *The identity relation satisfies the following*

1. $id_{domR} \subseteq R \circ R^{-1}$;

2. $id_{co-domR} \subseteq R^{-1} \circ R$;

3. $id_{domR} \circ R = R$;

4. $R \circ id_{co-domR} = R$.

Indeed, in case of Property 1, for $x \in domR$, there is $y \in Y$ with $<x,y>\in R$ hence $<y,x>\in R^{-1}$ so $<x,x>\in R \circ R^{-1}$. The proof for Property 2 is on the same lines. Properties 3 and 4 are obvious.

One more operation one may perform on relations is *restriction*. Given a relation $R \subseteq X \times Y$, and a subset $A \subseteq X$, the restriction $R|A$ is $R \cap A \times Y$.

Functions are a special class of relations singled out by the uniqueness property: a relation $R \subseteq X \times Y$ is a function if and only if the following property is observed

$$\forall x \forall y \forall z (xRy \wedge xRz \Rightarrow y = z) \tag{1.38}$$

In this case R is said to be a *function from the set X into the set Y* and the notation $R : X \to Y$ is in use. To denote functions, usually small letters $f, g, h, k, ...$ are used and in place of the formula xfy we write $y = f(x)$ calling *y the value of f at the argument x.*

A function $f : X \to Y$ is said to be *injective* in case the following is satisfied

$$\forall x \forall x' (f(x) = f(x') \Rightarrow x = x') \tag{1.39}$$

For a subset $A \subseteq X$, the *image* fA of A by a function $f : X \to Y$ is the set

$$\{y \in Y : \exists x \in A(y = f(x))\} \tag{1.40}$$

Similarly, for $B \subseteq Y$, the *inverse image* $f^{-1}B$ is the set

$$\{x \in X : \exists y \in B(y = f(x))\} \tag{1.41}$$

Proposition 1.6. *Operations of image and inverse image do satisfy the following rules*

1. $f(A \cup B) = fA \cup fB$;

2. $f(A \cap B) \subseteq fA \cap fB$;

3. $f^{-1}(A \cup B) = f^{-1}A \cup f^{-1}B$;

4. $f^{-1}(A \cap B) = f^{-1}A \cap f^{-1}B$;

5. $f^{-1}(A \setminus B) = f^{-1}A \setminus f^{-1}B$.

These facts are easily established; let us note the difference between Properties 2 and 4: in case of Property 2 we have only inclusion as for some f even disjoint sets may be mapped onto the same image, whereas in Property 4 we have identity by the uniqueness of the value of a function.

A function $f : X \to Y$ maps X *onto* Y (f *is a surjection*) if and only if $fX = Y$. A function f which is injective and onto is said to be a *bijection* between X and Y. For a bijection f, the relation f^{-1} is again a function $f^{-1} : Y \to X$ called the *inverse* of f. Let us observe that the function f^{-1} is a bijection between Y and X. Clearly, for a bijection f, it may not happen that disjoint sets map to the same value so in 2 above we have the identity: $f(A \cap B) = fA \cap fB$.

1.4 Ordering Relations

Relations on a Cartesian product $X \times X$ with $X = dom_R$ are called *relations on the set* X; they are classified according to some properties,viz.,

1. (*Reflexivity*) $\forall x(xRx)$;

2. (*Symmetry*) $\forall x \forall y(xRy \Rightarrow yRx)$;

3. (*Weak anti–symmetry*) $\forall x \forall y(xRy \wedge yRx \Rightarrow x = y)$;

4. (*Linearity*) $\forall x \forall y(xRy \vee x = y \vee yRx)$;

5. (*Transitivity*) $\forall x \forall y \forall z(xRy \wedge yRz \Rightarrow xRz)$.

A relation R on a set X is an *ordering* on X if and only if R is reflexive, weak anti–symmetric and transitive. In this case we write $x \leq_R y$ in place of xRy.

An ordering $\leq_R$ on the set X is *linear* in case R is linear; we say also that X is *linearly ordered* by R. More generally, we say that a subset $Y \subseteq X$ of a set X ordered by an ordering R is a *chain* in X when the restriction $R|Y \times Y$ is linear.

An ordering does stratify elements of its domain; $x, y \in X$ are *comparable* if $x \leq_R y \vee y \leq_R x$. In case $x \leq_R y$ we say that x is *less or equal to* y.

An element x is an *upper bound* (respectively a *lower bound*) of a set $A \subseteq X$ in case $a \leq_R x$ for each a in A (respectively, $x \leq_R a$ for each $a \in A$); an upper bound x of A is the *least upper bound* (*the supremum of A*) in case $x \leq_R y$ for each upper bound y of A and we denote x by the symbol $supA$.

Similarly, we call a lower bound x of A the *greatest lower bound* (*the infimum of A*) when $y \leq_R x$ for each lower bound y of A. We denote an infimum of A with the symbol $inf A$. In case infimum respectively supremum are admitted i.e. $inf A \in A$ respectively $sup A \in A$ we call $inf A$ the *least element in A* respectively we call $sup A$ *the greatest element in A*.

We denote also by A^+ the set of all upper bounds of the set A and by A^- the set of all lower bounds of the set A. Thus, $sup A = inf A^+$ and $inf A = sup A^-$ whenever $sup A$, respectively $inf A$ exists.

A set $Y \subseteq X$ is *bounded from above* (respectively, *bounded from below*) when there exists an upper bound for Y (respectively, a lower bound for Y).

In search of archetypical orderings, we may turn to inclusion $\subseteq$ on the power set of a given set X. It is clearly an ordering on any family of subsets of the set X. It is however not linear. To find suprema of families of sets with respect to inclusion, we have to turn to the sum axiom: given a family F of subsets of X we have $sup F = \bigcup F$. Similarly, $inf F = \bigcap F$. A family F is *closed with respect to unions* when $sup F' \in F$ for any $F' \subseteq F$; by analogy, F is *closed with respect to intersections* when $inf F' \in F$ for any $F' \subseteq F$.

A set X is *completely ordered* by an ordering R on X when for any subset $A \subseteq X$ there exist $sup A$ and $inf A$. The ordering R is said to be *complete* in this case. An example of a complete ordering is inclusion on the power set 2^X for any set X and for any family of sets $F \subseteq 2^X$ its supremum is $sup F = \bigcup F$ and its infimum is $inf F = \bigcap F$.

Relations on various sets may be compared to each other by means of functions satisfying adequate conditions. In the general case of a relation R on the product $X \times X$ and a relation $S \subseteq Y \times Y$, we say that a function $f : X \rightarrow Y$ *agrees with R, S* when

$$\forall x \forall y (xRy \Rightarrow f(x)Sf(y)) \tag{1.42}$$

In case f is an injection, we say that f is an *embedding* of R into S and when f is a bijection, then we say that f is an *isomorphism* between R and S.

In the particular case when R, S are orderings on respectively sets X, Y, we call a function $f : X \rightarrow Y$ which agrees with R, S an *isotone* function. Thus we have for the isotone function f

$$x \leq_R y \Rightarrow f(x) \leq_S f(y) \tag{1.43}$$

In general there is little one may say about isotone functions; however in the case when $R = S$ is a complete ordering, isotone functions between R and S have an important property of having a *fixed point* which means that $f(x) = x$ for some $x \in X$ called a *fixed point of f*.

Proposition 1.7. *(Knaster [11]–Tarski) If $f : X \rightarrow X$ is an isotone function on a completely ordered set X then f has a fixed point.*

Proof. Clearly, the set $A = \{x \in X : x \leq_R f(x)\}$ has a supremum $sup A = a$. As for $x \in A$ we have $x \leq_R a$, also $x \leq_R f(x) \leq_R f(a)$ hence $a \leq_R f(a)$.

It follows that $f(a) \leq_R f(f(a))$ so $f(a) \in A$ and hence $f(a) \leq a$ implying finally that $a = f(a)$ so a is a fixed point of f □

Given an ordered set X with an ordering $\leq$ one may construct a canonical isotone embedding of X into a completely ordered by inclusion family of sets. To this end, we make use of sets A^-, A^+ defined above for any set $A \subseteq X$. We state important properties of operations $(.)^{+,-}$.

Proposition 1.8. *The following properties are observed by operations* A^-, A^+

1. $A \subseteq B \Rightarrow B^- \subseteq A^-, B^+ \subseteq A^+$;

2. $A \subseteq (A^+)^-$;

3. $B = (A^+)^-$ *satisfies* $B = (B^+)^-$.

Proof. Property 1 is obvious: anything less (greater) than all elements of B is also less (greater) than all elements of A. Property 2 is also evident: for $x \in A$ we have $x \leq a$ for any $a \in A^+$ hence $x \in (A^+)^-$. For Property 3, we first notice that as $A \subseteq B$ by Property 2, we have by Property 1 that $B^+ \subseteq A^+$. But also $A^+ \subseteq B^+$ by transitivity of $\leq$. Since $A^+ = B^+$, we have $B = (A^+)^- = (B^+)^-$ □

Consider now for $x \in X$ the set $\{x\}^-$ denoted also $(\leftarrow, x]$ and called the *left interval* of x. We have $((\leftarrow, x]^+)^- = (\leftarrow, x]$.

We may consider now the function f which assigns to $x \in X$ its left interval $(\leftarrow, x]$. We may consider f as a function from X into the set $I(X) = \{A \subseteq X : A = (A^+)^-\}$ ordered by inclusion. We have

Proposition 1.9. *(Dedekind [7], MacNeille [17]) The function* f *is an isotone function from* X *into completely ordered by inclusion set* $I(X)$. *Moreover,* f *preserves suprema and infima.*

Proof. Clearly, $x \leq y$ implies $(\leftarrow, x] \subseteq (\leftarrow, y]$. For any $A \subseteq I(X)$, we have $inf A = (\bigcap A^+)^-$, $sup A = (\bigcup A^+)^-$ so $I(X)$ is completely ordered. Finally, given $A \subseteq X$, with $a = sup A$, we have easily that

$$(\leftarrow, a] = (\bigcup \{(\leftarrow, x] : x \in A\}^+)^-$$

i.e., $f(sup A) = sup f(A)$, and in case of infima we have for $b = inf A$ that $(\leftarrow, b] = \bigcap \{f(x) : x \in A\}$, i.e., $f(inf A) = inf f(A)$ □

We may by this result embed any ordered set into a completely ordered by inclusion family of sets. Let us observe that the Dedekind–MacNeille embedding is the least one: it embeds X into any completely ordered set into which X may be embedded isotonically.

A relation between two ordered sets, X ordered by $\leq_X$ and Y ordered by $\leq_Y$, can be established by a pair of isotone mappings, $f : X \to Y, g : Y \to X$, which satisfy the postulate

$$f(x) \leq_Y y \Leftrightarrow x \leq_X g(y) \tag{1.44}$$

which does establish a *Galois connection* between $(X, \leq_X)$ and $(Y, \leq_Y)$.

In the pair (f, g), f is the *lower adjoint*, g is the *upper adjoint*. From the definition (1.44), the following properties stem.

Proposition 1.10. *Essential properties of Galois connections are*

1. $x \leq_X g(f(x))$; $f(g(y)) \leq_Y y$;

2. $g(f(x)) = g(f(g(f(x))))$; $f(g(y)) = f(g(f(g(y))))$;

3. $x \leq_X z \Rightarrow g(f(x)) \leq_X g(f)z))$; $y \leq_Y w \Rightarrow f(g(y)) \leq_Y f(g(w))$.

The mapping $g \circ f : X \Rightarrow X$, is according to Proposition 1.10, *extensive* (Property 1), *idempotent* (Property 2), *isotone* (Property 3), i.e., it is a *closure mapping* on X. Dually, the mapping $f \circ g : Y \to Y$ is *co–extensive, idempotent* and *isotone*, i.e., it is an *interior mapping* on Y.

1.5   Lattices and Boolean Algebras

An ordered set $(X, \leq)$ in which for each pair x, y there exist the least upper bound (l.u.b.) $sup\{x, y\}$ as well as the greatest lower bound (g.l.b.) $inf\{x, y\}$ is called a *lattice*. Usually, $sup\{x, y\}$ is denoted $x \cup y$ and it is called the *join* of x, y while $inf\{x, y\}$ is denoted $x \cap y$ and it is called the *meet* of x, y.

When additionally X is endowed with a unary operation $-$ of *complementation* and the join and the meet satisfy the following conditions

1. $x \cup y = y \cup x, x \cap y = y \cap x$;

2. $x \cup (y \cup z) = (x \cup y) \cup z, x \cap (y \cap z) = (x \cap y) \cap z$;

3. $(x \cap y) \cup y = y, (x \cup y) \cap y = y$;

4. $x \cap (y \cup z) = (x \cap y) \cup (x \cap z), x \cup (y \cap z) = (x \cup y) \cap (x \cup z)$;

5. $(x \cap -x) \cup y = y, (x \cup -x) \cap y = y$.

then the lattice X is called a *Boolean algebra*. We may observe duality here: replacing $\cup/\cap, \cap/\cup$ we obtain a dual condition; we may also not confuse the symbols for the join and the meet used in general with the particular symbols of set union and intersection. The latter clearly satisfy 1–4. Let us observe that any field of sets is a Boolean algebra with operations of set union, intersection and complementation. The reader may consult Rasiowa [21] for a review of lattices, also from the point of view of their roles as models for logics.

1.6   Infinite Sets

We have witnessed a good deal of standard set theory having basically only finite sets (in an informal sense of the word) at our disposal. To introduce infinite sets, we need a new axiom.

A9 (*The infinity axiom*) *There exists a set X such that* (i) $\emptyset \in X$ (ii) *if $A \in X$ then $A \cup \{A\} \in X$*

Taking the intersection of the family of all sets satisfying axiom A9, we obtain the set N which contains elements $0 = \emptyset$, $1 = \emptyset \cup \{\emptyset\}$, $2 = 1 \cup \{1\}$,..., $n + 1 = n \cup \{n\}$,... and only those elements. N is the set of *natural numbers*. Having N we may attempt at defining precisely *finite* and *infinite sets*. A standard way of arguing with natural numbers employs the *principle of mathematical induction*.

Proposition 1.11. *Assume $A \subseteq N$ is such that (i) $0 \in A$ (ii) $\forall n(n \in A \Rightarrow n + 1 \in A)$. Then $A = N$.*

Indeed, by (i), (ii), A does satisfy requirements of axiom A9, hence, $N \subseteq A$ and finally $A = N$.

The principle of mathematical induction will serve us as a tool in proving basic facts about natural numbers.

Proposition 1.12. *For all natural numbers n, m the following statements are true*

1. $n \in m \Leftrightarrow n \subseteq m \wedge n \neq m$;

2. $n \in m \vee n = m \vee m \in n$;

3. $n \subseteq N \wedge n \neq N$;

4. $n \notin n$.

Proof. For Property 1: we consider $A = \{m \in N : \forall n(n \in m \Rightarrow n \subseteq m)\}$. Clearly, $0 \in A$ by default. Assuming $m \in A$, consider $m + 1$ along with $n \in m + 1$. Then either $n \in m$ or $n = m$; in the former case $n \subseteq m$ hence $n \subseteq m + 1$ and in the latter case again $n \subseteq m$ whence $n \subseteq m + 1$. Thus $m + 1 \in A$. By the principle of mathematical induction, $A = N$. This proves the part (i) $n \in m \Rightarrow n \subseteq m$.

We now prove Property 4; clearly, $0 \notin 0$. Assume $n \notin n$ and $n + 1 \in n + 1$. Then either $n + 1 \in n$ or $n + 1 = n$. In the former case by (i) $n + 1 \subseteq n$ hence $n \in n$, a contradiction. So only $n + 1 = n$ may hold but then again $n \in n$, a contradiction. It follows that we have $n + 1 \notin n + 1$ and by the principle of mathematical induction Property 4 holds.

Returning to Property 1, we infer from Property 4 that (ii) $n \in m \Rightarrow n \subseteq m \wedge n \neq m$. It remains to prove the converse. Again we apply induction. We use $\subset$ instead of $\subseteq \wedge \neq$. So we are going to check that $n \subset m$ implies $n \in m$. For $m = 0$ this is true. Assume this is true with m and consider $n \subset m + 1$. Was $m \in n$ then $m \subset n$ and $m + 1 \subseteq n$, a contradiction with $n \subset m + 1$. Hence $m \notin n$ and thus $n \subseteq m$. We have either $n \subset m$ in which case by assumption $n \in m$ and a fortiori $n \in m + 1$ or $n = m$ in which case $n \in m + 1$ so Property 1 is proved.

Property 3 follows easily: by Property 4, we have $n \neq N$, and by induction $n \subset N$ follows.

It remains to prove trichotomy 2. We apply induction proving that for each $n \in N$ the set $A_n = \{m \in N : n \in m \vee n = m \vee m \in n\}$ is N. Clearly, $A_0 = N$ as $0 \in m$ for each m. For A_n, it is clear that $0 \in A_n$. Assume that $m \in A_n$ and consider $m + 1$. By assumption either $m \in n$ or $m = n$ or $n \in m$. In the first case $m \subset n$ by 1 hence either $m + 1 \subset n$ and thus $m + 1 \in n$ or $m + 1 = n$. In both cases $m + 1 \in A_n$. In the second case when $m = n$ clearly $n \in m + 1$ and again $m + 1 \in A_n$. Finally, when $n \in m$ a fortiori $n \in m + 1$ so $m + 1 \in A_n$. It follows by the induction principle that $A_n = N$ for each $n \in N$ so Property 2 is proved $\qquad\square$

It follows that N is ordered linearly by the membership relation $\in$; using the symbol $<$ in place of $\in$, we have $0 < 1 < 2 < 3 < \ldots\ldots < n < n + 1 < \ldots$. We have used here the symbol $<$ of a *strict ordering* related to the ordering $\leq$ by the identity $\leq = < \cup =$ i.e. $n \leq m$ if and only if $n < m$ or $n = m$. The relation $<$ has one more important property.

Proposition 1.13. *In any non–empty subset $A \subseteq N$, there exists the least element with respect to $<$.*

Indeed, there is $n \in A$ and clearly it suffices to consider the set $A|n = A \cap (\leftarrow, n]$. If $0 \in A|n$ then 0 is the least element in A; otherwise we may check whether $1 \in A|n$ etc. Proceeding in this way, after at most $n - 1$ steps, we find k such that $\{0, 1, ..., k - 1\} \cap A = \emptyset$ and $k \in A$, Thus k is the least element in A.

A relation R of linear ordering on a set X with the property expressed in the last proposition is said to *well–order* the set X and it is called a *well–ordering* of X. Thus $\in$ well–orders N.

A question may arise at this point whether a well–ordering may exist for an arbitrary set X. We do not have by now any tool which would enable us to prove such a result. We need a new axiom which would guarantee us the existence of a well–ordering on any set.

1.7 Well–Ordered Sets

The axiomatic statement which would guarantee the existence of a well-ordering on each set is

A10 (the *Axiom of well–ordering*) *For each set X there exists a relation R on X which well–orders X*

We may be aware of the fact that this axiom is non–effective and it does not give us any procedure for constructing a well–ordering on a given set; its value is ontological, allowing us to explore consequences of its content. We first state one of these consequences known as the *axiom of choice* of Zermelo.

Proposition 1.14. *Axiom A10 implies the following: for any family $\mathcal{F}$ of non–empty sets, there exists a function $f : \mathcal{F} \to \bigcup \mathcal{F}$ with the property that $f(X) \in X$ for each $X \in \mathcal{F}$.*

Indeed, by A10, the set $\bigcup \mathcal{F}$ may be well–ordered by a relation $<$. To construct f with the desired property, it is sufficient to define $f(X)$ to be the least element in X with respect to $<$.

A function f with the property stated in the last proposition is called a *choice function* for the family $\mathcal{F}$ (or, a *selector* for $\mathcal{F}$). From the axiom of choice (hence from the axiom of well–ordering) one may derive some very important consequences. We now present some of them. Recall that a subset A of an ordered by $\leq$ set X is a *chain* if it is linearly ordered by the restriction $\leq |A = \leq \cap A \times A$. An element c of X is *maximal* (respectively *minimal*) in X in case there is in X no element $d \neq c$ with $c \leq d$ (respectively $d \leq c$).

We may give a formulation of the *maximum principle* known also as the *Zorn lemma* due independently to Hausdorff [10], Kuratowski [12] and Zorn [32].

Proposition 1.15. *(the Maximum Principle) Assume that a set X ordered by a relation $\leq$ has the property that any chain in X is bounded from above. Then in X there exists a maximal element.*

Proof. (Balcar–Štěpánek [2]) Let f be a selector on the set $2^X \setminus \{\emptyset\}$ of all non–empty subsets of X. Assume, to the contrary, that in X there is no maximal element i.e. for any $x \in X$ the set $M(x) = \{y \in X : x < y\}$ is non–empty and thus $s(x) = f(M(x))$ exists.

We will start with an element a in X and the set $S(a) = [a, \rightarrow)$, and we observe properties of $Y \subseteq S(a)$

$$\begin{cases} (i) \ a \in Y \\ (ii) \ y \in Y \to s(y) \in Y \\ (iii) \ if \ L \ is \ a \ chain \ in \ Y, \ then \ supL \in Y \end{cases} \quad (1.45)$$

For the family $\mathcal{F}(a)$ of all $Y \subseteq S(a)$ which satisfy (1.45), we form the intersection $\bigcap \mathcal{F}(a) = L$. Clearly, L satisfies (1.45) being a minimal set with this property. We would like to show that L is a chain.

To this end, first, we exploit minimality of L by looking at the set

$$K = \{y \in L : \forall x(x \in L \wedge x < y \Rightarrow s(x) \leq y)\}$$

Claim 1. *For $x \in K, y \in L$: $x \leq y \vee y \leq x$.*

To check this claim, we take $x \in K$ and we consider the set

$$K_x = \{y \in L : x \leq y \vee y \leq x\}$$

We check that K_x does satisfy (1.45). Clearly, $a \in K_x$ so (i) holds. Assume $y \in K_x$. In case $x \leq y$ we have $x < s(y)$; in case $x > y$, by definition of K, we have $s(y) \leq x$. In either case, $s(y) \in K_x$ so (ii) is satisfied. For a chain C in K_x, either $x < c$ for some $c \in C$ hence $x < supC$ or $c \leq x$ for every $c \in C$ hence $supC \leq x$. In either case $supC \in K_x$ witnessing (iii). By minimality of L we must have $K_x = L$ so Claim 1 is verified.

Claim 2. *For $x \in K, y \in L$: $x < y \Rightarrow s(x) \leq y$.*

We apply the same technique: given $x \in K$, we look at the set

$$K^x = \{y \in L : x < y \Rightarrow s(x) \leq y\}.$$

Again (i) holds obviously, for (ii) we look at $y \in K^x$. By Claim 1, either $x \leq y$ or $y \leq x$. The latter case impossible, we are left with $x \leq y$ so either $x = y$ hence $s(x) = s(y)$ or $x < y$ hence $s(x) \leq s(y)$ as $y \in K^x$. Thus in all cases $s(y) \in K^x$ proving (ii). In case (iii), for a chain C in K^x, if $x < supC$, then $x < c$ for some $c \in C$ hence $s(x) \leq c$ and thus $s(x) \leq supC$. It follows that $supC \in K^x$ and (iii) holds. By minimality of L, we have $K^x = L$ which proves Claim 2.

Claim 3. $K = L$.

We check that K satisfies (1.45) so Claim 3 would follow by minimality of L. Clearly, $a \in K$ so (i) holds. For $x \in K$, let $y < s(x)$. We have by Claim 1 that $x \leq y$ or $y \leq x$. But $x \leq y$ cannot hold by Claim 2 so only $y < x$ remains. Then $s(y) \leq x$ hence $s(y) < s(x)$ witnessing $s(y) \in K$ so (ii) holds. Finally, consider a chain C in K with $y < supC$ for some $y \in L$. Then $y < c$ for some $c \in C$ hence $s(y) \leq c$ and $s(y) \leq supC$ implying that $supC \in K$ so (iii) holds. By minimality of L, $K = L$ and it follows that L is a chain. By (iii), $supL \in L$ and by (ii) $s(supL) \in L$ which is a contradiction with $supL < s(supL)$. Thus in X there exists a maximal element $\square$

It is much easier to show that the axiom of well–ordering follows from the maximum principle.

Proposition 1.16. *The maximum principle implies the axiom of well–ordering.*

Proof. Consider a set X assuming the maximum principle. Now, consider the set $\mathcal{F}$ of all pairs of the form (a subset Y of X, a relation R well–ordering Y), i.e.,

$$(Y, R) \in \mathcal{F} \Leftrightarrow Y \subseteq X \wedge R \; well-orders \; Y$$

We introduce an ordering $\sqsubseteq$ on $\mathcal{F}$ by letting $(Y, R) \sqsubseteq (Z, S)$ if and only if $Y \subseteq Z$ and $S|Y = R$. Let us observe that any chain C in $\mathcal{F}$ with respect to $\sqsubseteq$ has a supremum, viz., $(\bigcup F, \bigcup\{R : (Y, R) \in C\})$.

By the maximum principle (Proposition 1.15) there exists a maximal element (Y_0, R_0) in $\mathcal{F}$. If it was $Y_0 \neq X$, we could pick an element $a \in X \setminus Y_0$ and extend Y_0 to $Y_1 = Y_0 \cup \{a\}$, extending also R_0 to a new ordering R_1 by declaring a greater than all elements in Y_0, i.e., $R_1 = R_0 \cup \{(y, a) : y \in Y_0\}$. Then R_1 well–orders Y_1 and $(Y_0, R_0) \sqsubseteq (Y_1, R_1)$, a contradiction. Hence $Y_0 = X$ and R_0 well–orders X $\qquad\square$

It follows that the axiom of choice, the axiom of well–ordering and the maximum principle are equivalent. There are a number of other statements which turn out to be equivalent to those three. For instance the following are often used in various branches of mathematics.

1. (Birkhoff [3]) *In any ordered set X, any chain is contained in a chain maximal with respect to inclusion*;

2. (Vaught [27]) *Any family $\mathcal{F}$ of sets contains a maximal with respect to inclusion sub–family consisting of pair–wise disjoint sets.*

Indeed, the Birkhoff principle follows from the maximum principle by the fact that the union of a chain $\mathcal{L}$ of chains is a chain being the supremum of L hence in a set of all chains over X ordered by inclusion there exists a maximal chain. By analogy, the Vaught principle follows from the maximum principle by the fact that the union of a linearly ordered by inclusion family $\mathcal{L}$ of sub–families of $\mathcal{F}$ consisting of pair–wise disjoint sets consists of pair–wise sets and constitutes the supremum of $\mathcal{L}$.

1.8   Finite versus Infinite Sets

We now are in position to define formally the notion of *finiteness* and to contrast it with the notion of *being infinite*.

For a set X, we say that X is a *finite set* (in the Dedekind sense) in case there exist: a natural number $n \in N$ and a bijection $f : X \to n$. In this case we say that the *cardinality* of X is n, in symbols $|X| = n$. Let us observe that given a singleton k, we have $|X| = |X \times \{k\}|$. From this observation we may derive

Proposition 1.17. *Finite sets have properties*

1. *Any subset of a finite set is finite;*
2. *The union $\bigcup_i^m X_i$ of finite sets $X_1, X_2, ..., X_m$ is a finite set.*

Proof. Indeed, if $f : X \to n$ witnesses $|X| = n$ and $Y \subseteq X$ then $f(Y) \subseteq n$ hence either $f(Y) = n$ and thus $|Y| = n$ or $f(Y) \subset n$ hence $|Y| = m < n$ by Proposition 1.12. This proves Property 1.

Given $X_1, X_2, ..., X_m$ with $|X_i| = m_i$ for $i = 1, 2, ..., m$ we may observe that (a) $\bigcup_i^m X_i$ embeds as a subset into $\oplus_i X_i = \bigcup_i X_i \times \{i\}$ and as the latter is the union of pair–wise disjoint sets we have $| \oplus_i X_i | = \sum_i m_i$ hence it is a finite set. By Property 1, $\bigcup_i^m X_i$ is finite □

Apparently distinct is the *finiteness in the Tarski sense*, viz., a set X is finite in the sense of Tarski [25], when the following *Tarski property* is observed

TF *In any non–empty subset $\mathcal{A} \subseteq 2^X$ there exists a set maximal in $\mathcal{A}$ with respect to inclusion*

One may try to relate the two notions of finiteness; it turns out that they are equivalent under the axiom of choice (or, for that matter under anyone of its equivalent statements). To this end we prove first that

Proposition 1.18. *Every natural number n is finite in the Tarski sense TF, hence, each set finite in the Dedekind sense is finite in the Tarski sense.*

Proof. We introduce a predicate *Tfin(k)* read k *is finite in the Tarski sense*. We check that the set $A = \{n \in N : Tfin(n)\}$ satisfies premises of the principle of mathematical induction.

Clearly, $0 \in A$. Assuming that $n \in A$, we prove that $n + 1 \in A$. We have $2^{n+1} = 2^n \cup \{X \cup \{n+1\} : X \in 2^n\}$. Now, for a set $Y \subseteq 2^{n+1}$, either $Y \subseteq 2^n$ in which case in Y there exists a maximal element or it is not true that $Y \subseteq 2^n$ in which case in the set $Y_1 = \{B \in 2^n : B \cup \{n + 1\} \in Y\}$ there exists a maximal set B_0. Then $B_0 \cup \{n + 1\}$ is a maximal set in Y. It follows that $n+1 \in A$ and $A = N$ by the principle of mathematical induction (Proposition 1.11) so each natural number is finite in the Tarski sense.

Now, if X is finite in the Dedekind sense, $|X| = n$ being witnessed by $f : X \to n$, the sets $2^X, 2^n$ are in bijective correspondence via $F : Y \to f(Y)$, hence 2^X has the Tarski property □

The converse fact that any set finite in the Tarski sense is finite in the Dedekind sense requires the use of axiom of choice. First, we have to say what an infinite set is. We say that X is an *infinite set* in either sense, when it is not finite in this sense. We begin with

Proposition 1.19. *Under axiom of choice: if a set X is not finite in the sense of Dedekind, then X contains a copy of the set N of natural numbers.*

Proof. Assume that X is not Dedekind–finite so no function $f : n \to X$ can be onto (otherwise, we could pick, by the axiom of choice, a point c_x from any non–empty set $f^{-1}(x)$ and then the function $g : X \to n$ defined via the formula $g(x) = c_x$ would be an embedding of X into n so $|X| = m < n$ by Proposition 1.12 , a contradiction to infinity of X. Let s be a selector on the set $2^X \setminus \{\emptyset\}$ of all non–empty subsets of X. We define a function $h : N \to X$ defining the value $h(n)$ by induction on n. We assign $h(0) = x_0$ where x_0 is an arbitrarily chosen element in X. Assume that we have defined values $h(0), h(1), ..., h(n)$ in such a way that $h(i) \neq h(j)$ whenever $i \neq j$. Then we let $h(n+1) = s(X \setminus \{h(0), ..., h(n)\})$ so $h(n+1)$ is distinct from any of $h(0), .., h(n)$. By the principle of mathematical induction (Proposition 1.11), a function $h : N \to X$ is defined with $h(i) \neq h(j)$ when $i \neq j$ i.e. h embeds N into X and $h(N)$ is a copy of the set N of natural numbers   □

The proposition follows stating that each Tarski–finite set is Dedekind–finite.

Proposition 1.20. *Under axiom of choice, each set finite in the Tarski sense is finite in the Dedekind sense.*

Proof. We prove this fact by showing that any set which is Dedekind–infinite is also Tarski–infinite. Assume thus that X is Dedekind infinite. By Proposition 1.19, X contains a copy Y of the set N of natural numbers so $Y = \{x_1, x_2, ..., x_n, ...\}$. Consider the family $Y_n = \{\{x_1, x_2, ..., x_n\} : n \in N\}$ of subsets of X. Clearly, this family has no maximal element hence X is no Tarski–finite set   □

Proposition 1.21. *Under axiom of choice both kinds of finiteness coincide.*

We will from now on simply discuss *finite sets* without referring to the kind of finiteness. It also follows from our discussion that any infinite set has to contain a copy of the set of natural numbers. As a particular case, the set N itself is infinite.

1.9 Equipotency

We have defined finite sets as sets that map bijectively onto a natural number. This notion may be extended, viz., for two sets X, Y we say that X and Y are *equipotent* whenever there exists a bijection $f : X \to Y$. We are justified in such statement as clearly the relation of equipotency is *symmetric*: if $f : X \to Y$ is a bijection of X onto Y then $f^{-1} : Y \to X$ is a bijection of Y onto X. It is also evident that any set X is equipotent with itself, and that the relation of equipotency is *transitive*: if X, Y are equipotent and Y, Z are equipotent then X, Z are equipotent because of the evident fact that if $f : X \to Y, g : Y \to Z$ are bijections then the composition $g \circ f : X \to Z$ is also a bijection. In case X, Y are equipotent, we say that their *cardinalities* coincide, in symbols $|X| = |Y|$.

We may attempt at ordering a given set of cardinalities: we may say that $|X| \leq |Y|$ if there exists an injection $f : X \rightarrow Y$. Then clearly, $|X| \leq |X|$, and from $|X| \leq |Y|, |Y| \leq |Z|$ it follows that $|X| \leq |Z|$. So, the only question that remains, if we would like to claim that the relation $\leq$ is an ordering on any family of cardinalities of sets, is that whether this relation is weakly symmetrical. It turns out to be such and this fact is by no means trivial and requires a proof.

Proposition 1.22. *(the Cantor [6]–Bernstein quoted in Borel [5] theorem) If $|X| \leq |Y|$ and $|Y| \leq |X|$, then $|X| = |Y|$.*

Proof. We apply in the proof the Knaster–Tarski fixed point theorem (Proposition 1.7). Assume that $|X| \leq |Y|$ and $|Y| \leq |X|$ so injections $f : X \rightarrow Y$ and $g : Y \rightarrow X$ exist.

Consider the function $h : 2^X \rightarrow 2^X$ defined as follows: for $A \subseteq X$

$$h(A) = g[Y \setminus f(X \setminus A)] \tag{1.46}$$

It is easy to check that h is isotone: if $A \subseteq B$ then $X \setminus B \subseteq X \setminus A$ hence $f(X \setminus B) \subseteq f(X \setminus A)$ so

$$Y \setminus f(X \setminus A) \subseteq Y \setminus f(X \setminus B)$$

and hence

$$g[Y \setminus f(X \setminus A)] \subseteq g[Y \setminus f(X \setminus B)]$$

As the power set 2^X is completely ordered by inclusion, by the Knaster–Tarski theorem it follows that for some $C \subseteq X$ we have $h(C) = C$, i.e.,

$$g[Y \setminus f(X \setminus C)] = C$$

This means that f maps $X \setminus C$ onto the set B such that $g(Y \setminus B) = C$.

We may define then a function $k : X \rightarrow Y$ by the recipe

$$k(x) = \begin{cases} f(x) \text{ when } x \in X \setminus C \\ g^{-1}(x) \text{ when } x \in C \end{cases} \tag{1.47}$$

where the inverse g^{-1} is defined for the restriction $g|B$ of the function g to the set B. As both f, g are injections, it is clear that k is a bijection of X onto Y so X, Y are equipotent, i.e., $|X| = |Y|$ ☐

We now have proved that the relation $\leq$ is an ordering on any set of cardinalities. We denote in particular the cardinality $|N|$ of the set of natural numbers N by the symbol ω. One may ask whether there exist sets whose cardinalities exceed that of ω i.e. about *degrees of infinity*. The classical result which has told us that there are more potent sets that N and at the same time has posed very difficult questions about cardinal numbers is the *Cantor diagonal theorem*.

Proposition 1.23. *(Cantor) For no set X there exists a function $f : X \to 2^X$ from X onto 2^X.*

Indeed, it is sufficient to consider the subset $Y = \{x \in X : x \notin f(x)\}$. Assume that $Y = f(y)$ for some $y \in X$. Then: if $y \in Y$ then $y \notin f(y)$ i.e. $y \notin Y$ and vice versa, if $y \notin Y$ then $y \in f(y)$ i.e. $y \in Y$. Finally we get: $y \in Y$ if and only if $y \notin Y$, a contradiction. Thus, Y cannot be the value of f and f cannot map X onto 2^X.

The upshot of this result is that no set X can be equipotent with its power set 2^X. In particular we have $|2^N| > \omega$ as clearly N embeds into 2^N via the function $f(n) = \{n\}$. We may form higher and higher cardinalities by looking at sets $2^N, 2^{2^N}, ...$ For our purposes, it will be enough to stay with finite sets or sets equipotent with ω which all together are called *countable sets*.

1.10 Countable Sets

To make the realm of countable sets more familiar to us, we should give some examples of countable sets besides already known to us finite sets and N itself. In particular, a question may arise about the cardinality of Cartesian products $N \times N$, $N \times N \times N$, and in general, about the cardinality of powers $N^k = N_1 \times N_2 \times ... \times N_k$ where $N_i = N$ for $i = 1, 2, .., k$. It turns out that all these sets are countable being equipotent with N.

To demonstrate this one should exhibit some bijections among these sets and N. We begin with the square N^2.

Proposition 1.24. *The function $f(m,n) = \frac{(m+n)\cdot(m+n+1)}{2} + m$ when $(m,n) \neq (0,0)$ and $f(0,0) = 0$ is a bijection from N^2 onto N.*

Proof. The function f is injective: consider points of the form (p, q) where $p, q \in N$ in the Euclidean plane: each (p, q) lies in the line $x + y = p + q$ and each line $x + y = c$ contains $c + 1$ such integer points, viz., from $(c, 0)$ to $(0, c)$. It follows that $(p+q) \cdot (p+q+1)/2 + p$ is the number of segments connecting neighboring integer points on lines of the form $x + y = c$ where $c = 0, 1, 2, ...$ necessary to reach the point (p, q) from $(0, 0)$. So if $f(m, n) = f(p, q)$ then necessarily $m = p, n = q$.

The function f is onto: by the above interpretation, for any k, going k segments the way as described above brings us to some (p, q). One may give an explicit formula for the inverse of f: for a natural number r, we have $f^{-1}(r) = (p, q)$ where p, q follow uniquely from the two equations

$$\begin{cases} p + q = \lfloor (\lfloor (8r + 1)^{\frac{1}{2}} \rfloor + 1)/2 \rfloor - 1 \\ 3p + q = 2r - (\lfloor (\lfloor (8r + 1)^{\frac{1}{2}} \rfloor + 1)/2 \rfloor - 1)^2 \end{cases} \tag{1.48}$$

To get these equations, it suffices to let $r = \frac{(p+q)\cdot(p+q+1)}{2}$ and observe that $8r + 1 = (2p+2q+1)^2 + 8p$ from which it follows that $2p + 2q + 1 \leq (8r+1)^{\frac{1}{2}} \leq 2p + 2q + 3$ implying (1.48).

It follows that f does establish the equipotency of N^2 with N. We may write down $f^{-1}(r) = (k(r), l(r))$ where k, l are functions which satisfy $k(r) = p, l(r) = q$ in the above equations.

We may iterate the above bijection: given the set N^k of ordered k–tuples of natural numbers we may define inductively a bijection f_k from N^k onto N, viz., assuming $f_1, ..., f_k$ defined, we define

$$f_{k+1}(n_1, n_2, .., n_{k+1}) = f(f_k(n_1, n_2, .., n_k), n_{k+1})$$

where f is the bijection of Proposition 1.24. One may prove by induction that all f_k are bijections. Thus, for any $k \in N$, the set N^k of all k–tuples of natural numbers is countable $\qquad\qquad\qquad\qquad\qquad\qquad\qquad\qquad\qquad\qquad\quad\square$

This result may be carried yet further: we may consider the set

$$N^* = \bigcup \{N^k : k \in N\}$$

of all finite tuples of natural numbers (sometimes called from the grammatical point of view *strings*, or *words*). As for each k the function $f_k : N^k \rightarrow N$ is a bijection, and sets N^k are pair–wise disjoint, we may construct a bijection h of N^* onto N by first defining a function $g : N^* \rightarrow N \times N$, by letting $g(n_1, ..., n_k) = (f_k(n_1, ..., n_k), k)$ for each $k \in N$ and each string $(n_1, ..., n_k) \in N^k$ and then defining $h(n_1, ..., n_k) = f(g(n_1, ..., n_k))$.

Thus, the set of all finite tuples (strings) of natural numbers is countable.

A step further in this direction would be the ultimate result that a union of countably many countable sets is countable. It is indeed so, but the proof will require the axiom of choice.

Proposition 1.25. *Under axiom of choice, the union of a countable set X whose elements are countable sets is countable.*

As X is countable, we may enumerate its elements as $\{X_n : n \in N\}$ (where in case X be finite, $X_n = X_{n+1} = X_{n+2} = ...$ for some n). For each n, the set B_n of all injections from X_n onto N is non–empty so by the axiom of choice we may select a function $k_n \in B_n$ for each n. For $S = \bigcup X$, we look for each $x \in S$ at the least $n(x)$ such that $x \in X_{n(x)}$. We define a function $h : S \rightarrow N^2$ by letting $h(x) = (k_{n(x)}(x), n(x))$; clearly, h is an injection so the composition $f \circ h$ embeds S into N.

1.11 Filters and Ideals

In ordered sets, one may single out certain subsets regular with respect to the ordering relation. Among them of primary interest are *ideals* and *filters*.

Here we restrict ourselves to ideals and filters in fields of sets ordered by inclusion. Our primary field of sets will be that of 2^X for a non–empty set X. The notions of a *filter* and of an *ideal* are dual to each other in the sense that if a family $\mathcal{F}$ is a filter then the family $\mathcal{I}_\mathcal{F} = \{X \setminus A : A \in \mathcal{F}\}$ is an ideal (and vice versa). Hence, we restrict ourselves to filters as the exposition of ideals follows by duality.

We consider a field of sets $\mathcal{B} \subseteq 2^X$. A non–empty family $\mathcal{F} \subseteq \mathcal{B}$ is a *filter* when

F1 $A, B \in \mathcal{F} \Rightarrow A \cap B \in \mathcal{F}$

F2 $A \subseteq B \wedge A \in \mathcal{F} \Rightarrow B \in \mathcal{F}$

F3 $\emptyset \notin \mathcal{F}$

A filter is then a family which is closed with respect to the intersection and taking a superset and is distinct from $\mathcal{B}$ as it cannot have $\emptyset$ as an element. In particular, $X \in \mathcal{F}$ by F2. Thus, by duality, an ideal is a family of sets in $\mathcal{B}$ closed with respect to unions and taking subsets so any ideal contains $\emptyset$ as an element and does not contain X.

As an example of a filter, we may take (in case $\mathcal{B} = 2^X$) for any $x \in X$ the set $\mathcal{F}_x = \{A \in 2^X : x \in A\}$ which clearly satisfies F1, F2, and F3 called the *principal filter induced by* x. By duality, *the principal ideal induced by* x is the ideal $\mathcal{I}_x = \{A \in \mathcal{B} : x \notin A\}$.

We may notice an important property of a principal filter (respective of a principal ideal): given any filter $\mathcal{F}$ with $\mathcal{F}_x \subseteq \mathcal{F}$ (respectively any ideal $\mathcal{I}$ with $\mathcal{I}_x \subseteq \mathcal{I}$), we have $\mathcal{F}_x = \mathcal{F}$ (respectively $\mathcal{I}_x = \mathcal{I}$).

Indeed, assuming that $\mathcal{F}_x \subset \mathcal{F}$ we may pick a set $A \in \mathcal{F} \setminus \mathcal{F}_x$ hence $x \notin A$ but then $\{x\} \cap A = \emptyset$ contrary to F3. Thus, $\mathcal{F}_x$ is a maximal with respect to inclusion filter; maximal filters are called also *ultrafilters*. Similarly, the ideal $\mathcal{I}_x$ is a *maximal ideal*. Principal filters and ideals are the only examples of maximal filters respectively ideals which we may construct effectively. In all other cases, to justify the existence of an ultrafilter or a maximal ideal we have to resort to the maximum principle.

Proposition 1.26. *By the maximum principle, any filter $\mathcal{F}$ may be extended to an ultrafilter $\mathcal{F}^* \supseteq \mathcal{F}$.*

Proof. Consider the set $\mathbf{F}$, of all filters on a field of sets $\mathbf{B}$ containing the filter $\mathcal{F}$ as a subset, ordered by inclusion. It remains to check that the union $\mathcal{S} = \bigcup \mathbf{L}$ of any chain $\mathbf{L}$ of filters in $\mathbf{F}$ is again a filter. To this end, consider $A, B \in \mathcal{S}$: there are filters $F, G \in \mathbf{L}$ such that $A \in F, B \in G$. By linearity of ordering on $\mathbf{L}$, either $F \subseteq G$ or $G \subseteq F$. In the former case $A, B \in F$, in the latter case $A, B \in G$ so either $A \cap B \in F$ or $A \cap B \in G$ and in either case $A \cap B \in \mathcal{S}$ proving the condition (i). Similarly (ii, (iii) may be proved so we conclude that $\mathcal{S}$ is a filter. Clearly, $\mathcal{S} = sup\mathbf{L}$. Thus, any chain in $\mathbf{F}$ has a supremum and the maximum principle implies the existence of a maximal element in $\mathbf{F}$ which is clearly an ultrafilter extending $\mathcal{F}$ □

The dual result on the existence of a maximal ideal extending a given ideal follows by duality.

As one more important example of a filter, we may consider the set N of natural numbers and the family $cof N = \{A \subseteq N : N \setminus A \text{ is } finite\}$ i.e. the family of all co–finite subsets in N. That $cof N$ is a filter follows easily as the union of any two finite sets is finite and any subset of a finite set is finite. The filter $cof N$ is called *the Fréchet filter*. By the last proposition, there exists on N an ultrafilter $cof N^*$ extending the Fréchet filter.

We give now a characterization of ultrafilters in terms of set operations.

Proposition 1.27. *For a filter $\mathcal{F}$, the following are equivalent*

1. $\mathcal{F}$ is an ultrafilter;

2. If a subset $Y \subseteq X$ is such that $Y \cap A \neq \emptyset$ for each $A \in \mathcal{F}$, then $Y \in \mathcal{F}$;

3. For any A, B, if $A \cup B \in \mathcal{F}$, then either $A \in \mathcal{F}$ or $B \in \mathcal{F}$;

4. For any A either $A \in \mathcal{F}$ or $X \setminus A \in \mathcal{F}$.

Proof. Property 1 implies Property 2: assume $\mathcal{F}$ an ultrafilter and $Y \cap A \neq \emptyset$ for each $A \in \mathcal{F}$. Was $Y \notin \mathcal{F}$, we could define $\mathcal{F}^*$ by adding to $\mathcal{F}$ all sets containing as a subset an intersection of the form $Y \cap A$ where $A \in \mathcal{F}$. Then clearly, $\mathcal{F}^*$ does satisfy F1, F2 and F3, so $\mathcal{F}^*$ is a filter. Also $\mathcal{F} \subset \mathcal{F}^*$ as $Y \in \mathcal{F}^* \setminus \mathcal{F}$ but this is a contradiction as $\mathcal{F}$ is an ultrafilter.

Property 2 implies Property 3: assume $A \cup B \in \mathcal{F}$ but neither $A \in \mathcal{F}$ nor $B \in \mathcal{F}$. By Property 2 there exist $C, D \in \mathcal{F}$ with $C \cap A = \emptyset = D \cap B$. But then $(A \cup B) \cap (C \cap D) = \emptyset$ and $C \cap D \in \mathcal{F}$, a contradiction with F1.

Property 3 implies Property 4: as $A \cup X \setminus A = X \in \mathcal{F}$ either $A \in \mathcal{F}$ or $X \setminus A \in \mathcal{F}$ by Property 3.

Property 4 implies Property 1: assume to the contrary that $\mathcal{F}$ is not an ultra-filter so $\mathcal{F} \subset \mathcal{G}$ for some filter $\mathcal{G}$ hence there is $A \in \mathcal{G} \setminus \mathcal{F}$ i.e. $A \notin \mathcal{F}$. But also $X \setminus A \notin \mathcal{F}$: otherwise, we would have $X \setminus A \in \mathcal{G}$ and $A \cap X \setminus A = \emptyset \in \mathcal{G}$, a contradiction with F3 □

1.12 Equivalence Relations

Tasks of classification require that elements in a universe be clustered into aggregates within which the elements are identical or similar with respect to some properties whereas elements in distinct aggregates are clearly discernible with respect to some of those properties.

A simplest class of classifying relations are *equivalence relations*; we call a relation R on a set X an *equivalence relation* if and only if R is reflexive, symmetric and transitive.

We recall the basic property of equivalence relations concerned with clustering of elements of X into classes of indiscernible objects. For an element $x \in X$, we let

$$[x]_R = \{y \in X : xRy\} \tag{1.49}$$

and we call $[x]_R$ the *equivalence class of x with respect to R*. Then we have

Proposition 1.28. *For each pair $x, y \in X$, classes $[x]_R, [y]_R$ are either disjoint or equal.*

We include a proof of this well–known fact. If $z \in [x]_R \cap [y]_R$, then by symmetry xRz, yRz, hence, xRz, zRy and by transitivity xRy; for $w \in [x]_R$, we have wRx, hence, wRy and it follows that $[x]_R \subseteq [y]_R$. By symmetry, $[y]_R \subseteq [x]_R$ and finally $[x]_R = [y]_R$.

We will denote the set $\{[x]_R : x \in X\}$ by the symbol X/R and we will call it the *quotient set* of the set X by the relation R. There is a function $q_R : X \rightarrow X/R$ defined by letting $q_R(x) = [x]_R$, and called the *quotient function* which maps X onto the quotient set X/R.

A dual way of introducing equivalence relations is by means of *partitions*. A *partition* on a set X is a family $\mathcal{P}$ of its subsets such that

P1 $\mathcal{P}$ *is non–empty*

P2 $\bigcup \mathcal{P} = X$

P3 *If $A, B \in \mathcal{P}$, then $A \cap B = \emptyset$*

By the last Proposition and the fact implied by reflexivity of R that $x \in [x]_R$, it follows that equivalence classes of the relation R form a partition $\mathbf{P}_R$ of the set X.

On the other hand, any partition $\mathbf{P}$ of X induces an equivalence relation R_P, viz., we let xR_Py if and only if there exists a set $A \in \mathbf{P}$ such that $x, y \in A$. The correspondence between equivalence relations and partitions is canonical, i.e.,

$$\mathbf{P}_{R_P} = \mathbf{P}, R_{\mathbf{P}_R} = R \tag{1.50}$$

We may also compare equivalence relations on the set X: for equivalence relations R, S on X, we say that R is *finer* than S in symbols $R \prec S$ if and only if xRy implies xSy for any pair x, y of elements of X. We have obvious

Proposition 1.29. $R \prec S$ *if and only if $[x]_R \subseteq [x]_S$ for each $x \in X$.*

We say in case $R \prec S$ that the partition $\mathbf{P}_R = \{[x]_R : x \in X\}$ is a *refinement* of the partition $\mathbf{P}_S = \{[x]_S : x \in X\}$.

The relation $\prec$ is a complete ordering of the set $Eq(X)$ of all equivalence relations on X.

Proposition 1.30. *For any family $\mathcal{R}$ of equivalence relations on a set X, there exist equivalence relations $inf\mathcal{R}, sup\mathcal{R}$ with respect to $\prec$, i.e., the family $Eq(X)$ of all equivalence relations on the set X is completely ordered by $\prec$.*

Clearly, the intersection $\bigcap \mathcal{R}$ of all relations in $\mathcal{R}$ is an equivalence relation hence it is $inf\mathcal{R}$. For supremum, let us consider the relation $S = \bigcup\{R \in Eq(X) : \forall A : (A \in \mathcal{R} \Rightarrow A \subseteq R)\}$. By the part already proven, S is an equivalence relation and for any relation $R \in Eq(X)$ with $A \subseteq R$ for each $A \in \mathcal{R}$ we have $S \subseteq R$ so $S = sup\mathcal{R}$.

Assuming that $R \prec S$, we may define on the quotient set X/R a new relation S/R by letting $[x]_R S/R[x']_R$ if and only if xSx'. It is evident that this definition does not depend on the choice of elements in $[x]_R, [x']_R$. It is also easy to see that $[[x_R]]_{S/R}=[x]_S$. We may write down this fact in a concise form as

$$(X/R)/(S/R) = X/S \qquad (1.51)$$

Given equivalence relations $R \subseteq X \times X$ and $S \subseteq Y \times Y$, a function $f : X \to Y$ is an $R, S-morphism$ if the condition $xRy \Rightarrow f(x)Sf(y)$ holds. Then we have a function $\overline{f} : X/R \to Y/S$ with $\overline{f}([x]_R) = [f(x)]_S$; indeed, if zRx then $f(z)Sf(x)$ hence $f([x]_R) \subseteq [f(x)]_S$ and $\overline{f}$ is defined uniquely. Let us observe that $q_S \circ f = \overline{f} \circ q_R$.

1.13   Tolerance Relations

The idea of similarity is a weakening of the idea of equivalence; a similarity may be rendered in many ways leading to various formal notions. One of these ways is related to problems of visual perception in which a similarity of objects may be interpreted as inability to distinguish between them, e.g., the distance between the objects may be smaller then a given discernibility threshold: if points x, y in the real line are regarded as *similar* when their distance is less than δ then we may have x, y, and y, z similar while x, z need not be similar, see Poincaré [19].

It follows that such a relation of similarity lacks the transitivity property. The idea of a *tolerance relation* was introduced formally in Zeeman [29] to capture this intuitive understanding of similarity, cf., also, Menger [16], cf., Nieminen [18], Polkowski et al. [20]. Following Zeeman, we call a relation R on a set X a *tolerance relation* when R is (i) reflexive (ii) symmetric. Thus a tolerance relation lacks the transitivity property of equivalence relations. This makes the analysis of tolerance relations more difficult and their structure more intricate.

If we have tried to mimic the case of equivalence relations, then we would look at the sets $R(x)=\{y \in X : xRy\}$. However, due to the lack of transitivity, we cannot say for instance whether two elements $y, z \in R(x)$ are in the relation R and a fortiori there is nothing essential we may say about the structure of $R(x)$. So, we should with equivalence classes in mind, consider sets with the property that any two of its elements are in the relation R. Such sets are called *tolerance pre–classes*.

Formally , a set $A \subseteq X$ is a tolerance pre–class when xRy for any pair $x, y \in A$. In particular, any singleton $\{x\}$ is a tolerance pre–class, and if xRy then $\{x, y\}$ is a tolerance pre–class.

We may easily see that any union of a linearly ordered by inclusion family of tolerance pre—classes is a tolerance pre–class. From the maximum principle it follows that

Proposition 1.31. *Any tolerance pre–class A is a subset of a maximal with respect to inclusion tolerance pre–class.*

Maximal tolerance pre–classes are called *tolerance classes*. Within a tolerance class C, any two elements are in the relation R, and, for each element $y \notin C$, there is $x \in C$ such that $\neg yRx$.

An example of a tolerance relation is the following: given a family of non–empty sets $\mathcal{F}$, define a relation $\tau_\cap$ on $\mathcal{F}$ as

$$A\tau_\cap B \ if \ and \ only \ if \ A \cap B \neq \emptyset \qquad (1.52)$$

$\tau_\cap$ is clearly a tolerance relation. Its classes are maximal *2–centered* sub–families of $\mathcal{F}$, where a sub–family $F' \subseteq F$ is *centered* if for any finite $G \subseteq F'$ the intersection $\bigcap G \neq \emptyset$ and being *k–centered* means the specialization of the preceding definition to G consisting of at most k elements.

The relation $\tau_\cap$ is archetypical for tolerance relations: it turns out that any tolerance relation can be represented in that form. To see this, for a tolerance relation R on X, we denote by the symbol $\mathcal{C}$ the family of all tolerance classes of R and we let

$$\mathcal{C}_x = \{C \in \mathcal{C} : x \in C\}$$

We have, see Shreider [23]

Proposition 1.32. *For $x, y \in X$, we have xRy if and only if $\mathcal{C}_x \cap \mathcal{C}_y \neq \emptyset$.*

Proof. Indeed, in case xRy, the pre–class $\{x, y\}$ extends to a tolerance class $C \in \mathcal{C}$ and $C \in \mathcal{C}_x \cap \mathcal{C}_y$. Conversely, in case $\mathcal{C}_x \cap \mathcal{C}_y \neq \emptyset$, there exists a tolerance class C with $x, y \in C$ and thus xRy by the definition of a tolerance class   □

We may now consider a way to embed a tolerance relation into an equivalence relation. To this end, given a relation R on a set X, we define the *n–th power R^n* of R as the composition $R \circ R \circ R \circ .... \circ R$ of n copies of R. Clearly, xR^ny holds if and only if there exist $x = x_0, x_1, ..., x_n = y$ with x_iRx_{i+1} for $i = 0, .., n - 1$. Then, we define the *transitive closure R^+* of the relation R as the union $\bigcup\{R^n : n \in N\}$. Clearly, $R \subseteq R^+$ and R^+ is transitive: given xR^+y, yR^+z, we have xR^ny, yR^mz for some $n, m \in N$ and then $xR^{n+m}z$. It follows from the construction of R^+ that it is the least transitive relation extending R.

We may notice that

Proposition 1.33. *For a tolerance relation R, the transitive closure R^+ is an equivalence relation. For any $x \in X$, the union of tolerance classes of R containing x is a subset of the equivalence class of x by R^+: $\bigcup C_x \subseteq [x]_{R^+}$.*

Proof. As R^+ is reflexive since $R \subseteq R^+$ and symmetric by the symmetry of R, R^+ is an equivalence relation because we have checked its transitivity. For any tolerance class $C \ni x$, $C \times C \subseteq R$ so $C \times C \subseteq R^+$ i.e. $C \subseteq [x]_{R^+}$. It follows that $\bigcup C_x \subseteq [x]_{R^+}$ □

The relation R^+ is constructed by inflating tolerance classes until they become disjoint or identical; we may think also of a way of constructing an equivalence by restricting tolerance classes. To this end, we first introduce a new notion: given a family $\mathcal{F}$ of subsets of a set X, we consider a new family of sets constructed in the following way. For any sub–family $\mathcal{G} \subseteq \mathcal{F}$, we let $A_G = \bigcap \mathcal{G} \setminus \bigcup (\mathcal{F} \setminus \mathcal{G})$. We let $\mathcal{C}(\mathcal{F}) = \{A_G : \mathcal{G} \subseteq \mathcal{F}\}$. Sets of the form A_G are called *components* of $\mathcal{F}$. Let us observe that

Proposition 1.34. *For $\mathcal{G}, \mathcal{G}' \subseteq \mathcal{F}$, if $\mathcal{G} \neq \mathcal{G}'$ then $A_G \cap A_{G'} = \emptyset$. Thus the family $\mathcal{C}(\mathcal{F})$ of components of $\mathcal{F}$ is a partition of $Y = \bigcup \mathcal{F}$.*

Indeed, for $\mathcal{G} \neq \mathcal{G}'$, let e.g. $A \in \mathcal{G} \setminus \mathcal{G}'$. Then $A_G \subseteq A$ and $A_{G'} \cap A = \emptyset$. Hence $A_G \cap A_{G'} = \emptyset$.

We now apply the idea of components to the family $\mathcal{C}_R$ of tolerance classes of a tolerance relation R on a set X.

Proposition 1.35. *Components in the family $\mathcal{C}(\mathcal{C}_R)$ are equivalence classes of the equivalence relation E defined by the condition xEy if and only if $zRx \Leftrightarrow zRy$ for each z.*

Proof. The relation E is clearly an equivalence (we have xEy if and only if $R(x) = R(y)$). Assume xEy. Then for any tolerance class $C \ni x$, $y \in C$ as yRz for any $z \in C$. Obviously the converse holds by symmetry of R: for any tolerance class $C \ni y$, we have $x \in C$. Hence for $\mathcal{G} = \{C \in \mathcal{C}_R : x \in C\}$ we have $\mathcal{G} = \{C \in \mathcal{C}_R : y \in C\}$. It follows that x, y belong to the same component A_G. Conversely, if x, y belong to the same component A_G, then given xRz we have a tolerance class C with $x, z \in C$; but $y \in C$ hence yRz. By symmetry, if yRz then xRz. It follows that xEy □

1.14 A Deeper Insight into Lattices and Algebras

We know from section 1.5 that a lattice is a set X with two operations, of the *meet*, $x \cap y$, and the *join*, $x \cup y$, which satisfy the postulates

L1 $x \cup y = y \cup x$

L2 $x \cup (y \cup z) = (x \cup y) \cup z$

L3 $x \cap y = y \cap z$

L4 $x \cap (y \cap z) = (x \cap y) \cap z$

L5 $x \cap (y \cup z) = x = x \cup (x \cap y)$

We know from section 1.5 that such structures arise in ordered sets $(X, \leq)$ with the property that $g.l.b.(x,y)$ and $l.u.b.(x,y)$ exist for each pair x, y. Conversely, in each lattice $(X, \cup, \cap)$, an ordering $\leq_L$ can be introduced by means of

$$x \leq_L \; \Leftrightarrow \; x \cup y = y \; \Leftrightarrow \; x \cap y = x \qquad (1.53)$$

and, moreover, in the canonical way, i.e., the meet $\cap_L$ and the join $\cup_L$ introduced by $\leq_L$ coincide with $\cap$ and $\cup$, respectively.

The *unit* element, 1, is defined by the property that

$$x \cup 1 = 1 \; for \; each \; x \in X \qquad (1.54)$$

i.e, $x \leq_L 1$ for each $x \in X$.

Dually, the *zero* element 0 is introduced by means of the postulate that

$$x \cap 0 = 0 \; for \; each \; x \in X \qquad (1.55)$$

i.e., $0 \leq_L x$ for each $x \in X$.

A lattice with 0 and 1 will be called a 0,1–lattice.

The notions of a filter and an ideal, known for fields of sets, can be defined in the more general realm of lattices.

A *filter* on a lattice $(X, \cup, \cap)$ is a subset $\mathcal{F} \subseteq X$, satisfying postulates

F1 $x, y \in \mathcal{F} \Rightarrow x \cap y \in \mathcal{F}$

F2 $x \in \mathcal{F}, x \leq_L y \Rightarrow y \in \mathcal{F}.$

In each 0,1–lattice, $1 \in \mathcal{F}$. If $0 \in \mathcal{F}$, then $\mathcal{F} = X$, and $\mathcal{F}$ is *improper*.

The corresponding dual postulates with replacement of $\cap$ by $\cup$ and $\leq_l$ by $\geq_L$ define an ideal. The theory of ideals is then dual to theory of filters: every statement about filters becomes a statement about ideals after replacement of $\cap$ by $\cup$, $\cup$ by $\cap$, $\leq_L$ by $\geq_L$, and 1 by 0.

The notion of a *maximal filter* is defined as in sect. 1.5.

A filter $\mathcal{F}$ is *prime* if and only if it satisfies the postulate

$$x \cup y \in \mathcal{F} \Rightarrow x \in \mathcal{F} \; \vee \; y \in \mathcal{F} \qquad (1.56)$$

A lattice $(X, \cup, \cap)$ is *distributive* if and only if it satisfies the postulate

$$x \cap (y \cup z) = (x \cap y) \cup (x \cap z) \qquad (1.57)$$

and then obviously its dual: $x \cup (y \cap z) = (x \cup y) \cap (x \cup z)$.

In distributive lattices

Proposition 1.36. *Every maximal filter is prime.*

Proof. Assume a maximal filter $\mathcal{F}$ is not prime. There are x, y with $x \cup y \in \mathcal{F}$ and $x \notin \mathcal{F}$, and $y \notin \mathcal{F}$. For each $z \in x \cap z$, $(x \cup y) \cap z = (x \cap z) \cup (y \cap z) \in \mathcal{F}$, hence, $x \cap z \in \mathcal{F}$ or $y \cap z \in \mathcal{F}$.

Consider $\mathcal{G} = \{w \in X : x \cap z \leq_l w \vee y \cap z \leq_L w\}$. $\mathcal{G}$ is a filter, $x, y \in \mathcal{G}$, and $\mathcal{F} \subseteq \mathcal{G}$, contrary to maximality of $\mathcal{F}$ □

A subclass of distributive lattices are relatively pseudo–complemented lattices; a lattice X is *relatively pseudo–complemented* if for each pair $x, y \in X$, there exists an element $x \Rightarrow y$ called the *pseudo–complement of x relative to y* which satisfies the postulate

$$x \Rightarrow y \geq_L z \Leftrightarrow x \cap z \leq_L y \qquad (1.58)$$

for each $z \in X$. Let us observe that (1.58) defines a Galois connection $(x \cap z, x \Rightarrow y)$, see sect. 1.4.

The existence of pseudo–complement implies some important features of a lattice

Proposition 1.37. *Assume that a lattice X is relatively pseudo–complemented. Then*

1. X has a unit element 1;

2. X is distributive.

Proof. By (1.58), $z \leq_L x \Rightarrow x$ for each z, hence, $x \Rightarrow x = 1$. Distributivity follows by (1.58): $x \cap y, x \cap z \leq_L x \cap (y \cup z)$ imply $x \cap y \cup x \cap z \leq_L x \cap (y \cup z)$. To prove the converse, observe that $x \cap y, x \cap z \leq_L x \cap y \cup x \cap z$, hence, by (1.58), $y, z \leq_L x \Rightarrow x \cap y \cup x \cap z$, which implies that $y \cup z \leq_L x \Rightarrow x \cap y \cup x \cap z$ and again (1.58) implies that $x \cap (y \cup z) \leq_L x \cap y \cup x \cap z$ □

A complement to $x \in X$ is an element $-x \in X$ such that

$$(x \cup (-x)) \cap y = y; \quad (x \cap (-x)) \cup y = y \qquad (1.59)$$

for every $y \in X$.

Thus, $x \cup (-x) = 1$ and $x \cap (-x) = 0$.

A lattice is *complemented* if and only if each element has a complement.

A Boolean algebra is a complemented $0, 1$–distributive lattice. The complement acts in Boolean algebras as dualizer: it turns every true statement into its dual by replacements: $\cup/\cap$, $\cap/\cup$, $x/-x$, $0/1$.

In Boolean algebras, filters have the same properties as in fields of sets of sect. 1.5.

Proposition 1.38. *In every Boolean algebra, each prime filter $\mathcal{F}$ is maximal.*

Indeed, as $1 = x \cup (-x) \in \mathcal{F}$, either x or $-x$ belongs in $\mathcal{F}$ for each $x \in X$; but this characterizes maximal filters.

We let $x \Rightarrow y$ to stand for $-x \cup y$. Let us observe a far analogy of this definition with the tautology $p \Rightarrow q \Leftrightarrow \neg p \vee q$ of the propositional calculus. We define the relation $\sim_F$ for a maximal filter $\mathcal{F}$ by letting,

$$x \sim_F y \Leftrightarrow\ x \Rightarrow y \in \mathcal{F} \wedge y \Rightarrow x \in \mathcal{F} \qquad (1.60)$$

The relation $\sim_F$ is an equivalence: the condition on the right–hand side of (1.60) is equivalent to the condition

C *either* $x, y \in \mathcal{F}$ *or* $-x, -y \in \mathcal{F}$

Hence, $x \sim_F x$ holds for every $x \in X$ and $\sim_F$ is symmetric. Transitivity follows easily by C.

Clearly, $x \in \mathcal{F}$ if and only if $x \sim_F 1$ and $x \notin \mathcal{F}$ if and only if $x \sim_F 0$. Thus, the quotient set $X/ \sim_F$ contains only classes $[1]_{\sim_F}$ and $[0]_{\sim_F}$.

An *atom* in a Boolean algebra X is an element $a \neq 0$ with the property

$$y \leq_L a \Rightarrow y = 0 \vee y = a \qquad (1.61)$$

A Boolean algebra X is *atomic* if for every $y \in X$ there exists an atom a with $a \leq y$, see Tarski [26].
By $At(X)$, we denote the set of all atoms in a Boolean algebra X. It is obvious that if $a, b \in At(X)$, then $a \cap b = 0$.

For $x, y \in X$, if $x \neq y$, then, e.g., $x \cap -y \neq 0$; if X is atomic, then there is an atom a with $a \leq_L x \cap -y$, hence $a \leq x$ and $a \cap y = 0$. It follows that distinct elements in an atomic Boolean algebra can be separated by an atom. Moreover, for each element x, $x = sup\{a \in At(X) : a \leq_L x\}$.

Hence, the mapping $A : X \to 2^{At(X)}$ embeds the atomic Boolean algebra X into the field of sets $2^{At(X)}$.

A subset $D \subseteq X \setminus \{0\}$ is *dense* in X if and only if for each element $x \in X$ there exists $d \in D$ such that $d \leq_L x$. Clearly, $At(X) \subseteq D$ for each dense D.

Proposition 1.39. *(Rasiowa and Sikorski [22]) If $D_n : n = 1, 2, ...$ is a countable sequence of dense sets in a Boolean algebra X, and $x \neq 0$ is an element of X, then there exists an ultrafilter $\mathcal{F}$ such that $x \in \mathcal{F}$ and $\mathcal{F} \cap D_n \neq \emptyset$ for each n.*

Proof. By definition of a dense set, by induction, one can find a sequence $(x_n : n = 1, 2, ...)$ such that $x_1 = x$ and $x_{n+1} \leq_L x_n$ for $n = 1, 2, ...$ $Y = \{x_n : n = 1, 2, ...\}$ extends to an ultrafilter $\mathcal{F}$ which is what is needed   □

The reader will find a deeper discussion of classical aspects of set theory in a monograph by Kuratowski and Mostowski [13]. Modern aspects are treated in Balcar and Štěpánek [2]. For a lattice–based theory of concepts, see Wille [28].

References

1. Aristotle.: Prior Analytics. Hackett Publ. Co., Indianapolis (1989)
2. Balcar, B., Štěpánek, P.: Teorie Množin. Academia, Praha (1986)
3. Birkhoff, G.: Lattice Theory, 3rd edn. AMS, Providence (1967)
4. Bocheński, I.M.: Die Zeitgönossischen Denkmethoden. A. Francke AG, Bern (1954)
5. Borel, E.: Leçons sur la Théorie des Fonctions. Paris (1898)
6. Cantor, G.: Gesammelte Abhandlungen. Julius Springer, Berlin (1932)
7. Dedekind, R.: Was sind und was sollen die Zahlen. Braunschweig (1881)
8. Fraenkel, A.: Zu den Grundlagen der Cantor–Zermeloschen Mengenlehre. Math. Annalen 86, 230–237 (1922)
9. Frege, G.: Grundgesetze der Arithmetik II. Verlag Hermann Pohle, Jena (1903)
10. Hausdorff, F.: Grundzüge der Mengenlehre (1914); von Veit, Leipzig, Hausdorff, F.: Set Theory. Chelsea, New York (1962)
11. Knaster, B.: Un théoréme sur les fonctions d'ensembles. Ann. Soc. Polon. Math. 6, 133–134 (1928)
12. Kuratowski, C.: Une méthode d'élimination des nombres transfinis des raisonnement mathematiques. Fundamenta Mathematicae 3, 89 (1922)
13. Kuratowski, C., Mostowski, A.: Set Theory. North Holland, Amsterdam (1968)
14. Łukasiewicz, J.: On Aristotle's Syllogistic (in Polish). Compt. Rend. Acad. Polon. Lettr. Cracovie 44 (1939)
15. Łukasiewicz, J.: Aristotle's Syllogistic from the Standpoint of Modern Formal Logic, 2nd edn. Oxford University Press, Oxford (1957)
16. Menger, K.: Statistical metrics. Proc. Natl. Acad. Sci. USA 28, 535–537 (1942)
17. Mac Neille, H.M.: Partially ordered sets. Trans. Amer. Math. Soc. 42, 416–460 (1937)
18. Nieminen, J.: Rough tolerance equality and tolerance black boxes. Fundamenta Informaticae 11, 289–296 (1988)
19. Poincaré, H.: La Science et l'Hypothése. Flammarion, Paris (1902)
20. Polkowski, L., Skowron, A., Żytkow, J.: Tolerance based rough sets. In: Lin, T.Y., Wildberger, M. (eds.) Soft Computing: Rough Sets, Fuzzy Logic, Neural Networks, Uncertainty Management, Knowledge Discovery, pp. 55–58. Simulation Councils Inc., San Diego (1994)
21. Rasiowa, H.: Algebraic Models of Logic. Warsaw University Press, Warszawa (2001)
22. Rasiowa, H., Sikorski, R.: A proof of the completeness theorem of Gödel. Fundamenta Mathematicae 37, 193–200 (1950)
23. Shreider, Y.: Equality, Resemblance, Order. Mir Publishers, Moscow (1960)
24. Słupecki, J.: On Aristotle's Syllogistic. Studia Philosophica (Poznań) 4, 275–300 (1949–1950)

25. Tarski, A.: Sur les ensembles finis. Fundamenta Mathematicae 6, 45–95 (1924)
26. Tarski, A.: Zur Grundlegung der Booleschen Algebra. I. Fundamenta Mathematicae 24, 177–198 (1935)
27. Vaught, R.L.: On the equivalence of the axiom of choice and the maximal principle. Bull. Amer. Math. Soc. 58, 66 (1952)
28. Wille, R.: Restructuring lattice theory: An approach based on hierarchies of concepts. In: Rival, I. (ed.) Ordered Sets, pp. 445–470. Reidel, Dordrecht (1982)
29. Zeeman, E.C.: The topology of the brain and the visual perception. In: Fort, K.M. (ed.) Topology of 3-manifolds and Selected Topics, pp. 240–256. Prentice Hall, Englewood Cliffs (1965)
30. Zermelo, E.: Beweiss das jede Menge wohlgeordnet werden kann. Math. Annalen 59, 514–516 (1904)
31. Zermelo, E.: Untersuchungenüber die Grundlagen der Mengenlehre I. Math. Annalen 65, 261–281 (1908)
32. Zorn, M.: A remark on method in transfinite algebra. Bull. Amer. Math. Soc. 41, 667–670 (1935)

Chapter 2
Topology of Concepts

Topology is a theory of certain set structures which have been motivated by attempts to generalize geometric reasoning based on Euclidean distance invariants and replace it by more flexible schemes. As the notion of closeness, or, distance, permeates a plethora of reasoning schemes, topological structures are often in focus of a reasoner. In many schemes of reasoning one resorts to the idea of a *neighbor* with the assumption that reasonably selected neighbors of a given object preserve its properties to a satisfactory degree (look at methods based on the notion of the *nearest neighbor*, for instance). The notion of a neighbor as well as a collective notion of a *neighborhood* are studied by topology.

Basic topological notions were originally defined by means of a notion of *distance* which bridges geometry to more general topological structures. The analysis of properties of distance falls into theory of *metric spaces*, see Hausdorff [12] and Fréchet [9], with which we begin our exposition of fundamentals of topology.

Generalizations to abstract structures are discussed in following sections, with emphasis on basic properties of compactness, completeness, continuity, relations to algebraic structures, quotient structures, and hyperspaces.

As the last topic, an important notion of a Čech topology is discussed, which arises often in applications, where the underlying structure cannot satisfy all requirements for a topology; such is , e.g., the case of mereological structures in general.

2.1 Metric Spaces

Consider a copy $\mathbf{R}^1$ of the set of *real numbers*. The distance between real numbers is defined usually by means of the *natural metric* $|x - y|$. Denoting $|x - y|$ by $\rho(x, y)$, we may write down the essential properties of the natural metric well–known from elementary mathematics:

L. Polkowski: Approximate Reasoning by Parts, ISRL 20, pp. 45–78.
springerlink.com   © Springer-Verlag Berlin Heidelberg 2011

1. $\rho(x, y) \geq 0$ and $\rho(x, y) = 0$ if and only if $x = y$;

2. $\rho(x, y) = \rho(y, x)$;

3. $\rho(x, z) \leq \rho(x, y) + \rho(y, z)$ (the triangle inequality).

Properties 1–3 are taken as characterizing any distance function (called also a *metric*). Thus, a *metric* ρ on a set X is a function $\rho : X^2 \to R^1$ which satisfies the conditions 1–3. A set X endowed with a metric ρ is called a *metric space*. Elements of a metric space are usually called *points*.

The natural metric is an example of a metric. Other examples are

1. *The function $\rho = \sqrt{\sum_{i=1}^{k}(x_i^2 - y_i^2)}$ on the vector space $\mathbf{R}^k$(the Euclidean metric);*

2. *The function $\rho(x, y) = max_i|x_i - y_i|$ on $\mathbf{R}^k$ (the Manhattan metric);*

3. *The function $\rho(x, y) = \sum_i |x_i - y_i|$ on $\mathbf{R}^k$.*

An example of a metric on an arbitrary set X may be the *the discrete metric* $\rho(x, y)$ defined as $\rho(x, y) = 1$ when $x \neq y$ and $\rho(x, y) = 0$ when $x = y$.

In general, in order to construct a non–trivial metric we would need a more refined set–theoretic context.

Now, we will examine set structures induced by a metric. So assume ρ is a metric on a set X. The elementary notion of topology is the notion of a *neighborhood*, due to Hausdorff [12]. To derive this notion, we first introduce the notion of an *open ball*. For $x \in X$ and a positive $r \in \mathbf{R}^1$, the *open ball* of *radius r about x* is the set $B(x, r) = \{y \in X : \rho(x, y) < r\}$.

Let us record the basic property of open balls

Proposition 2.1. *Assume that $y \in B(x, r) \cap B(z, s)$. Then, there exists $t \in R^1$ with the property that $B(y, t) \subseteq B(x, r) \cap B(z, s)$.*

Proof. Let $t = min\{r - \rho(x, y), s - \rho(z, y)\}$ and assume that $w \in B(y, t)$. By the triangle inequality, $\rho(w, x) \leq \rho(w, y) + \rho(y, x) < r - \rho(x, y) + \rho(x, y) = r$ and similarly $\rho(w, z) \leq \rho(w, y) + \rho(y, z) < s - \rho(z, y) + \rho(z, y) = s$. It follows that $w \in B(x, r) \cap B(z, s)$ and it follows further that $B(y, t) \subseteq B(x, r) \cap B(z, s)$ □

An *open* set in the metric space (X, ρ) is a union of a family of open balls; formally $U \subseteq X$ is *open* when $U = \bigcup \mathcal{B}$, where $\mathcal{B}$ is a family of open balls. We have

Proposition 2.2. *In any metric space (X, ρ),*

1. *The intersection of any two open balls is an open set;*

2. *The intersection of any finite family of open sets is an open set;*

3. *The union of any family of open sets is an open set.*

Proof. Property 1 follows from Proposition 2.1 as for any point in the intersection of two balls there exists an open ball about that point contained in the intersection so the intersection is a union of open balls. For Property 2, consider open sets U, W with $U = \bigcup \mathcal{B}$, $W = \bigcup \mathcal{C}$ where $\mathcal{B}, \mathcal{C}$ are families of open balls. Then $U \cap W = \bigcup \mathcal{B} \cap \bigcup \mathcal{C} = \bigcup \{B \cap C : B \in \mathcal{B}, C \in \mathcal{C}\}$ and as any intersection $B \cap C$ of open balls is a union of open balls by Property 1, it follows that $U \cap W$ is a union of a family of open balls hence it is an open set. By induction on n one can then prove that any intersection $U_1 \cap U_2 \cap ... \cap U_n$ of open sets is an open set. For Property 3, if $T = \bigcup \mathcal{U}$ where $\mathcal{U}$ is a family of open sets so each $U \in \mathcal{U}$ is a union of a family of open balls, $U = \bigcup \mathcal{B}_U$, then $T = \bigcup \{B : \exists U \in \mathcal{U} (B \in \mathcal{B}_U)\}$ i.e. T is the union of a family of open balls and so T is an open set $\qquad \square$

We denote by the symbol $\mathcal{O}(X)$ the family of all open sets in a metric space (X, ρ). By the last proposition this family is closed with respect to finite intersections and all unions. Clearly, $\emptyset$ and X are open sets, the former as the union of the empty family of open balls, the latter as, e.g., the intersection of the empty family of open balls.

A dual notion of a *closed set*, due to Cantor [6], is also of principal interest. A set $K \subseteq X$ is *closed* in case its complement $X \setminus K$ is open. Thus the family $\mathcal{C}(X)$ of all closed sets in X is closed with respect to finite unions and arbitrary intersections.

As not any subset of X is open or closed, one may try to characterize the position of sets in X with respect to open or closed sets. To this end, we introduce the operation of *interior* of a set as follows.

For a set $A \subseteq X$, we let

$$Int A = \bigcup \{B(x, r) : B(x, r) \subseteq A\} \tag{2.1}$$

i.e., $Int A$ is the union of all open balls contained in A; clearly, $Int A$ is an open set, moreover by its definition it is the largest open subset of A. The operator Int is the *interior* operator and its value $Int A$ is the *interior of the set A*. This operation has the following properties

Proposition 2.3. *In any metric space (X, ρ),*

1. $Int \emptyset = \emptyset$;

2. $Int A \subseteq A$;

3. $Int(A \cap B) = IntA \cap IntB$;

4. $Int(IntA) = IntA$.

Proof. As $\emptyset$ is the only open subset of itself, so Property 1 holds. Clearly $intA \subseteq A$ by definition i.e. Property 2 is fulfilled. For Property 3, $x \in IntA \cap IntB$ if and only if $B(x,r) \subseteq A$ and $B(x,s) \subseteq B$ for some $r, s > 0$ if and only if $B(x,t) \subseteq A \cap B$ where $t = min\{r,s\}$ if and only if $x \in Int(A \cap B)$. Finally, Property 4 follows as $IntA$ is an open set and thus $Int(IntA) = IntA$ □

The dual operator Cl of *closure*, introduced in Riesz [22], cf., Kuratowski [13], is defined via the formula

$$ClA = X \setminus Int(X \setminus A) \tag{2.2}$$

Hence, by duality, ClA is a closed set containing A and it is the smallest among such sets. From the previous proposition, the following properties of Cl follow

Proposition 2.4. *For any metric space* (X, ρ),

1. $ClX = X$;

2. $A \subseteq ClA$;

3. $Cl(A \cup B) = ClA \cup ClB$;

4. $Cl(ClA) = ClA$.

Proof. For Property 1 we have $ClX = X \setminus Int(X \setminus X) = X \setminus Int\emptyset = X \setminus \emptyset = X$. In case of Property 2, as $Int(X \setminus A) \subseteq X \setminus A$, we have $A = X \setminus (X \setminus A) \subseteq X \setminus Int(X \setminus A) = ClA$. Similarly Property 3 follows: $Cl(A \cup B) = X \setminus Int(X \setminus (A \cup B)) = X \setminus Int((X \setminus A) \cap (X \setminus B)) = X \setminus (Int(X \setminus A) \cap Int(X \setminus B)) = X \setminus (Int(X \setminus A)) \cup X \setminus (Int(X \setminus B)) = ClA \cup ClB$. Finally, ClA is a closed set hence it is identical to its closure: $Cl(ClA) = ClA$ □

The difference $ClA \setminus IntA = BdA$ is a closed set called the *boundary* of A. In case $BdA = \emptyset$, we say that A is *closed–open* (or, *clopen* for short).

     For any $x \in X$, a set $A \subseteq X$ is called a *neighborhood* of x if and only if $x \in IntA$. It is easy to see that the intersection of two neighborhoods of x is a neighborhood of x and that any set containing a neighborhood of x is itself a neighborhood of x, i.e., the family $N(x)$ of all neighborhoods of x is a *filter*, see Ch. 1, sect. 11. Usually one does not need to look at all neighborhoods of x. It is sufficient to look at a *co–final* sub–family of $N(x)$ viz. we call a sub–family $B(x)$ of $N(x)$ a *neighborhood basis* if and only if for any $A \in N(x)$ there exists $B \in B(x)$ with the property that $B \subseteq A$. Thus, $B(x)$ contains *arbitrarily small* neighborhoods of x.

Proposition 2.5. *In any metric space* (X, ρ), *any point* x *has a countable neighborhood basis.*

Proof. It suffices to select for each $n \in N$ the open ball $B(x, 1/n)$ and let $B(x) = \{B(x, 1/n) : n \in N\}$. Indeed, given any $A \in N(x)$, as $x \in IntA$, there exists an open ball $B(x, r)$ with $r > 0$ and $B(x, r) \subseteq A$. For $0 < 1/n < r$, we have $B(x, 1/n) \subseteq B(x, r)$ $\qquad\qquad\qquad\qquad\qquad\qquad\qquad\qquad\qquad\qquad\qquad\quad$ □

The fact that neighborhood bases of all points in a metric space may be chosen countable, makes it possible to describe metric topology by means of *sequences*. A *sequence* in a set X is the image $f(N)$ of the set N of natural numbers under a function $f : N \to X$. Usually, we write down a sequence as the set $\{f(n) : n \in N\}$ or in a concise form as $(x_n)_n$ disregarding the function f (which after all may be fully restored from the latter notation as $f(n) = x_n$, each $n \in N$).

A sequence $(x_n)_n$ is *converging to a limit* $x \in X$ if and only if each neighborhood A of x contains *almost all* (meaning: all but a finite number) members x_n of the sequence. In this case we write $x = limx_n$. More formally, $x = limx_n$ if and only if for each $r > 0$ there exists $n(r) \in N$ such that $x_m \in A$ for each $m \geq n(r)$. A canonical example of a sequence converging to x may be obtained via axiom of choice by selecting a point $x_n \in B(x, 1/n)$ for each $n \in N$.

We may now characterize interiors and closures in a metric space by means of neighborhoods as well as by means of sequences.

Proposition 2.6. *In any metric space* (X, ρ)

1. $x \in IntA$ *if and only if there exists a neighborhood* $P \in N(x)$ *such that* $P \subseteq A$;

2. $x \in IntA$ *if and only if for any sequence* $(x_n)_n$ *with* $x = limx_n$ *there exists* $n_0 \in N$ *such that* $x_m \in A$ *for each* $m \geq n_0$;

3. $x \in ClA$ *if and only if for every* $P \in N(x)$ *we have* $P \cap A \neq \emptyset$;

4. $x \in ClA$ *if and only if there exists a sequence* $(x_n)_n \subseteq A$ *with* $x = limx_n$.

Proof. Property 1 follows by definition of $IntA$. Property 2 follows by definition of a limit of a sequence: if $x \in IntA$ and $x = limx_n$ then almost every $x_n \in IntA$ a fortiori almost every $x_n \in A$. Assume conversely that $x \notin IntA$; then for any $n \in N$, we have that it is not true that $B(x, 1/n) \subseteq A$ so we may choose $x_n \in B(x, 1/n) \setminus A$ for $n \in N$. Obviously, $x = limx_n$ and $\{x_n : n \in N\} \cap A = \emptyset$ proving Property 2.

Property 3 follows from Property 1 by duality : $x \in ClA$ if and only if $x \notin Int(X \setminus A)$. As for Property 4, if $x \in ClA$ then $B(x, 1/n) \cap A \neq \emptyset$ so selecting $x_n \in B(x, 1/n) \cap A$ for each n, we produce a sequence $(x_n)_n \subseteq A$ with $x = limx_n$.

Conversely if $x = lim x_n$ for a sequence $(x_n)_n \subseteq A$, then for any neighborhood $P \in N(x)$ we have $P \cap \{x_n : n \in N\} \neq \emptyset$ a fortiori $P \cap A \neq \emptyset$ so $x \in ClA$ in virtue of Property 3   □

Now we show some methods for constructing new spaces.

2.2   Products of Metric Spaces

It may be desirable at this point to introduce some techniques for constructing more intricate metric spaces of general importance. We now begin with a discussion of topologies on Cartesian products. For a family $\mathcal{F} = \{F_i : i \in I\}$ of sets, the *Cartesian product* $\Pi_i F_i$ is the set of all functions $f : I \to \bigcup \mathcal{F}$ with the property that $f(i) \in F_i$ for each $i \in I$. We refer to a function $f \in \Pi_i F_i$ as to a *thread* and we usually write down this thread in the form $(f_i)_i$, i.e., as a *generalized sequence*.

Assume now that each F_i is a metric space with a metric ρ_i; we consider a question: is there a way to introduce a metric on the Cartesian product $\Pi_i F_i$ so it be compatible in a sense with metrics ρ_i? It turns out that in order to answer this question positively we should restrict ourselves to *countable* products in which case the set I is countable.

In case we have a Cartesian product $\Pi_n F_n$ of a countable family $\mathcal{F} = \{F_n : n \in N\}$ of metric spaces (F_n, ρ_n), we first *normalize* metrics ρ_n by letting $\eta_n(x, y) = min\{1, \rho_n(x, y)\}$ for $x, y \in F_n$, each n. Then

Proposition 2.7. *For each n, metrics ρ_n, η_n induce the same topology in the sense that families of open sets are identical in both metrics. Hence, convergence of sequences is identical with respect to both metrics.*

Indeed, for any set $A \subseteq F_n$, if A is open with respect to ρ_n i.e. $B_\rho(x, r(x)) \subseteq A$ for each $x \in A$ then $B_\eta(x, min\{1, r(x)\}) \subseteq A$ for each $x \in A$ i.e. A is open with respect to η. The converse follows on same lines.

We say that metrics ρ, η are *equivalent*. We may assume that ρ_n is already bounded by 1 for each n. Then, for $f = (f_i)_i, g = (g_i)_i \in \Pi_i F_i$, we let

$$\rho(f, g) = \sum_{n=0}^{\infty} \frac{1}{2^n} \cdot \rho_n(f_n, g_n) \qquad (2.3)$$

and we declare ρ to be a *product metric* on the Cartesian product $\Pi_n F_n$. It is evident that ρ is a metric.

A classical example of a metric Cartesian product is the *Cantor cube* $\mathcal{C} = \Pi_n \{0, 1\}_n$ where $\{0, 1\}_n = \{0, 1\}$ for each n. Using the symbol **2** for the set $\{0, 1\}$, we may use for the Cantor cube the symbol $\mathbf{2}^N$ as any thread $f \in \mathcal{C}$ may be regarded as the *characteristic function* of a set $A \subseteq N$, i.e., $A = \{n \in N : f(n) = 1\}$ and thus $\mathcal{C}$ is the set of function codes for subsets of the set of natural numbers.

Regarding the set $\mathbf{2} = \{0, 1\}$ as a metric space with the discrete metric, we have the product metric

$$\rho(f, g) = \sum_{n=0}^{\infty} \frac{1}{2^n} \cdot |f_n - g_n| \tag{2.4}$$

on $\mathbf{2}^N$.

We may be interested in basic neighborhoods in this metric space. Let us observe that any set of the form

$$B(i_1, i_2, ..., i_n) = \{f \in \mathbf{2}^N : f_k = i_k, k = 1, 2, .., n\} \tag{2.5}$$

where $i_1, i_2, ..., i_n \in \mathbf{2}$ is open: given $f \in B(i_1, i_2, ..., i_n)$ we have $B(f, \frac{1}{2^{n+1}}) \subseteq B(i_1, i_2, ..., i_n)$; indeed, $g \in B(f, \frac{1}{2^{n+1}})$ implies $g_i = f_i$ for $i = 1, 2, ..., n$. Conversely, any open ball $B(f, r)$ contains a neighborhood of f of the form $B(i_1, i_2, .. ., i_n)$: it suffices to take the least n with the property that $\sum_{j=n+1}^{\infty} \frac{1}{2^j} < r$ and then $B(f_1, ..., f_n) \subseteq B(f, r)$. We have verified

Proposition 2.8. *Open sets of the form $B(i_1, i_2, ..., i_n)$ form a basis for the product metric topology of the Cantor cube: each open set is the union of a family of sets of the form $B(i_1, i_2, ..., i_n)$.*

In particular, for any $f \in \mathbf{2}^N$, we have $N(f) = \{B(f_1, f_2, ..., f_n) : n \in N\}$.

Using the Cantor cube as an example, we may exhibit two important properties of topological, in particular metric, spaces, viz., *completeness* and *compactness*.

To begin with, let us look at an arbitrary sequence $(x_n)_n$ in the Cantor cube. As each x_n may take at each coordinate i only one of two values $0, 1$, there exists $y \in \mathbf{2}^N$ with the property that for each $j \in N$ the set $A_j = \{x_n : x_{n_i} = y_i, i = 1, 2, ..., j\}$ is infinite. Let n_j be the first natural number n such that $x_n \in A_j$. Then the *sub-sequence* $(x_{n_j})_j$ converges to y: for any basic neighborhood $B(y_1, y_2, .., y_j)$ of y we have $x_n \in B(y_1, y_2, .., y_j)$ for $n \geq n_j$. We have proved that

Proposition 2.9. *Any sequence in the Cantor cube contains a convergent subsequence.*

Of a metric space with this property we say that it is *compact*. Let us establish other characterizations of compactness, more convenient in case of general topological spaces.

2.3   Compact Metric Spaces

We say that a family $\mathcal{U}$ of open sets in a metric space X is an *open covering* of X if and only if $X = \bigcup \mathcal{U}$. A family $\mathcal{U}' \subseteq \mathcal{U}$ is a *sub-covering* in case $X = \bigcup \mathcal{U}'$.

A family of sets $\mathcal{F}$ is *centered* if and only if for any finite sub–family $\mathcal{F}' \subseteq \mathcal{F}$, the intersection $\bigcap \mathcal{F}'$ is non–empty.

Let us observe that in a compact metric space X there exists a countable base for open sets: to this end we observe that compact metric spaces have the property of having *finite nets*.

Proposition 2.10. *In each compact metric space* (X, ρ), *for each* $r > 0$, *there exists a finite subset* $X_r = \{x_1, ..., x_{k_r}\}$ *with the property that for each* $y \in X$ *there exists* $j \leq k_r$ *with* $\rho(y, x_j) < r$.

Proof. Assume to the contrary that for some $r > 0$ no finite subset $Y \subseteq X$ satisfies the condition in the proposition. Let us select $x_0 \in X$. There exists $x_1 \in X$ with $\rho(x_0, x_1) \geq r$ and in general once we find $x_0, x_1, .., x_m$ with $\rho(x_i, x_j) \geq r$ for $i \neq j, i, j \leq m$ then we can find $x_{m+1} \in X$ with $\rho(x_{m+1}, x_j) \geq r$ for $j = 1, 2, ..., m$. By the principle of mathematical induction we define thus a sequence $(x_n)_n$ in X with $\rho(x_i, x_j) \geq r$ for $i \neq j$ which clearly cannot have any convergent subsequence   □

A set X_r is called an *r–net*; thus, in compact metric spaces there exist r-nets for each $r > 0$. Now, let us select for each $n \in N$ a $1/n$–net X_n. Let $Y = \bigcup \{X_n : n \in N\}$. Then for each $x \in X$ and any $r > 0$, there exists $y \in Y$ with $\rho(x, y) < r$.

It follows that the family $\{B(y, 1/m) : y \in Y, m \in N\}$ is a *countable basis for open sets*: given an open ball $B(x, r)$ and $z \in B(x, r)$, we may find $y \in Y$ and $m \in N$ such that $2/m < r - \rho(x, z)$ and $\rho(z, y) < 1/m$ and then $z \in B(y, 1/m) \subseteq B(x, r)$. We have proved

Proposition 2.11. *In each compact metric space* (X, ρ) *there exists a countable basis for open sets.*

It follows that in case of any compact metric space X we may restrict ourselves to countable coverings: any open covering contains a countable sub–covering; indeed, given an open covering $\mathcal{U}$ of X and a countable base $\mathcal{B}$ for open sets, the sub–family $\mathcal{U}'$, defined by selecting for each $B \in \mathcal{B}$ a $U \in \mathcal{U}$ such that $B \subseteq U$, is a countable covering of X. Now we may turn to promised characterizations.

Proposition 2.12. *For a metric space* (X, ρ), *the following statements are equivalent*

1. *X is compact;*

2. *Each countable centered family of closed sets in X has a non–empty intersection;*

3. *Each countable open covering of X has a finite sub–covering.*

Proof. Assume Property 1. Let $\mathcal{F} = \{F_n : n \in N\}$ be a centered family of closed sets. Hence each intersection $F_1 \cap ... \cap F_n$ is non–empty and we may select a point $x_n \in F_1 \cap ... \cap F_n$ for each n. There exists a convergent sub–sequence $(x_{n_k})_k$ with $x = \lim x_{n_k}$. Then $x \in F_n$ for each n and thus $x \in \bigcap \mathcal{F}$. This proves Property 2.

Now, Property 3 follows from Property 2 by duality: given a countable open covering $\{U_n : n \in N\}$ without any finite sub–covering, the family $\mathcal{F} = \{F_n = X \setminus U_n : n \in N\}$ is a centered family of closed sets so by 2 there exists $x_0 \in \bigcap \mathcal{F}$. But then $x_0 \notin \bigcup \mathcal{U}$, a contradiction.

Finally, Property 3 implies Property 1, as if we had a sequence $(x_n)_n$ without any convergent sub–sequence then we would have for each $x \in X$ an open neighborhood U_x with the property that $x_n \in U_x$ for finitely many n only. Then the open covering $\{U_x : x \in X\}$ contains a finite sub–covering $U_{x_1} \cup ... \cup U_{x_k} = X$ which misses infinitely many elements of the sequence $(x_n)_n$, a contradiction $\qquad\qquad\qquad\qquad\qquad\qquad\qquad\qquad\qquad\qquad\qquad\qquad\square$

The last proposition indicates that compactness may be defined in purely topological terms of open, respectively closed, sets regardless of a specific metric from the class of equivalent metrics inducing the given topology. Compactness allows for existential arguments based on the convergent sub–sequences.

We have begun with an example of the Cantor cube as a compact metric space. To give more examples, let us observe that the unit interval $I = [0,1]$ is a compact metric space in natural metric topology induced by the metric $\rho(x,y) = |x - y|$; indeed, assume that $\mathcal{U}$ is an open covering of I (we may also assume that elements of $\mathcal{U}$ are open intervals of the form $(a,b), [0,a)$, or $(a,1]$). Was I not compact, we could consider a subset $C \subseteq I$ defined as follows: $x \in C$ if and only if the interval $[0,x]$ may be covered by a finite number of elements of $\mathcal{U}$.

As, under our assumption, $1 \notin C$, there exists $s = \sup C < 1$. We may pick an element $s \in (a,b) \in \mathcal{U}$ along with some $a < t < s$. Then the interval $[0,t]$ may be covered by finitely many elements $U_1, .., U_k$ of $\mathcal{U}$ hence $U_1, .., U_k, (a,b)$ cover any interval of the form $[0,w]$ for $s < w < b$, contrary to the definition of s. It follows that $\sup C = 1$ and a fortiori $[0,1]$ is compact. Same argument will show that each interval $[a,b]$ in the real line is compact.

2.4 Complete Metric Spaces

A weakening of compactness is the condition of *completeness*. This notion depends on the notion of a *fundamental sequence*. A sequence $(x_n)_n$ is *fundamental* if and only if for each $r > 0$ there exists $n_r \in N$ such that $\rho(x_{n_r+k}, x_{n_r}) < r$ for each k. Let us observe that each convergent sequence is fundamental.

A metric space (X, ρ) is *complete* if and only if each fundamental sequence converges in X.

Proposition 2.13. *If a fundamental sequence* $(x_n)_n$ *contains a convergent sub–sequence, then the sequence* $(x_n)_n$ *converges. Hence any compact metric space is complete.*

Indeed, assuming a sub–sequence $(x_{n_k})_k$ to be convergent to a point x, we have $\rho(x, x_n) \leq \rho(x, x_{n_k}) + \rho(x_{n_k}, x_n)$ and it is sufficient to observe that $\rho(x, x_{n_k})$, $\rho(x_{n_k}, x_n)$ tend to 0 as n, n_k tend to ∞. Hence $\rho(x, x_n)$ approaches 0 as n tends to ∞ and this implies $x = lim x_n$. Thus, in a compact metric space any fundamental sequence converges implying completeness.

As any fundamental sequence in the real line $\mathbf{R}^1$ is bounded, i.e., it is contained in a closed interval of the form $[a, b]$ which is compact, it contains a convergent subsequence and it does converge itself by Proposition 2.13. Thus the real line in natural metric is a complete metric space. One may prove that the set of irrational real numbers is also a complete metric space under the natural metric.

As with compactness, we may attempt at characterizing completeness in topological terms. Contrary to the case of compactness, we will not get rid of the metric.

For a subset $A \subseteq X$ of a metric space (X, ρ), we define the *diameter of A*, $diam A$ as follows: $diam A = sup\{\rho(x, y) : x, y \in A\}$. We say that a family $\mathcal{A}$ of subsets of X contains *arbitrarily small sets* if and only if for each $r > 0$ there exists $A \in \mathcal{A}$ with $diam A \leq r$.

Proposition 2.14. *(Cantor [5]) A metric space* (X, ρ) *is complete if and only if each centered countable family* $\mathcal{F}$ *of closed non–empty sets containing arbitrarily small sets has a non–empty intersection.*

Proof. Assume (X, ρ) is complete. We may also assume that $\mathcal{F} = \{F_n : n \in N\}$ satisfies the condition $F_{n+1} \subseteq F_n$ for each n. Indeed, otherwise we could consider the family $\mathcal{G} = \{F_1 \cap ... \cap F_n : F_i \in \mathcal{F}\}$ as $\bigcap \mathcal{G} = \bigcap \mathcal{F}$. Selecting $x_n \in F_n$ for each n, we produce a sequence $(x_n)_n$ which is fundamental: $\rho(x_{n+k}, x_n) \leq diam F_n$ and $lim diam F_n = 0$. Hence there exists a convergent sub–sequence $(x_{n_k})_k$ with the limit $x \in X$. Clearly, $x \in Cl F_n = F_n$ for each n so $x \in \bigcap \mathcal{F}$.

Now, we assume that the condition on $\mathcal{F}$ is satisfied and we prove completeness. Given a fundamental sequence $(x_n)_n$ in X, we consider $F_n = Cl\{x_k : k \geq n\}$ and $\mathcal{F} = \{F_n : n \in N\}$. Clearly, $F_{n+1} \subseteq F_n$ for each n and $lim diam F_n = 0$. Thus, there exists $x \in \bigcap \mathcal{F}$, a fortiori, $x = lim x_n$: for each ball $B(x, r)$ we have $F_n \subseteq B(x, r)$ for almost every n hence $x_n \in B(x, r)$ for almost every n □

Complete metric spaces proved to be very important due to their properties of which we recall here the two most fundamental and very often recalled in applications, viz., the *Baire category theorem* and the *Banach fixed point theorem*.

The Baire category theorem basically states that a complete metric space cannot be covered by countably many "small" subsets. We should define the meaning of 'small'.

We say that a subset $A \subseteq X$ is *meager* when $Int A = \emptyset$ and A is *nowhere dense* in case $Int Cl A = \emptyset$. When $B = \bigcup\{A_n : n \in N\}$ and each A_n is nowhere dense then we call B a *set of 1st category*. An archetypical example of a meager set is $Cl A \setminus A$ for any A; indeed, was $Int(Cl A \setminus A) \neq \emptyset$ we would have $G \subseteq Cl A \setminus A$ for some open non–empty G implying $G \cap A = \emptyset$ on one hand and $G \cap A \neq \emptyset$ on the other.

Proposition 2.15. *(the Baire category theorem, Baire [3]) In any complete metric space (X, ρ), any set of the 1st category is meager.*

Proof. Consider a set $Y = \bigcup\{Y_n : n \in N\}$ of the 1st category. For an arbitrary $x_0 \in X$, we define inductively closed sets F_n for $n \in N$ such that $x_0 \in Int F_n$ and

$$F_{n+1} \subseteq Int F_n$$

for each n and $\mathcal{F} = \{F_n : n \in N\}$ contains arbitrarily small sets.

To begin with, as $Int Cl Y_0 = \emptyset$, taking an arbitrary open ball $B(x_0, s_0)$ we find

$$B(y_0, r_0) \subseteq Cl B(y_0, r_0) \subseteq B(x_0, s_0) \setminus Cl Y_0 \qquad (2.6)$$

with $r_0 < 1/2$. Assuming that open balls $B(y_0, r_0), ..., B(y_k, r_k)$ with

$$\begin{cases} (i) \ Cl B(y_{i+1}, r_{i+1}) \subseteq B(y_i, r_i) \ for \ i = 1, 2, .., k-1 \\ (ii) \ Cl B(y_i, r_i) \cap Cl Y_i = \emptyset \ for \ i \leq \ k \\ (iii) \ r_i < \frac{1}{2^{i+1}} \ for \ i \leq \ k \end{cases} \qquad (2.7)$$

have been defined, we repeat our argument with Y_{k+1} in place of Y_0 and $B(y_k, r_k)$ in place of $B(x_0, s_0)$ to produce

$$B(y_{k+1}, r_{k+1}) \subseteq Cl B(y_{k+1}, r_{k+1}) \subseteq B(y_k, r_k) \setminus Cl Y_{k+1}$$

with $r_{k+1} < \frac{1}{2^{k+2}}$.

By induction, we arrive at the sequence $(F_k = Cl B(y_k, r_k))_k$ satisfying (2.7) and thus $F_{k+1} \subseteq F_k$, $F_k \subseteq X \setminus Cl Y_k$, each k, and $lim diam F_k = 0$. It follows by completeness that there exists $x \in \bigcap\{F_k : k \in N\}$ and clearly $x \notin Y$. As $x \in B(x_0, s_0)$, we have proved that $\bigcup\{Y_k : k \in N\}$ contains no non–empty open set, i.e., it is meager $\qquad \square$

It follows from the proof that the difference $X \setminus Y$ is *dense* in X: any open non–empty ball contains a point from $X \setminus Y$. In other form: $Cl(X \setminus Y) = X$.

The Baire category theorem is an important tool in proving the existence (although non–effectively) of certain objects and moreover in proving that these objects form a large (not of the 1st category) set (like, e.g., continuous nowhere differentiable real functions).

The Banach fixed point theorem also allows for proving existential facts. It has to do with *contracting functions*. A function $f : X \to X$ on a metric space (X, ρ) is a *contracting function* with a *contraction factor* $0 < \lambda < 1$ if and only if $\rho(f(x), f(y)) \leq \lambda \cdot \rho(x, y)$. Then we claim that f has a *fixed point* c, i.e., $f(c) = c$.

Proposition 2.16. *(the Banach fixed point theorem, Banach [4]) Each contracting function on a complete metric space (X, ρ) has a (unique) fixed point.*

Proof. Let us begin with an arbitrary $x_0 \in X$. We define inductively a sequence $(x_n)_n$ by letting $x_{n+1} = f(x_n)$ for each n. This sequence is fundamental: letting $K = \rho(x_0, x_1) = \rho(x_0, f(x_0))$, we have $\rho(x_1, x_2) = \rho(f(x_0), f(x_1)) \leq \lambda \rho(x_0, x_1) = \lambda \cdot K$. Analogously, $\rho(x_2, x_3) \leq \lambda^2 \cdot K$, and in general $\rho(x_n, x_{n+1}) \leq \lambda^n \cdot K$. Thus, we have

$$\rho(x_{n+k}, x_n) \leq K \cdot \sum_{i=n}^{i=n+k-1} \lambda^i = K \cdot \lambda^n \frac{1 - \lambda^k}{1 - \lambda} \tag{2.8}$$

and hence $\rho(x_{n+k}, x_n)$ tends to 0 as $n, n+k$ tend to ∞. By completeness, there exists $c = \lim x_n$. As $\rho(f(c), f(x_n)) \leq \lambda \cdot \rho(c, x_n)$ we have $f(c) = \lim f(x_n) = \lim x_{n+1} = c$ so c is a fixed point of f. Observe that f has a unique fixed point: was d a fixed point other than c, we would have $\rho(c, d) = \rho(f(c), f(d)) \leq \lambda \cdot \rho(c, d)$, a contradiction $\square$

Passing with k to ∞ in the formula (2.8) and noticing that $c = \lim x_{n+k}$, we have

$$\rho(c, x_n) \leq K \cdot \lambda^n \frac{1}{1 - \lambda} \tag{2.9}$$

which allows for an estimate of the depth of procedure necessary to approximate c with x_n with a given accuracy. In order to have $\rho(c, x_n) \leq \delta$ it suffices by (2.9) to have

$$K \cdot \lambda^n \frac{1}{1 - \lambda} \leq \delta \tag{2.10}$$

hence it suffices that

$$n = \lfloor \frac{log K^{-1} \cdot (\delta \cdot (1 - \lambda))}{log \lambda} \rfloor \tag{2.11}$$

We now pass to general topological spaces.

2.5   General Topological Spaces

By a *topological space*, we mean a set X along with a *topology* τ defined usually as a family $\tau \subseteq 2^X$ of sets in X which satisfies the conditions for open sets familiar to us from Proposition 2.2, viz.,

1. *For any finite sub–family $\tau' \subseteq \tau$, we have $\bigcap \tau' \in \tau$;*

2. *For any sub–family $\tau' \subseteq \tau$, we have $\bigcup \tau' \in \tau$.*

Hence, τ is closed with respect to operations of finite unions and arbitrary intersections; let us notice that $\emptyset \in \tau$ as the union of the empty family and $X \in \tau$ as the intersection of the empty family.

Elements of τ are *open sets*, Alexandroff [1], Fréchet [10]. As with metric spaces, we introduce the operator Int of *interior* defined as follows,

$$Int A = \bigcup \{U \in \tau : U \subseteq A\} \tag{2.12}$$

$IntA$ is the union of all open subsets of A and a fortiori it is the greatest open subset of A. By closeness of τ on finite intersections and unions, we obtain basic properties of Int.

Proposition 2.17. *The following are properties of the operation of interior*

I1 $Int\emptyset = \emptyset$

I2 $IntA \subseteq A$

I3 $Int(A \cap B) = IntA \cap IntB$

I4 $Int(IntA) = IntA$

Indeed, Properties I1, I2 follow by remarks following the definition of Int, Property I3 follows by the finite intersection property of τ and Property I4 does express obvious fact that any open set is its interior. Let us observe that by Property I4, we have also,

I5 $A \subseteq B$ *implies* $IntA \subseteq IntB$

Conversely, open sets may be introduced via the interior operator Int satisfying Properties I1–I4.

Proposition 2.18. *If an operator* $I : 2^X \to 2^X$ *does satisfy Properties I1–I4, then* $\tau = \{A \subseteq X : A = IA\}$ *is a topology on* X *and* I *is the interior operator* Int *with respect to* τ.

Proof. We call A such that $A = IA$ open. As $I(A \cap B) = IA \cap IB$ by Property I3, the intersection of two open sets is open and by induction it follows that τ is closed with respect to finite intersections. For a family $\{A_i : i \in I\}$ where $IA_i = A_i$, we have by Property I5, that $A_i = IA_i \subseteq I\bigcup_i A_i$ for each i hence $\bigcup_i A_i \subseteq I\bigcup_i A_i$ and finally by Property I2, $\bigcup_i A_i = I\bigcup_i A_i$, i.e., $\bigcup_i A_i \in \tau$ ☐

By duality, we may define the *closure* operator Cl via

$$Cl A = X \setminus Int(X \setminus A) \tag{2.13}$$

The closure operator does satisfy dual conditions

C1 $ClX = X$

C2 $A \subseteq ClA$

C3 $Cl(A \cup B) = ClA \cup ClB$

C4 $Cl(ClA) = ClA$

and we may define the family $\mathcal{C}(X)$ of *closed sets* in X by letting $\mathcal{C}(X) = \{A \subseteq X : ClA = A\}$. By duality, this family is closed with respect to finite unions and arbitrary intersections. The family $\tau = \{X \setminus A : A \in \mathcal{C}(X)\}$ is clearly the family of open sets with respect to topology induced by the operator Cl.

Operators Int and Cl give rise to a topological calculus on sets. Let us observe that both these operators are monotone: for $A \subseteq B$, we have $IntA \subseteq IntB$ by Property I5, and

C5 $ClA \subseteq ClB$

by the dual to Property I5. We begin with a useful lemma of this calculus.

Proposition 2.19. *For an open set P and a set A we have:* $P \cap ClA \subseteq Cl(P \cap A)$.

Indeed, for $x \in P \cap ClA$ and any open set $Q \ni x$, we have $(Q \cap P) \cap A \neq \emptyset$ as $Q \cap P$ is a neighborhood of x and $x \in ClA$, hence $Q \cap (P \cap A) \neq \emptyset$, i.e., $x \in Cl(P \cap A)$.

Various algebras of sets may be defined with help of these operations. We begin with the two most important.

2.6   Regular Open and Regular Closed Sets

We say that a set $A \subseteq X$ is *regular open* if and only if $A = Int(ClA)$, i.e., A is the interior of its closure.

Similarly, we say that a set $A \subseteq X$ is *regular closed* if and only if $A = Cl(IntA)$, i.e., A is the closure of its interior.

Both types of sets are related to each other by duality: A is regular open if and only if $X \setminus A$ is regular closed. Indeed, if $A = IntClA$ then $X \setminus A = X \setminus IntClA = X \setminus Int(X \setminus (X \setminus ClA)) = Cl(X \setminus ClA) = Cl(Int(X \setminus A))$.

We now prove a basic fact that *regular open sets* form a Boolean algebra, denoted $RO(X)$. In order to prove this fact, we introduce a new auxiliary symbol, and we let

$$A^{\perp} = X \setminus ClA$$

The set A is regular open if and only if $A = A^{\perp\perp}$. Indeed, $A^{\perp\perp} = X \setminus Cl(X \setminus ClA) = IntClA$. We sum up properties of the operation $A^{\perp}$

Proposition 2.20. *The following are properties of the operation* $A^{\perp}$

1. If $A \subseteq B$, *then* $B^{\perp} \subseteq A^{\perp}$;

2. If A *is an open set, then* $A \subseteq A^{\perp\perp}$;

3. If A *is an open set, then* $A^{\perp} = A^{\perp\perp\perp}$, *hence,* $A^{\perp\perp} = A^{\perp\perp\perp\perp}$;

4. If A, B *are open sets, then* $(A \cap B)^{\perp\perp} = A^{\perp\perp} \cap B^{\perp\perp}$;

5. $(A \cup B)^{\perp} = A^{\perp} \cap B^{\perp}$;

6. If A *is an open set, then* $(A \cup A^{\perp})^{\perp\perp} = X$.

Proof. Property 1 follows immediately: $A \subseteq B$ implies $ClA \subseteq ClB$ implies $X \setminus ClB \subseteq X \setminus ClA$. For Property 2, A open implies $A \subseteq IntClA = A^{\perp\perp}$ as $A \subseteq ClA$. In case of Property 3, we get $A^{\perp} \subseteq A^{\perp\perp\perp}$ from Property 2 by substituting $A^{\perp}$ for A; also from Property 2 we get by applying Property 1: $A^{\perp\perp\perp} \subseteq A^{\perp}$, hence, Property 3 follows.

For Property 4, we have by applying Property 1 twice to the inclusion $A \cap B \subseteq A$ that $(A \cap B)^{\perp\perp} \subseteq A^{\perp\perp}$ and similarly $(A \cap B)^{\perp\perp} \subseteq B^{\perp\perp}$ hence $(A \cap B)^{\perp\perp} \subseteq A^{\perp\perp} \cap B^{\perp\perp}$.

For the converse, we apply Proposition 2.19 and we observe that its statement $A \cap ClB \subseteq Cl(A \cap B)$ may be paraphrased as

$$A^{\perp\perp} \cap B \subseteq (A \cap B)^{\perp\perp} \tag{2.14}$$

Hence, for A, B open, we have $A \cap ClB \subseteq Cl(A \cap B)$, i.e.,

$$(A \cap B)^{\perp} = X \setminus Cl(A \cap B) \subseteq X \setminus (A \cap ClB) = (X \setminus A) \cup B^{\perp} \tag{2.15}$$

From $(A \cap B)^{\perp} \subseteq (X \setminus A) \cup B^{\perp}$, we get

$$A \cap B^{\perp\perp} = ((X \setminus A) \cup B^{\perp})^{\perp} \subseteq (A \cap B)^{\perp\perp} \tag{2.16}$$

Substituting in

$$A \cap B^{\perp\perp} \subseteq (A \cap B)^{\perp\perp} \tag{2.17}$$

$A^{\perp\perp}$ for A, and applying the second statement in Property 3, we come at

$$A^{\perp\perp} \cap B^{\perp\perp} \subseteq (A^{\perp\perp} \cap B)^{\perp\perp} \subseteq (A \cap B)^{\perp\perp\perp\perp} = (A \cap B)^{\perp\perp} \tag{2.18}$$

which proves Property 4.

Property 5 follows immediately by duality. To check Property 6, let us observe that in case A open we have $A \cup A^{\perp} = ClA \setminus A$ hence

$$(A \cup A^{\perp})^{\perp} = IntCl(ClA \setminus A) = Int(ClA \setminus A) = \emptyset \qquad (2.19)$$

as $ClA \setminus A$ is meager. Thus $(A \cup A^{\perp})^{\perp\perp} = X$ $\qquad\qquad$ $\square$

Let us observe that from the second part of Property 3 above: $A^{\perp\perp} = A^{\perp\perp\perp\perp}$ it follows that $IntClIntClA = IntClA$ and $ClIntClIntA = ClIntA$ for any A. A consequence of this is that one may obtain from a given set A by applying $Int, Cl, \setminus$ only at most 14 distinct sets. The reader will easily write them down. From this property also follows immediately that any set of the form $A^{\perp\perp}$ is regular open, i.e., any set of the form $IntClA$ is regular open. By duality, any set of the form $ClIntB$ is regular closed.

Now, we pass to the family $RO(X)$ of regular open sets.

Proposition 2.21. $RO(X)$ *is a Boolean algebra under operations* $\wedge, \vee, \prime$, *defined as follows*

1. $A \vee B = (A \cup B)^{\perp\perp} = IntCl(A \cup B)$;

2. $A \wedge B = A \cap B$;

3. $A' = A^{\perp} = X \setminus ClA$.

and with constants $\mathbf{0} = \emptyset, \mathbf{1} = X$.

Proof. All Boolean operations listed above give regular open sets by properties of $(.)^{\perp}$ listed above in Proposition 2.20. It remains to check that axioms of a Boolean algebra are satisfied. Commutativity laws $A \vee B = B \vee A, A \wedge B = B \wedge A$ are satisfied evidently. The laws $A \vee \mathbf{0} = A, A \wedge \mathbf{1} = A$ are also manifest. We have $A \wedge A' = A \cap A^{\perp} = A \setminus ClA = \emptyset = \mathbf{0}$ as well as $A \vee A' = (A \cup A^{\perp\perp})^{\perp\perp}) = X = \mathbf{1}$. The distributive laws $A \vee (B \wedge C) = (A \vee B) \wedge (A \vee C)$ as well as $A \vee (B \wedge C) = (A \vee C) \wedge (A \vee C)$ hold by Property 5 and dualization $\qquad\qquad$ $\square$

A particular sub–algebra of $RO(X)$ is the algebra $CO(X)$ of *clopen sets* in X. In case of $CO(X)$ boolean operations $\vee, \wedge, \prime$ specialize to usual set–theoretic operations $\cup, \cap, \setminus$ i.e. $CO(X)$ is a field of sets.

The basic distinction between $RO(X)$ and $CO(X)$ is the fact that $RO(X)$ is a *complete Boolean algebra* for any X whereas $CO(X)$ is not always such. We prove this important fact.

Proposition 2.22. *The boolean algebra* $RO(X)$ *is complete for any topological space* X.

Proof. Let us observe that the boolean ordering relation $\leq$ is in this case the inclusion $\subseteq$. Consider $\mathcal{A} \subseteq RO(X)$. Let $s(\mathcal{A}) = (\bigcup \mathcal{A})^{\perp\perp}$; we check that $s(\mathcal{A})$ is the supremum of $\mathcal{A}$. Indeed, for $A \in \mathcal{A}$, we have $A \in \bigcup \mathcal{A}$ hence

$A = A^{\perp\perp} \subseteq (\bigcup \mathcal{A})^{\perp\perp}$ i.e. $A \leq s(\mathcal{A})$. It follows that $s(\mathcal{A})$ is an upper bound for $\mathcal{A}$. Now, assume that $B \in RO(X)$ is an upper bound for $\mathcal{A}$ i.e. $A \subseteq B$ for each $A \in \mathcal{A}$. Hence $\bigcup(\mathcal{A}) \subseteq B$ and thus $(\bigcup \mathcal{A})^{\perp\perp} \subseteq B^{\perp\perp} = B$ i.e. $s(\mathcal{A}) \leq B$ proving that $s(\mathcal{A})$ is the supremum of $\mathcal{A}$. Finally, by duality it follows that $i(\mathcal{A}) = (\bigcap \mathcal{A})^{\perp\perp}$ is the infimum of $\mathcal{A}$ □

By duality applied to the family $RC(X)$ of regular closed sets in X, we obtain a dual proposal

Proposition 2.23. $RC(X)$ *is a boolean algebra under operations* $\wedge, \vee, \prime$ *defined as follows*

1. $A \vee B = A \cup B$.
2. $A \wedge B = ClInt(A \cap B)$.
3. $A' = X \setminus IntA$.

and with constants $\mathbf{0} = \emptyset, \mathbf{1} = X$. *The algebra* $RC(X)$ *is complete.*

We now pass to a study of compactness in general spaces.

2.7 Compactness in General Spaces

Now it is time for exploring notions we have met with in the metric case in the general topological context. We begin with the notion of *compactness* as the most important by far and yielding most easily to generalizations.

As a definition of compactness in general spaces, we adopt the open covering condition proved for metric spaces in Proposition 12(3). Thus, we say that a topological space (X, τ) is *compact*, after Alexandroff–Urysohn [2], if and only if any open covering $\mathcal{U}$ of X contains a finite sub–covering $\{U_1, .., U_k\}$. The following characterizations follow easily.

For a filter $\mathcal{F}$, we say that $x \in X$ is a *cluster point* of $\mathcal{F}$ if $P \cap F \neq \emptyset$ for any $F \in \mathcal{F}$ and any neighborhood P of x. Let us observe that in case $\mathcal{F}$ is an ultrafilter, and x its cluster point, we have $N(x) \subseteq \mathcal{F}$, i.e., $\mathcal{F}$ has as elements all neighborhoods of x. In this case we say that x is a *limit* of $\mathcal{F}$, in symbols $x = lim\mathcal{F}$.

Proposition 2.24. *For a topological space* (X, τ) *the following are equivalent*

1. X *is compact;*

2. *Each centered family of closed subsets of* X *has a non–empty intersection;*

3. *Each filter on* X *has a cluster point;*

4. *Each ultrafilter on* X *has a limit point.*

Proof. Properties 1 and 2 are equivalent as they are dual statements with respect to duality open – closed. Consider a filter $\mathcal{F}$. As any finite intersection of elements of $\mathcal{F}$ is non–empty, it follows that

$$\overline{\mathcal{F}} = \{ClF : F \in \mathcal{F}\} \tag{2.20}$$

is a centered family of closed sets hence

$$\bigcap \overline{\mathcal{F}} \neq \emptyset \tag{2.21}$$

Any $x \in \bigcap \overline{\mathcal{F}}$ is a cluster point of $\mathcal{F}$ proving Property 3. Property 4 follows from Property 3 as for any ultrafilter its any cluster point is a limit point. Now assume Property 4 and consider a centered family $\mathcal{F}$ of closed sets. Letting

$$\mathcal{F}^* = \{A \subseteq X : \exists F \in \mathcal{F}(F \subseteq A)\} \tag{2.22}$$

we define a filter $\mathcal{F}^*$ which extends to an ultrafilter $\mathcal{G}$. By Property 4, there exists $x = lim\mathcal{G}$; hence, any neighborhood of x does intersect any set in $\mathcal{G}$ and moreover any set in $\mathcal{F}$ thus $x \in ClF = F$ for any $F \in \mathcal{F}$, hence, $x \in \bigcap \mathcal{F}$ proving Property 2 which as we know is equivalent to Property 1   □

Compact spaces have the Baire property: any countable union of nowhere–dense sets is meager. The proof of this fact mimics the proof of the corresponding result for complete metric spaces.

When discussing compactness, in general context, two important findings come to the fore. The first is the *Tikhonov theorem*, due to Tikhonov [25], stating that any Cartesian product of compact spaces is a compact space and the second is the *Stone duality theorem* of Stone [24] allowing to represent any Boolean algebra **B** as the field $CO(X)$ of clopen sets in a compact space X. We give the proofs of these both facts of fundamental importance.

First, we have to define topology of a Cartesian product $\Pi_i X_i$ of a family $\{X_i : i \in I\}$ of topological spaces (X_i, τ_i). To this end, we call a *box defined by coordinates $i_1, .., i_k$ and open sets $U_i \in \tau_i, i = 1, 2, .., k$*, the set

$$O(i_1, .., i_k, U_1, .., U_k) = \{f \in \Pi_i X_i : f_i \in U_i, i = 1, 2, .., k\} \tag{2.23}$$

We define a topology τ_Π on $\Pi_i X_i$ by accepting the family of all boxes as an open basis for τ_Π, i.e., $U \in \tau_\Pi$ if and only if there exists a family $\{B_i : i \in I\}$ of boxes with $U = \bigcup\{B_i : i \in I\}$.

Proposition 2.25. *(the Tikhonov theorem) A Cartesian product $X = \Pi_i X_i$ of a family $\{X_i : i \in I\}$ of compact spaces with the topology τ_Π is a compact space.*

Proof. Consider projections $p_i : X \to X_i$ defined as follows: $p_i(f) = f_i$ for $f \in X, i \in I$. For an ultrafilter $\mathcal{F}$ on X, consider filters $\mathcal{F}_i$ induced by $p_i(\mathcal{F})$ in each X_i.

By compactness, there exists a cluster point x_i of $\mathcal{F}_i$ in X_i and we may define $x = (x_i)_i$ in X. Now for any neighborhood G of x in X there exists a box $x \in O(i_1, .., i_k, U_1, .., U_k) \subseteq G$ hence x_i in U_i for $i = i_1, .., i_k$. It follows that $U_i \cap F \neq \emptyset$ for each $F \in \mathcal{F}_i$ and thus $G \cap F \neq \emptyset$ for any $F \in \mathcal{F}$. Thus x is a cluster point hence a limit point for $\mathcal{F}$ □

By the Tikhonov theorem, e.g., cubes $[0,1]^n$ in Euclidean n–spaces ($n = 2, 3, ...$) as well as the *Hilbert cube* $[0,1]^N$ are compact metric spaces.

Now we will consider a Boolean algebra $\mathbf{B}$ and we define a set $S(\mathbf{B})$ as follows. Elements of $S(\mathbf{B})$ are ultrafilters on $\mathbf{B}$. We define a topology σ on $S(\mathbf{B})$ by letting the sets of the form

$$F_a = \{f \in S(\mathbf{B}) : a \in f\} \qquad (2.24)$$

for $a \in \mathbf{B}$ to make a basis for open sets. Thus,

$$\sigma = \left\{ \bigcup \{F_a : a \in A\} : A \subseteq \mathbf{B} \right\} \qquad (2.25)$$

We observe that $F_{a \wedge b} = F_a \cap F_b$ hence indeed σ is a topology.

We consider an ultrafilter Φ on $S(\mathbf{B})$ and along with it we look at the set

$$A_\Phi = \{a \in S(\mathbf{B}) : F_a \in \Phi\} \qquad (2.26)$$

Then A_Φ is a filter as $F_a \subseteq F_b$ when $a \leq b$ and $F_{a \wedge b} = F_a \cap F_b$. Actually, A_Φ is an ultrafilter as, dually, for a filter $\mathcal{G}$ on $\mathbf{B}$ the family $\{F_a : a \in \mathcal{G}\}$ is a filter on $S(\mathbf{B})$. We have duality

$$A_\Phi \in F_a \leftrightarrow a \in A_\Phi \leftrightarrow F_a \in \Phi \qquad (2.27)$$

i.e., $A_\Phi = lim\Phi$. Thus any ultrafilter on $S(\mathbf{B})$ converges proving this space to be compact.

The compact space $S(\mathbf{B})$ is called the *Stone space*, after Stone [24], of the Boolean algebra $\mathbf{B}$.

Let us observe that we have $F_a = S(\mathbf{B}) \setminus F_{a'}$ for each a thus proving that each base set F_a is closed so finally it is clopen. It follows that the function $h : \mathbf{B} \to S(\mathbf{B})$ which satisfies already established conditions,

$$\begin{cases} h(a \wedge b) = h(a) \cap h(b) \\ h(a \vee b) = h(a) \cup h(b) \\ h(a') = S(\mathbf{B}) \setminus h(a) \end{cases} \qquad (2.28)$$

does establish an isomorphism between $\mathbf{B}$ and $CO(S(\mathbf{B}))$.Hence, each Boolean algebra may be represented as a field of clopen sets in a compact space with a clopen base for open sets (such a space is called 0–*dimensional*).

One may ask about the Stone space in case of a complete Boolean algebra, e.g., $RO(X)$: does it have some additional properties? It turns out that completeness of the Boolean algebra **B** is rendered in Stone spaces by the property of *extremal disconnectedness*: this means that closure of any open set is open, in symbols, $IntClIntA = ClIntA$ for each $A \subseteq S(\mathbf{B})$.

Proposition 2.26. *For any complete Boolean algebra* **B**, *the Stone space* $S(\mathbf{B})$ *is extremally disconnected.*

Proof. Consider $U = \bigcup \{U_a : a \in A\}$. For $A \subseteq \mathbf{B}$, there exists $c = supA$. Then $U_a \subseteq U_c$ for each $a \in A$ so $U \subseteq U_c$ and thus $ClU \subseteq U_c$. Assume $f \in U_c \setminus ClU$ for some ultrafilter f. Then there exists U_b with $f \in U_b$, $U_b \cap U_a = \emptyset$ for $a \in A$. It follows that $a \wedge b = 0$ for $a \in A$ hence $a \leq b'$ for $a \in A$ implying that $supA = c \leq b'$. But then $f \in U_{b'}$, contradicting $f \in U_b$. It follows that $ClU = U_c$ i.e. ClU is clopen   □

This fact allows for a *completion* of a Boolean algebra via topological arguments: given a Boolean algebra **B**, we have an isomorphism $h : \mathbf{B} \to CO(S(\mathbf{B}))$. Now, we embed $CO(S(\mathbf{B}))$ into complete Boolean algebra $RO(S(\mathbf{B}))$ of regular open sets in the Stone space via identity embedding $i : A \to A$. The composition $i \circ h$ embeds the algebra **B** into $RO(S(\mathbf{B}))$. Thus any Boolean algebra **B** has a completion $\mathbf{B}^*$ i.e. a complete Boolean algebra into which it can be embedded as a dense set (it is so as any open set contains a basic clopen set).

As the Stone space of a Boolean algebra is compact it does satisfy the Baire theorem. A counterpart of this theorem was formulated in the language of Boolean algebras as the Rasiowa–Sikorski theorem, Rasiowa and Sikorski [21], cf., Ch.1, sect. 14

RS *In any Boolean algebra* **B**, *given a countable sequence* $(A_n)_n$ *of dense sets and* $0 \neq a \in \mathbf{B}$, *there exists an ultrafilter* f *with the properties that* $a \in f$ *and* $f \cap A_n \neq \emptyset$ *for each* n.

We translate the assumptions of RS into the language of topological spaces. For each A_n, the set $W_n = \bigcup \{U_a : a \in A_n\}$ is open dense in the Stone space $S(\mathbf{B})$. Hence the intersection $W = \bigcap \{W_n : n \in N\}$ is a dense set by the Baire theorem (Proposition 2.15) hence there exists $f \in W \cap U_a$. This f is as desired.

2.8 Continuity

To this point, we have been interested in the inner properties of topological spaces, without any attempt to relate spaces one to another. In order to compare or relate topological structures, we have to define an appropriate type of functions or *morphisms* among topological spaces. Of such a morphism we would require that it represents the topological structure of one space in the other. It has turned out that proper morphisms in topological case are *continuous functions.* A function $f : X \to Y$ of a topological space (X, τ) into a topological space (Y, σ) is said to be *continuous* if and only if the condition $f(ClA) \subseteq Cl(f(A))$ holds for every $A \subseteq X$.

It does mean that f maps closures of subsets in X into closures of their images in Y. It may be convenient to have some characterizations of continuity in terms of open or closed sets; to this end we have to look at inverse images in X of sets in Y.

Proposition 2.27. *For $f : X \to Y$, the following are equivalent*

1. f is continuous;

2. $Clf^{-1}(B) \subseteq f^{-1}(ClB)$ for every $B \subseteq Y$;

3. $f^{-1}(IntA) \subseteq Intf^{-1}(A)$ for every $A \subseteq Y$;

4. The set $f^{-1}(A)$ is open for each open set $A \subseteq Y$;

5. The set $f^{-1}(A)$ is closed for each closed set $A \subseteq Y$.

Proof. Property 1 implies Property 2. In the condition $f(ClA) \subseteq Cl(f(A))$, we substitute $f^{-1}(B)$ for A so we get $f(Clf^{-1}(B)) \subseteq Cl(B))$ so $Clf^{-1}(B) \subseteq f^{-1}(ClB)$, i.e., Property 2 holds. Assuming Property 2, we have $f^{-1}(IntA) = f^{-1}(Y \setminus Cl(Y \setminus A)) = X \setminus f^{-1}(Cl(Y \setminus A)) \subseteq X \setminus Clf^{-1}(Y \setminus A) = X \setminus Cl(X \setminus f^{-1}(A) = Intf^{-1}(A)$, i.e., Property 3 follows from Property 2.

Now, Property 4 follows from Property 3 as in case A is an open set in Y we have $f^{-1}(A) = f^{-1}(IntA) \subseteq Intf^{-1}(A)$ hence $f^{-1}(A) \subseteq Intf^{-1}(A)$ so $f^{-1}(A) = Intf^{-1}(A)$, i.e., $f^{-1}(A)$ is open. Property 5 follows from Property 4 by duality.

Finally, assume property 5. We have for $A \subseteq X$ that $A \subseteq f^{-1}(f(A))$ hence $A \subseteq f^{-1}(Clf(A))$. As the set $f^{-1}(Clf(A))$ is closed by Property 5, we have $ClA \subseteq f^{-1}(Clf(A))$ hence $f(ClA) \subseteq Clf(A)$, i.e., f is continuous ☐

In case $f : X \to Y$ is a bijection and f as well as f^{-1} are continuous, we say that f is a *homeomorphism* of X onto Y.

A homeomorphism establishes a $1-1$ correspondence between open sets in X and open sets in Y via $A \to f(A)$ as well as between closed sets in either space via the same assignment. Thus homeomorphic spaces are identical from general topological point of view.

Let us finally observe that the characterization of continuity in case of metric spaces X, Y may be given in terms of converging sequences. As $x \in ClA$ if and only if $x = limx_n$ for a sequence $(x_n)_n \subseteq A$, we may conjecture that the condition $f(limx_n) = limf(x_n)$ for any convergent sequence $(x_n)_n \subseteq X$ is equivalent to continuity as it does guarantee that $f(ClA) \subseteq Cl(f(A))$ for any $A \subseteq X$.

Proposition 2.28. *For metric spaces X, Y, a function $f : X \to Y$ is continuous if and only if the condition*

$$f(lim x_n) = lim f(x_n) \tag{2.29}$$

holds for any convergent sequence $(x_n)_n \subseteq X$.

Proof. As already observed, the condition (2.29) implies continuity. If f is continuous then for any convergent sequence $(x_n)_n$ with $x = lim x_n$ and any open $G \subseteq Y$ with $f(x) \in G$ we have $x \in f^{-1}(G)$ and as $f^{-1}(G)$ is open, we have $x_n \in f^{-1}(G)$ hence $f(x_n) \in G$ for almost every n i.e. $f(x) = lim f(x_n)$ □

In particular, any contracting map on a metric space X is continuous. An equivalent characterization of continuity may be given in the $\varepsilon - \delta$ language viz. $f : (X, \rho) \to (Y, \eta)$ is continuous at $x \in X$ if and only if

   CONT *for each $\varepsilon > 0$ there exists $\delta > 0$ with the property that $\rho(x, y) < \delta$ implies $\eta(f(x), f(y)) < \varepsilon$*

and f is continuous if and only if CONT holds with every $x \in X$.

2.9   Topologies on Subsets

Assume that (X, τ) is a topological space and $A \subseteq X$. One may introduce a topology on A in a natural way by declaring the natural injection $i : A \to X$ a continuous function. This implies that for any open set $G \subseteq X$, the intersection $A \cap G$ is open in A. Thus the family $\tau_A = \{A \cap G : G \in \tau\}$ is a topology on A. Operators Cl_A of closure and Int_A of interior with respect to the topology τ_A are related to operators Cl, Int with respect to the topology τ as described below

Proposition 2.29. *For any subset $Y \subseteq A$,*

1. $Cl_A Y = A \cap Cl Y$.
2. $A \cap Int Y \subseteq Int_A Y$.

Proof. For 1., clearly $Cl_A Y \subseteq A \cap Cl Y$ as the latter set is closed in A. Conversely, for $x \in A \cap Cl Y$, we have that $A \cap (G \cap Y) = (A \cap G) \cap Y \neq \emptyset$ hence $x \in Cl_A Y$ as $A \cap G$ runs over open neighborhoods of x in A. In case of 2., as $A \cap Int Y$ is open in A we have $A \cap Int Y \subseteq Int_A Y$ □

The converse to Property 2 does not need hold.

2.10 Quotient Spaces

We now examine how a topology may be induced on a quotient set X/R for an equivalence relation R on X. Let us assume that (X, τ) is a topological space and R an equivalence relation on X. We define a topology on the quotient set X/R of equivalence classes by requiring that the quotient function $q_R :$ $X \to X/R$ be continuous. This means that for each open set $G \subseteq X/R$ the inverse image $q_R^{-1}(G) = \bigcup G$ should be open in X. Thus we define the *quotient topology* τ_R on X/R by letting

$$\tau_R = \{G \subseteq X/R : \bigcup G \in \tau\}$$

Then we have

Proposition 2.30. *The following are true for each quotient topology*

1. *If $f : X \to Y$ is a continuous function defined on a space X with an equivalence relation R into a space Y with an equivalence relation S and xRy implies $f(x)Sf(y)$ then $\overline{f} : X/R \to Y/S$ defined as $\overline{f}[x]_R = [f(x)]_S$ is continuous;*

2. *If R, S are equivalence relations on X with $S \subseteq R$ then the quotient spaces $(X/S)/(R/S)$ and X/R are homeomorphic.*

Proof. In case of Property 1, for an open in Y/S set P of equivalence classes of S we have $q_R^{-1}(\overline{f}^{-1}(P)) = f^{-1}(q_S^{-1}(P))$ and as the latter set is open it follows that $q_R^{-1}(\overline{f}^{-1}(P))$ is an open set in X hence $\overline{f}^{-1}(P)$ is open in X/R witnessing that $\overline{f}$ is continuous.

For Property 2, we consider the function $q : (X/S)/(R/S) \to X/R$ defined by letting $q([[x]_S]_{R/S}) = [x]_R$. Then q is a bijection and q is continuous: for any open $P \subseteq X/R$ we have

$$q_S^{-1}(q_{R/S}^{-1}(q^{-1}(P))) = q_R^{-1}(P) \tag{2.30}$$

and the latter set is open in X hence $q_S^{-1}(q_{R/S}^{-1}(q^{-1}(P)))$ is open and finally $q^{-1}(P)$ is open witnessing continuity of q. We have to prove that q^{-1} is continuous which amounts to showing that q is *an open function* i.e. the image $q(Q)$ is open in X/R for each open in $(X/S)/(R/S)$ set Q. As we have

$$q_R^{-1}(q(Q)) = q_S^{-1}(q_{R/S}^{-1}(Q)) \tag{2.31}$$

and the latter set is open, it follows that $q_R^{-1}(q(Q))$ is open i.e. $q(Q)$ is open in X/R. Thus q is a homeomorphism □

2.11   Hyperspaces

In many applications, e.g., in *morphology, control theory, mathematical economy and game theory* we have to consider spaces whose points are subsets in some underlying space, see Michael [15]. Then we say of *hyperspaces*. The problem of inducing topologies on such new spaces will be our subject in this section. Consider a topological space (X, τ) and a family $\mathcal{F}$ of subsets of X. It is our aim to define a topology on $\mathcal{F}$ related to the topology τ.

In order to do this, we consider subsets of $\mathcal{F}$ of the form $C(P, \mathcal{V})$ where

$$C(P, \mathcal{V}) = \{A \in \mathcal{F} : A \subseteq P \wedge \forall V \in \mathcal{V}(A \cap V \neq \emptyset)\} \qquad (2.32)$$

Let us observe that

$$C(P, \mathcal{V}) \cap C(Q, \mathcal{W}) = C(P \cap Q, \mathcal{V} \cup \mathcal{W}) \qquad (2.33)$$

hence, if P, Q come from a family of subsets of X having the finite intersection property and $\mathcal{V}, \mathcal{W}$ are sub–families of a family of sets in X closed on arbitrary unions then sets of the form $C(P, V)$ have the finite intersection property and may be taken as a basis for open sets for a topology on $\mathcal{F}$. Clearly, the family $\mathcal{O}(X)$ of open sets in X poses itself as a candidate from whom sets P as well as elements of $\mathcal{V}$ could be taken.

Accepting that candidate we arrive at the *Vietoris topology*, after Vietoris [26], known in applications also as the *hit–or–miss topology* on $\mathcal{F}$ induced by an open base consisting of sets of the form $C(P, V)$ where P is an open set in X and $\mathcal{V}$ is a *finite* collection of open sets in X. Although the collection $\mathcal{F}$ may in principle consist of arbitrary sets in X yet the most important cases are when $\mathcal{F}$ is the family of all closed sets in X or it is the family of all compact sets in X.

We will look at topologies on closed subsets of X.

2.11.1   Topologies on Closed Sets

The *hit–or–miss topology*, see Polkowski [19] for a brief survey, on the family $\kappa(X)$ of closed sets in X is generated by the collection of sets of the form $C(P, \mathcal{V})$, where P is open and $\mathcal{V}$ is a finite collection of open sets, taken as an open basis hence each open set is of the form $\bigcup_{i \in I} C(P_i, \mathcal{V}_i)$. It is most important for applications to determine the case when this topology is metrizable, i.e., when it is identical to a metric topology induced by some metric.

We consider a metric space (X, ρ). We assume that ρ is *bounded by 1*, i.e., $\rho(x, y) = min\{1, \rho(x, y)\}$; as we already know this has no impact on the induced metric topology. Now, we define a metric δ_ρ on closed sets, called the *Hausdorff–Pompéïu metric*, see Hausdorff [12], Pompéïu [20]. For two closed sets M, N in X, we calculate subsequently the following

1. *For each $x \in M$, we compute $dist_\rho(x, N) = inf\{\rho(x, y) : y \in N\}$ called the distance from x to N;*

2. *We compute $sup\{\rho(x, N) : x \in M\}$;*

3. *For each $y \in N$, we compute $dist_\rho(y, M) = inf\{\rho(x, y) : x \in M\}$ (the distance from y to M);*

4. *We compute $sup\{\rho(y, M) : y \in N\}$;*

5. *We let $\delta_\rho(M, N) = max\{sup\{\rho(x, N) : x \in M\}, sup\{\rho(y, M) : y \in N\}\}$.*

The function δ_ρ is defined for all pairs M, N of closed sets due to our assumption that the metric ρ is bounded. We should check that δ_ρ is a metric. It is manifest that

$$\begin{cases} \delta_\rho(M, N) = 0 \Leftrightarrow M = N \\ \delta_\rho(M, N) = \delta_\rho(N, M) \end{cases} \tag{2.34}$$

so it remains to prove only that the triangle inequality holds for δ_ρ.

Consider closed sets M, R, N; we are going to prove that

$$\delta_\rho(M, N) \leq \delta_\rho(M, R) + \delta_\rho(R, N) \tag{2.35}$$

For $x \in M, y \in N$, we have by the triangle inequality that

$$\rho(x, y) \leq \rho(x, r) + \rho(r, y) \tag{2.36}$$

for any $r \in R$. Taking infima with respect to r gives us

$$\rho(x, y) \leq dist(x, R) + dist(y, R) \leq \delta_\rho(M, R) + \delta_\rho(R, N) \tag{2.37}$$

whence

$$sup\{\rho(x, N) : x \in M\} \leq \delta_\rho(M, R) + \delta_\rho(R, N) \tag{2.38}$$

follows. By symmetry, we get

$$sup\{\rho(y, M) : y \in N\} \leq \delta_\rho(M, R) + \delta_\rho(R, N) \tag{2.39}$$

and from the last two inequalities (2.35) follows. Thus δ_ρ is a metric, indeed.

It turns out that in case of a compact metric space the two topologies: the hit–or–miss topology and the metric topology induced by the Hausdorff–Pompëiu metric, coincide. We prove this important fact.

Proposition 2.31. *For each compact metric space (X, ρ), the hit–or–miss topology coincides with the metric topology induced by the Hausdorff–Pompëiu metric.*

Proof. Consider a basic open set $C(P, \mathcal{V})$ in the hit–or–miss topology where $\mathcal{V} = \{V_1, .., V_k\}$ along with $M \in C(P, \mathcal{V})$, M a closed set. As $M \subseteq P$, the closed sets $M, X \setminus P$ are disjoint. Hence, by compactness of X, there exists $\varepsilon > 0$ such that

$$\rho(x, y) > \varepsilon \tag{2.40}$$

for any $x \in X, y \in Y$. Letting

$$B(M, \varepsilon) = \{x \in X : dist(x, M) < \varepsilon\} \tag{2.41}$$

we have that $B(M, \varepsilon) \subseteq P$. Similarly, as $M \cap V_i \neq \emptyset$ for $i = 1, 2.., k$ there exist $x_1, .., x_k$ with $x_i \in M \cap V_i$ each $i \leq k$ and for each x_i we may find $\varepsilon_i > 0$ with

$$B(x_i, \varepsilon_i) \subseteq V_i \tag{2.42}$$

for each $i \leq k$.

Let $\varepsilon_0 = min\{\varepsilon, \varepsilon_1, ..., \varepsilon_k\}$ and consider a closed set N with

$$\delta_\rho(M, N) < \varepsilon_0 \tag{2.43}$$

Then $N \subseteq B(M, \varepsilon)$ hence $N \subseteq P$; also, for each $i \leq k$, we find $y_i \in N$ with $\rho(x_i, y_i) < \varepsilon_i$ implying $N \cap V_i \neq \emptyset$. It follows that $N \in C(P, \mathcal{V})$ hence

$$B_{\delta_\rho}(M, \varepsilon_0) \subseteq C(P, \mathcal{V}) \tag{2.44}$$

which does imply that $C(P, \mathcal{V})$ is open in the metric topology of δ_ρ.

For the converse, consider an open ball $B_{\delta_\rho}(M, \varepsilon)$ along with the set $P = B(M, \varepsilon)$ which is open in the topology of X. For any closed $N \subseteq P$ we have clearly $dist(y, M) < \varepsilon$ for each $y \in N$ which settles one half of the condition for $\delta_\rho < \varepsilon$. To settle the other half, i.e. to assure that $dist(x, N) < \varepsilon$ for $x \in X$, let us take in M an $\varepsilon/2$-net i.e. a set $x_1, ..., x_m$ with the property that for any $x \in M$ we find x_i such that $\rho(x, x_i) < \varepsilon/2$.

Letting $V_i = B(x_i, \varepsilon/2)$ for $i \leq k$ we observe that if $N \cap V_i \neq \emptyset$ for each $i \leq k$ then $dist(x, N) < \varepsilon$ for each $x \in M$. Thus for every $N \in C(P, \{V_1, V_2, .., V_k\})$ we have,

$$\delta_\rho(M, N) < \varepsilon \tag{2.45}$$

i.e.,

$$C(P, \{V_1, V_2, .., V_k\}) \subseteq B_{\delta_\rho}(M, \varepsilon) \tag{2.46}$$

proving that basic sets in the metric topology are open in the hit–or–miss topology thus showing both topologies to be identical and concluding the proof   □

This result tells us that the topology induced by the Hausdorff–Pompéïu metric is in fact independent of the metric on X in case X is metric compact.

As we know, a weaker but very important and useful substitute of compactness is completeness. It turns out that in case (X, ρ) is a complete metric space, the space of closed sets in X with the Hausdorff–Pompéïu metric is complete as well. To discuss this fact, we need some new notions related to sequences of sets.

For a sequence $(A_n)_n$ of sets in X, we define the *lower limit* LiA_n of this sequence by letting,

$$LiA_n = \{x \in X : limdist(x, A_n) = 0\}$$

i.e., $x \in LiA_n$ if and only if for each ball $B(x, \varepsilon)$ we have

$$B(x, \varepsilon) \cap A_n \neq \emptyset \qquad (2.47)$$

for all but finitely many A_n, i.e., if and only if every neighborhood of x does intersect almost all sets A_n. Let us observe that LiA_n is a closed set: indeed, if $y \notin LiA_n$ then some neighborhood $P \ni y$ misses infinitely many A_n's and then clearly $P \cap LiA_n = \emptyset$.

Similarly, we define the *upper limit* LsA_n of the sequence $(A_n)_n$ as

$$LsA_n = \{x \in X : liminfdist(x, A_n) = 0\}$$

i.e., $x \in LsA_n$ if and only if $limdist(x, A_{n_k}) = 0$ for a subsequence $(A_{n_k})_k$, i.e., if and only if every neighborhood of x does intersect infinitely many A_n's. As with LiA_n, the upper limit LsA_n is a closed set which may be shown by the same argument. It follows immediately that $LiA_n \subseteq LsA_n$.

In case $LiA_n = LsA_n$, we say that the sequence $(A_n)_n$ *converges* to the *limit* $LimA_n = LiA_n = LsA_n$. Now, we turn to the completeness property.

Let us first observe that in case a sequence $(A_n)_n$ converges in the metric δ_ρ, i.e.,

$$lim\delta_\rho(A_n, B) = 0 \qquad (2.48)$$

for some closed set B, we have $B = LimA_n$.

Indeed, given $\varepsilon > 0$, there is $n_\varepsilon \in N$ with $A_n \subseteq B(B, \varepsilon)$ for $n \geq n_\varepsilon$. Hence, for each $x \in B$ we have $B(x, \varepsilon) \cap A_n \neq \emptyset$ for $n \geq n_\varepsilon$. It follows that $B \subseteq LiA_n$. On the other hand, given $x \in LsA_n$, we may define inductively a sequence $(a_{n_k})_k$ with $a_{n_k} \in A_{n_k}$ and $x = lima_{n_k}$. Passing eventually to a subsequence, in virtue of (2.48), we have $\rho(a_{n_k}, b_k) < 1/k$ with some $b_k \in B$ for each k. Thus $x = limb_k$ hence $x \in B$ and finally $LsA_n \subseteq B$. Therefore $B = LimA_n$.

Proposition 2.32. *(Hahn [11]) For a complete metric space (X, ρ), the space $\mathcal{C}(X)$ of closed sets in X with the Hausdorff–Pompéiu metric δ_ρ is complete.*

Proof. Assume $(A_n)_n$ is a fundamental sequence with respect to δ_ρ. Once we exhibit B with (2.48) we have $B = LimA_n$. It is the easiest to consider $B = LsA_n$. Given $\varepsilon > 0$, we have $n(\varepsilon)$ such that,

$$\delta_\rho(A_{n(\varepsilon)}, A_n) < \varepsilon \qquad (2.49)$$

for $n \geq n(\varepsilon)$. We consider $x \in LsA_n$. Then $B(x, \varepsilon)$ intersects infinitely many A_n's hence for some $n > n(\varepsilon)$ and some $y \in A_n$ we have $\rho(x, y) < \varepsilon$ and thus $dist(x, A_n) < \varepsilon$. By (2.48), we have,

$$dist(x, A_{n(\varepsilon)}) \leq 2 \cdot \varepsilon \qquad (2.50)$$

hence,

$$sup_x dist(x, A_{n(\varepsilon)}) \leq 2 \cdot \varepsilon \qquad (2.51)$$

On the other hand, for $z \in A_{n(\varepsilon)}$, we may find a sequence $(z_{n_k})_k$ with

$$\begin{cases} (i) \ z_{n_k} \in A_{n_k} \\ (ii) \ n_k > n(\varepsilon) \\ (iii) \ \rho(z_{n_k}, z_{n_{k+1}}) < 1/2^k \cdot \varepsilon \end{cases} \qquad (2.52)$$

This sequence may be defined by induction as $(A_n)_n$ is a fundamental sequence. As the sequence $(z_{n_k})_k$ is fundamental with respect to ρ, there exists $w = lim z_{n_k}$. Clearly, $w \in LsA_n$ and $\rho(z, w) \leq 2 \cdot \varepsilon$, hence,

$$\begin{cases} (iv) \ dist(z, LsA_n) \leq 2 \cdot \varepsilon \\ (v) \ sup_z\{dist(z, LsA_n)\} \leq 2 \cdot \varepsilon \end{cases} \qquad (2.53)$$

Hence,

$$\delta_\rho(A_{n(\varepsilon)}, LsA_n) \leq 2 \cdot \varepsilon \qquad (2.54)$$

and, by the triangle inequality, we have

$$\delta_\rho(A_n, LsA_n) \leq 3 \cdot \varepsilon \ for \ n > n(\varepsilon) \qquad (2.55)$$

This proves that $lim \delta_\rho(A_n, LsA_n) = 0$. Hence $(A_n)_n$ converges to $LsA_n = LimA_n$. Completeness of the space $(\mathcal{C}(X), \delta_\rho)$ has been established $\qquad \square$

We comment also on the status of continuous mappings of the form $f : X \to 2^Y$ from a topological space X into the space of closed subsets of the space Y endowed with the Vietoris topology. Clearly, the continuity condition of f is

Proposition 2.33. *A mapping $f : X \to 2^Y$ is continuous if and only if the set $\{x \in X : f(x) \subseteq P, f(x) \cap V_i \neq \emptyset\}$, for $i = 1, 2, ..., k$, is open in X for each open basic set $C(P, V_1, ..., V_k)$ in the space 2^Y.*

The continuity notion ramifies for multi–valued mappings into two weaker *semi–continuity* notions, viz., a mapping $f : X \to 2^Y$ is *upper–semicontinuous* if and only if the set $\{x \in X : f(x) \subseteq P\}$ is open in X for each open set P in Y.

A mapping $f : X \to 2^Y$ is *lower–semicontinuous* if and only if the set $\{x \in X : f(x) \cap V \neq \emptyset\}$ is open in X for each open set V in Y.

2.12 Čech Topologies

In many cases that arise in real problems, we have to deal with a weaker form of topology. Such situation, for instance, takes place when we try to define closure operator with respect not to the full family of open sets but with respect to a certain sub–family, e.g., a covering of a space. Assume that we are given a covering $\mathcal{U}$ of a set X and for each $A \subseteq X$, we let

$$Cl_c A = \{x \in X : \forall G \in \mathcal{U}(x \in G \Rightarrow G \cap A \neq \emptyset)\}$$

We state properties of the operator Cl_c

Proposition 2.34. *The operator Cl_c does satisfy the following properties*

1. $Cl_c\emptyset = \emptyset$;

2. $A \subseteq Cl_c A$ for each $A \subseteq X$;

3. $A \subseteq B \Rightarrow Cl_c A \subseteq Cl_c B$ for each pair A, B of subsets of X.

All these properties follow immediately from the definition of Cl_c. Let us observe that the operator Cl_c does not need to observe the properties of topological closure operator, notably, in general the properties that $Cl_c(A \cup B) = Cl_c A \cup Cl_c B$ and $Cl_c Cl_c A = Cl_c A$ do not hold and are replaced by weaker consequences of 1-3, viz.,

4. $Cl_c(A \cap B) \subseteq Cl_c A \cap Cl_c B$;

5. $Cl_c A \cup Cl_c B \subseteq Cl_c(A \cup B)$;

6. $Cl_c A \subseteq Cl_c Cl_c A$.

We will call the operator Cl_c satisfying 1-3, a fortiori also 4–6, the *Čech closure operator*, see Čech [7]. A most important case when a Čech closure operator does arise is that of a *tolerance relation* when we consider, e.g., a covering formed by tolerance sets of the form $\tau(x) = \{y \in X : x\tau y\}$. We denote by $\mathcal{T}(\tau)$ the set of all tolerance sets of τ. Then for any $A \subseteq X$, we have,

$$Cl_c A = \{x \in X : \forall T \in \mathcal{T}(\tau)(x \in T \Rightarrow T \cap A \neq \emptyset)\} \qquad (2.56)$$

The *Čech topology* generated by the Čech closure operator is the family of all *open sets* with respect to Cl_c, viz., we declare a set $A \subseteq X$ to be *closed* in case $Cl_c A = A$ and we declare a set $A \subseteq X$ to be *open* in case the complement $X \setminus A$ is a closed set. Then we have

Proposition 2.35. *In any Čech topological space*

1. Each intersection of a family of closed sets is a closed set;

2. Each union of a family of open sets is an open set.

Proof. Assume that $\mathcal{F}$ is a family of closed sets in a Čech topological space X. Let $K = \bigcap \mathcal{F}$. As $K \subseteq F$ for any $F \in \mathcal{F}$, we have $Cl_c K \subseteq Cl_c F = F$ for any $F \in \mathcal{F}$ hence $Cl_c K \subseteq K$ and finally $Cl_c K = K$ i.e. K is closed. Property 2. follows immediately by duality   □

We now confine ourselves to Čech topologies generated from coverings by tolerance sets in the way described above. We may observe that

$$\begin{cases} (i) \ Cl_c A \ \text{need not be a closed set} \\ (ii) \ Cl_c(A \cap B) \subseteq Cl_c A \cap Cl_c B \\ (iii) \ Cl_c A \cup Cl_c B \subseteq Cl_c(A \cup B) \end{cases} \tag{2.57}$$

hold but converse statements to (ii), (iii) need not hold.

We may characterize $Cl_c A$ via open sets as in case of topological spaces. For a point $x \in X$, we say that a set $Q \subseteq X$ is a *neighborhood* of x if and only if $x \notin Cl_c(X \setminus Q)$. Clearly, $x \in Q$. Then, we have

Proposition 2.36. *For $x \in X$, $x \in Cl_c A$ if and only if $Q \cap A \neq \emptyset$ for each neighborhood Q of x.*

Proof. If for some neighborhood Q of x we have $Q \cap A = \emptyset$, then $A \subseteq X \setminus Q$ and we have $Cl_c A \subseteq Cl_c(X \setminus Q)$ hence $x \notin Cl_c A$. Thus, $x \in Cl_c A$ implies $Q \cap A \neq \emptyset$ for any neighborhood Q of x. On the other hand, $x \notin Cl_c A$ implies $x \notin Cl_c(X \setminus (X \setminus A))$ hence $X \setminus A$ is a neighborhood of x disjoint to A □

Let us consider a sequence of closure operators $(Cl_c^n)_n$ defined recurrently for any $A \subseteq X$ by

$$\begin{cases} Cl_c^0 A = A \\ Cl_c^{n+1} A = Cl_c(Cl_c^n A) \ for \ n \geq 1 \end{cases} \tag{2.58}$$

For any $A \subseteq X$, we have

$$Cl_c^0 A = A \subseteq Cl_c A \subseteq ... \subseteq Cl_c^n A \subseteq Cl_c^{n+1} A \subseteq .... \tag{2.59}$$

In our setting of a tolerance relation, subsequent values $Cl_c^n A$ correspond to saturation of A with tolerance sets: $S^0 A = A$, $S^{n+1} A = S(S^n A)$ where $SA = \bigcup \{\tau(x) : \tau(x) \cap A \neq \emptyset\}$, i.e., we have $Cl_c^n A = S^n A$, any n.

Let us define the closure operator Cl^* by letting

$$Cl^* A = \bigcup \{Cl_c^n A : n \in N\}$$

Then, we have

Proposition 2.37. *For any tolerance relation τ on a set X, Cl^*A is a topological closure operator. Moreover, Cl^* is identical to the Čech closure operator induced by the equivalence relation $\bar{\tau}$, the transitive closure of the relation τ.*

Proof. Clearly, $Cl^*\emptyset = \emptyset$, $A \subseteq Cl^*A$ hold by definition of Cl^*. To proceed further we observe that,
(i) $x \in Cl^*A$ if and only if there is a finite sequence $x_0 \in A, x_1, ..., x_k = x$ with

$$x_i \tau x_{i+1} \ for \ i = 0, 1, ..., k-1 \tag{2.60}$$

and we assume that $x \in Cl^*Cl^*A$. Then there is a finite sequence $y_0 \in Cl^*A$, $y_1, ..., y_m = x$ for some $m \in N$ with $y_i \tau y_{i+1}$ for $i = 0, .., m-1$ and thus we have a sequence $z_0 \in A, z_1, ..., z_p = y_0$ for some $p \in N$ with $z_i \tau z_{i+1}$ for $i = 0, ..., p-1$. It follows that $x \in Cl_c^{m+p}A$ hence $x \in Cl^*A$. Thus, $Cl^*Cl^*A \subseteq Cl^*A$ and finally $Cl^*Cl^*A = Cl^*A$.

It remains to prove that $Cl^*A \cup Cl^*B = Cl^*(A \cup B)$. Clearly, $Cl^*A \cup Cl^*B \subseteq Cl^*(A \cup B)$. Assume that $x \in Cl^*(A \cup B)$ hence there is a sequence $x_0 \in A \cup B, x_1, ..., x_k = x$ with $x_i \tau x_{i+1}$ for $i = 0, 1, ..., k-1$. Then either $x_0 \in A$ or $x_0 \in B$; in the former case, $x \in Cl^*A$ and in the latter case $x \in Cl^*B$. Thus $Cl^*(A \cup B) \subseteq Cl^*A \cup Cl^*B$ and finally the equality of both sides follows   □

We have proved that Cl^* is a topological closure operator. For the second part of the proposition, indeed, we have $x \in Cl^*A$ if and only if $A \cap \bar{\tau}(x) \neq \emptyset$ i.e. the closure operator Cl^* and the Čech closure operator $Cl_{\bar{\tau}}$ coincide.

Remaining for a while with the case of an equivalence relation, we may observe that in that case the tolerance sets are equivalence classes, and the resulting Čech topology is a topology induced by taking the set of equivalence classes as an open basis for this topology. Thus, any open set is a union of a family of equivalence classes a fortiori any closed set is a union of a family of equivalence classes. Hence, in this topology, any open set is closed as well (we call such topology a *clopen* topology). Moreover, any intersection of a family of open sets is an open set. Let us observe that, conversely, given any clopen topology with the property that the intersection of any family of open sets is an open set, we may introduce an equivalence relation which induces this topology: it is sufficient to let xRy if and only if x and y belong in the same open sets, i.e.,

$$xRy \leftrightarrow \forall Q = IntQ(x \in Q \leftrightarrow y \in Q)$$

It follows that each equivalence class $[x]_R$ is the intersection of all open sets containing x hence it is clopen. Clearly, any open set Q is the union of the equivalence classes $[x]_R$ for $x \in Q$.

Proposition 2.38. *The Čech topology is generated by the set of equivalence classes of an equivalence relation if and only if*

(ii) any open set is closed

(iii) the intersection of each family of open sets is an open set

An analysis of clopen topologies induced by indiscernibility relations in information systems, see Ch. 4, is given in Skowron [23]; in Wiweger [27], topological calculus is applied to characterize sets which are not clopen in this context. Analysis of topological properties of rough sets can be found in Polkowski [16]–[18], cf., Ch. 4. Readers wishing to extend for themselves the outline of topology of concepts given here may be advised to consult Engelking [8] or Kuratowski [14].

References

1. Alexandrov, P.S.: Zur Begründung der n–dimensionalen mengentheoretischen Topologie. Math. Ann. 94, 296–308 (1925)
2. Alexandrov, P.S., Urysohn, P.S.: Mémoire sur les espaces topologiques compacts. Verh. Konink. Akad. Amsterdam 14 (1929)
3. Baire, R.: Ann. di Math. 3 (1899)
4. Banach, S.: Sur les opérations dans les ensembles abstraits et leur application aux équations intégrales. Fundamenta Mathematicae 3, 133–181 (1922)
5. Cantor, G.: Math. Ann. 17 (1880)
6. Cantor, G.: Math. Ann. 21 (1883)
7. Čech, E.: Topologické prostory. In: Čech, E., Frolik, Z. (eds.) Topological Spaces. Academia, Praha (1966)
8. Engelking, R.: An Outline of General Topology. North–Holland, PWN–Polish Scientific Publishers, Amsterdam–Warszawa (1965)
9. Fréchet, M.: Sur quelques points du Calcul fonctionnel. Rend. Circ. Matem. di Palermo 22 (1906)
10. Fréchet, M.: Les espaces abstraits, Paris (1928)
11. Hahn, H.: Reelle Funktionen I. Leipzig (1932)
12. Hausdorff, F.: Grundzüge der Mengenlehre (1914); von Veit, Leipzig, Hausdorff F.: Set Theory. Chelsea, New York (1965)
13. Kuratowski, C.: Sur l'operation Ā de l'Analysis Situs. Fundamenta Mathematicae 3, 182–199 (1922)
14. Kuratowski, C.: Topology I, II. Academic Press – Polish Scientific Publishers (PWN), New York – Warszawa (1966)
15. Michael, E.: Topologies on spaces of subsets. Trans. Amer. Math. Soc. 71, 152–182 (1951)
16. Polkowski, L.T.: On convergence of rough sets. In: Słowiński, R. (ed.) Intelligent Decision Support. Handbook of Applications and Advances of Rough Sets Theory. Kluwer, Dordrecht (1992)
17. Polkowski, L.T.: Mathematical morphology of rough sets. Bull. Pol. Ac.: Math. 41(3), 241–273 (1993)
18. Polkowski, L.T.: Metric spaces of topological rough sets from countable knowledge bases. Foundations of Computing and Decision Sciences 18, 293–306 (1993)
19. Polkowski, L.T.: Hit–or–miss topology. In: Hazewinkel, M. (ed.) Encyclopaedia of Mathematics. Supplement 1, p. 293. Kluwer, Dordrecht (1998)
20. Pompéïu, D.: Ann. de Toulouse 7 (1905)
21. Rasiowa, H., Sikorski, R.: A proof of the completeness theorem of Gödel. Fundamenta Mathematicae 37, 193–200 (1950)

22. Riesz, F.: Stetigskeitbegriff und abstrakte Mengenlehre. Atti IV Congr. Int. Mat., Rome (1909)
23. Skowron, A.: On topology in information systems. Bull. Pol. Ac.: Math. 36, 477–479 (1988)
24. Stone, M.H.: The theory of representations for Boolean algebras. Trans. Amer. Math. Soc. 40, 37–111 (1936)
25. Tikhonov, A.N.: Über einen Funktionenraum. Math. Ann. 111 (1935)
26. Vietoris, L.: Monat. Math. Ph. 31, 173–204 (1921)
27. Wiweger, A.: On topological rough sets. Bull. Pol. Ac.: Math. 37, 89–93 (1988)

Chapter 3
Reasoning Patterns of Deductive Reasoning

Processes of reasoning are divided into two main types: deductive and reductive. Deductive reasoning begins with a set of premises and concludes with a set of inferences obtained by specified rules of deduction, whereas reductive reasoning tries to obtain a set of premises/causes for an observed set of facts. In this chapter, we present the reader with some basic schemes of deductive reasoning. We begin with sentential calculus (propositional logic) which sets the pattern of deductive reasoning and then we discuss many–valued propositional calculi, predicate calculus, and modal calculi. Analysis of these types of reasoning results in notions which are relevant also for reductive types of reasoning, in particular, for rough and fuzzy types of reasoning discussed in the next chapter as examples of reductive reasoning.

3.1   The Nature of Exact Reasoning

Reasoning processes (logics) are divided into two classes: *deductive* and *reductive* reasoning, see Bocheński [9], Łukasiewicz [30]. Logics implementing *deductive reasoning*, proceed with inference from premises to a conclusion according to a scheme: if premises $p, p \Rightarrow q$, then the conclusion q. The other type of reasoning, the *reductive reasoning*, carries inference from conclusions to a set of plausible premises, according to a scheme: if $p \Rightarrow q$ and q, then p. As pointed to by Łukasiewicz, reductive reasoning does encompass *inductive reasoning*, i.e., inference about the general property of a class from behavior of a sample from the class, a generalization of evidence to a hypothesis about a population. Clearly, whereas deductive reasoning is exact, reductive reasoning is more intuitive.

Deductive reasoning in its formalized version, is a process by which a certain definite set of *conclusions* is reached from a set of certain *premises* in a formalized manner, by means of a priori specified *reasoning rules*. All patterns of deductive reasoning are modeled after, or are modifications of, the classical reasoning with *sentences*, initiated with the Stoic school of Athens, ca. III–th

L. Polkowski: Approximate Reasoning by Parts, ISRL 20, pp. 79–143.
springerlink.com   © Springer-Verlag Berlin Heidelberg 2011

century B.C., cf., Bocheński [3] and initially developed in the modern form
in works of Frege [9] and Post [39].

Calculus of sentences had been extended to the *1st–order calculus*, initiated
by Frege [9], *many–valued calculi*, see Łukasiewicz [30]–[28], including *fuzzy
calculi*, see, e.g., Hájek [17].

In what follows, we give a nutshell account of these calculi in order to bring
together pieces of knowledge scattered through various sources and to give
the reader a perspective on this classical subject.

3.2   Propositional Calculus

We begin with a discussion of the basic tool of mathematical reasoning: propositional calculus. By a *proposition, sentence* we mean any statement about
reality of interest to us whose truth value can be established with certainty.
By a *truth value* of a proposition, we understand either of two possible values:
truth (T or 1), *falsity* (F or 0). A proposition p may be therefore either *true* or
false and exactly one of the two possibilities actually holds for p. For instance,
the statement *"if today is Monday then tomorrow is Tuesday"* is according
to our best knowledge true while the statement of ordinary arithmetic *" 2+2
= 3"* is false.

Sentential calculus deals with propositions considered in their symbolic
form: there is no reason to consider all possible propositions (apart from
the real possibility of listing all of them); instead, the aim of the theory
of sentential calculus is to reveal all properties of this system which hold
regardless of the nature of statements considered, i.e., which depend only on
truth values of these statements (one says also that such properties are *truth
functional*). Thus, we consider, instead of actual propositions, *propositional
variables* $p, q, r, s, t, \ldots$ ranging over propositions and we aim at establishing
general properties of the sentential calculus which are preserved when we
substitute for any of variables $p, q, r, \ldots$ a specific proposition.

As with any language, in propositional calculus we may construct complex
expressions from simpler ones. Passing from simpler to more complex propositions is effected by means of *proposition–forming functors* (or, *functors*,
shortly). In propositional calculus we have four basic functors with which to
form more complex expressions. Let us look at them now.

The functor of *negation* denoted N or $\neg$ is a unary functor acting on single
propositions: for a proposition p, the expression Np (or, $\neg p$) is a proposition.
To describe fully the action of N on propositions, we must describe its semantic impact, thus we have to give truth values of Np for respective values
of p. So, we let $Np = 0$ when $p = 1$ and $Np = 1$ when $p = 0$. Thus, Np is
true if and only if p is false. The proposition Np is read: *"it is not true that
p"* or, shortly, *"not p"*. To give a concise description of the semantic content
of the functor N, we use the following conditions

$$\begin{cases} N0 = 1 \\ N1 = 0 \end{cases} \tag{3.1}$$

The binary functor of *alternation,disjunction* denoted OR (or, $\vee$) acts on pairs of propositions p, q to produce the proposition $ORpq$ (denoted also $pORq$, $p \vee q$; the proposition $p \vee q$ is read: *either p or q*. Truth functional description of $\vee$ is given by the following conditions

$$\begin{cases} 0 \vee 0 = 0 \\ 0 \vee 1 = 1 \\ 1 \vee 0 = 1 \\ 1 \vee 1 = 1 \end{cases} \tag{3.2}$$

It follows that $p \vee q$ is true if and only if at least one of p, q is a true proposition.

The binary functor AND of *conjunction* (denoted also $\wedge$) acts on pairs p, q to produce the expression $ANDpq$ (denoted also $p \wedge q$). The following conditions determine the semantics of $\wedge$.

$$\begin{cases} 0 \wedge 0 = 0 \\ 0 \wedge 1 = 0 \\ 1 \wedge 0 = 0 \\ 1 \wedge 1 = 1 \end{cases} \tag{3.3}$$

Hence, the proposition $p \wedge q$ is true if and only if each of propositions p, q is true.

The binary functor of *implication* C (denoted also $\Rightarrow$) acts on pairs p, q of propositions to form the proposition $p \Rightarrow q$ defined as follows

$$\begin{cases} 0 \Rightarrow 0 = 1 \\ 0 \Rightarrow 1 = 1 \\ 1 \Rightarrow 0 = 0 \\ 1 \Rightarrow 1 = 1 \end{cases} \tag{3.4}$$

We may now introduce the notion of a *meaningful expression(a formula)* of propositional calculus by stating that

1. *Each propositional variable is a meaningful expression;*

2. *$N\alpha$ is a meaningful expression whenever α is a meaningful expression;*

3. *Expressions $\alpha \vee \beta$, $\alpha \wedge \beta$, $\alpha \Rightarrow \beta$ are meaningful when α, β are meaningful expressions;*

4. *No expression other than obtained from 1–3 is a meaningful expression.*

Hence, the meaningful expressions are the least set containing all propositional variables and closed under propositional functors $\neg, \Rightarrow, \vee, \wedge$.

We will be interested from now on in meaningful expressions formed in accordance with the rules given in this paragraph. For a meaningful expression α, we will sometimes write down α in the form $\alpha(p_1, p_2, ..., p_k)$ to denote the fact that propositional variables occurring in α are exactly $p_1, p_2, ..., p_k$. Clearly, any meaningful expression $\alpha(p_1, p_2, ..., p_k)$ takes a truth value either 0 or 1 on assuming that each of p_i ($i = 1, 2, ..., k$) takes a truth value 0 or 1 (a simple proof may be done by induction on number of connectives in α). Thus $\alpha(p_1, p_2, ..., p_k)$ induces a mapping (a Boolean function) $f_\alpha : 2^k \to \{0, 1\}$.

For instance, the meaningful expression $\neg p \Rightarrow q$ takes the value 0 for $p = 0 = q$, the value 1 for $p = 0, q = 1$, the value 1 for $p = 1, q = 0$ and the value 1 for $p = 1, q = 1$.

We will say that two meaningful expressions α, β are *equivalent* when they take the same truth value for each substitution of the same truth values for equiform propositional variables occurring in them. To give an important example, we observe that

Proposition 3.1. *The following expressions are equivalent*

1. Meaningful expressions $p \vee q$, $\neg p \Rightarrow q$ are equivalent;

2. Meaningful expressions $p \wedge q$, $\neg(p \Rightarrow \neg q)$ are equivalent.

Proof. To verify this proposition, it is sufficient to compare truth values of both expressions for each substitution of truth values for p, q.

We will call a *tautology* of the sentential calculus any meaningful expression $\alpha(p_1, ..., p_k)$ with the property that the truth value of α is 1 for every substitution of truth values for $p_1, ..., p_k$.

We will write $\alpha \Leftrightarrow \beta$ in case α, β are equivalent meaningful expressions $\square$

Let us derive two important observations from the above proof.

1. *For any meaningful expression α, the truth value of α for any substitution of truth values of propositional variables in α may be checked in a finite number of steps. In particular, it may be checked in a finite number of steps whether α is a tautology of propositional calculus, i.e., the truth value of α is 1 for any substitution of truth values for propositional variables in α;*

2. *Connectives $\vee$ and $\wedge$ may be defined by means of connectives $\Rightarrow, \neg$: in any meaningful expression, $p \vee q$ may be replaced with $\neg p \Rightarrow q$ and $p \wedge q$ may be replaced with $\neg(p \Rightarrow \neg q)$ without any effect on the truth value of that expression. It follows that in forming meaningful expressions of propositional calculus, we may apply connectives $\Rightarrow, \neg$ only.*

When discussing meaningful expressions of propositional calculus, we are interested above all in *tautologies* of this calculus, i.e., as defined above, in expressions which are true regardless of truth values of propositional variables that occur in these expressions. The reason for this lies primarily in the fact that tautologies of propositional calculus may be regarded as *laws of thought* leading always to valid (true) conclusions regardless of truth values of premises (i.e., propositional variables or propositions substituted for them). The notion of a tautology is a *semantic* one, relying on our checking of truth values. On the other hand, we have defined a *syntactic mechanism* for generating meaningful expressions. The importance of relations between the syntactic and semantic aspects of propositional calculus had been fully recognized in Frege [9].

It had been his idea to distinguish between meaningful expressions and *derivation rules*, i.e., rules for forming new meaningful expressions from given ones. The Frege program assumed the existence of certain apriorical expressions called *axioms* and the existence of certain rules by means of which meaningful expressions may be derived from axioms. Accepted axioms and derivation rules constitute a *deduction system* for propositional calculus and define the *consequence relation* for this system, see Rasiowa and Sikorski, see [42], p. 151. Meaningful expressions derived from axioms by means of specified rules are called *theorems* of the system.

Clearly, in our choice of axioms and derivation rules, we are guided by certain constraints. Above all, we would like to invent axioms and rules of derivation which would give as theorems of the system all tautologies of propositional calculus; also, we would like to have the situation when all tautologies of propositional calculus are theorems of our system. When both constraints are met, we have the case when syntactic and semantic aspects of propositional calculus are mutually expressible: theorems of the deductive system = tautologies of propositional calculus.

Many axiom systems for sentential calculus were proposed by various authors.

The axioms proposed by Jan Łukasiewicz, see [30], are the following

T1 $(p \Rightarrow q) \Rightarrow [(q \Rightarrow r) \Rightarrow (p \Rightarrow r)]$

T2 $(\neg p \Rightarrow p) \Rightarrow p$

T3 $p \Rightarrow (\neg p \Rightarrow q)$

As the derivation rule, the *detachment* (or, *modus ponens, MP*) is applied, which consists in the following

> *Detachment, MP For any pair of meaningful expressions α, β, if α and $\alpha \Rightarrow \beta$ are theorems of the system, then β is a theorem of the system.*

We also allow *substitution*, i.e., in a meaningful formula, in place of each occurrence of a variable, a meaningful expression α can be substituted.

Thus, axiom formulas given here are in fact, *axiom schemes*, as in each of them, a substitution can be effected leading to an axiom *instance*.

Given axioms and the derivation rule, one may derive from axioms new theorems of the system. We will give a formal definition of a *derivation* (a *proof*) of a theorem.

For a meaningful expression α, a *derivation* of α from axioms is a sequence

$$\alpha_0, \alpha_1, .., \alpha_n$$

of meaningful expressions such that

1. α_0 *is an axiom*;

2. α_n *is equiform with α*;

3. *Each α_i for $i = 1, 2, ..., n$ either is obtained from some α_j, α_k with $j, k < i$ by means of detachment or α_i is an axiom, or α_i is obtained by substitution from some α_j with $j < i$.*

A meaningful expression having a derivation from axioms is called *a theorem* of the axiomatic system. To denote the fact that α is a theorem, we use the acceptance symbol $\vdash$ so $\vdash \alpha$ reads "α is a theorem".

We denote with the symbol (L) the deductive system consisting of axioms T1–T3 and the rules of detachment as well as of substitution as derivation rules.

Checking whether a given meaningful expression is a theorem requires a trial–and–error procedure. It is customary to apply the deductive mechanism of propositional calculus to derive a number of theorems on basis of which one may establish fundamental properties of this system.

In particular, one can derive the following theorems of propositional calculus, see, e.g., Łukasiewicz [30]

T4 $p \Rightarrow p$

T5 $p \Rightarrow (p \vee q)$

T6 $q \Rightarrow (p \vee q)$

T7 $(p \Rightarrow q) \Rightarrow [(r \Rightarrow q) \Rightarrow (p \lor r \Rightarrow q)]$

T8 $(p \land q) \Rightarrow p$

T9 $(p \land q) \Rightarrow q$

T10 $(p \Rightarrow q) \Rightarrow [(p \Rightarrow r) \Rightarrow (p \Rightarrow q \land r)]$

T11 $p \Rightarrow (q \Rightarrow r) \Rightarrow (p \land q \Rightarrow r)$

T12 $(p \land q \Rightarrow r) \Rightarrow [p \Rightarrow (q \Rightarrow r)]$

T13 $(p \land \neg p) \Rightarrow q$

T14 $[p \Rightarrow (p \land \neg p)] \Rightarrow \neg p$

T15 $p \lor \neg p$

This list can be continued at will; the question is: to what purpose can it serve? To answer this question, we introduce an algebraic component in analysis of propositional calculus, known as the Lindenbaum–Tarski algebra, see Rasiowa and Sikorski [42], p. 245, footnote 1.

For meaningful expressions (formulas) α, β, we say that α, β are *equivalent*, in symbols $\alpha \approx \beta$, when $\alpha \Rightarrow \beta$ and $\beta \Rightarrow \alpha$ are theorems of propositional calculus.

The relation $\approx$ is an equivalence relation, and one denotes with the symbol $[[\alpha]]_\approx$ the equivalence class of $\approx$ containing α. On equivalence classes of $\approx$, one does introduce the following operations

1. $[[\alpha]]_\approx \cup [[\beta]]_\approx = [[\alpha \lor \beta]]_\approx$;

2. $[[\alpha]]_\approx \cap [[\beta]]_\approx = [[\alpha \land \beta]]_\approx$;

3. $-[[\alpha]]_\approx = [[\neg \alpha]]_\approx$.

One introduces the following constants

4. $1 = [[T1[]]_\approx$;

5. $0 = [[\neg T1]]_\approx$.

The ordering $\leq_L$ in the Lindenbaum–Tarski algebra is given by, see Rasiowa and Sikorski [42], p. 246, 10.2,

Proposition 3.2. $[[\alpha]]_\approx \leq_L [[\beta]]_\approx \Leftrightarrow \alpha \Rightarrow \beta$ *is a theorem.*

A principal result, see Rasiowa and Sikorski [42], 10.6., is

Proposition 3.3. *If T4–T15 are theorems of propositional calculus, then the Lindenbaum–Tarski algebra LT(Prop) of propositional calculus with operations* $\cup$, $\cap$, $-$, *and constants* 0, 1 *is a Boolean algebra.*

As T1–T15 are theorems of propositional calculus, $LT(Prop)$ is a Boolean algebra.

By means of truth matrices of functors of propositional calculus, one makes it certain that formulas T1–T15 are tautologies. Moreover, every formula derivable from T1–T15 is a tautology: the detachment preserves tautologies, and substitution does not change the logical value of a formula.

One says that propositional calculus is *sound*: every theorem is a tautology. The converse also holds but the proof is more difficult. The property in question is the *completeness*: every tautology is a theorem. There are a few proofs of completeness, begun with Łukasiewicz [30]; we follow the lines of Rasiowa and Sikorski [42] and their proof of completeness using algebraic structures.

The observation that a formula with k propositional variables induces a Boolean function from 2^k into $\{0, 1\}$ can be generalized: for a given Boolean algebra B, and a formula α, a mapping $f_\alpha : B^k \to B$ is induced by replacing propositional functors $\vee, \wedge, \neg$ with, respectively, boolean algebra operations $\cup, \cap, --$. A particular case is the *canonical mapping* into $LT(Prop)$, at which each variable p_i in α is mapped onto its class $[[p_i]]_\approx$. Then, see Rasiowa and Sikorski [42], 1.2, p. 258,

Proposition 3.4. *Under the canonical mapping, each formula* α *is mapped onto its class* $[[\alpha]]_\approx$.

It is important to establish the value of a class of theorem in the Lindenbaum–Tarski algebra $LT(Prop)$, cf., Rasiowa and Sikorski [42], p. 247, 10.4,

Proposition 3.5. *In the Lindenbaum–Tarski algebra* $LT(Prop)$, $[[\alpha]]_\approx = 1$ *if and only if* α *is a theorem.*

Proof. Clearly, if α is a theorem, then $[[\alpha]]_\approx = 1$. If $[[\alpha]]_\approx = 1$, then $[[\beta]]_\approx \leq_L [[\alpha]]_\approx$ for each formula β, in particular $[[\alpha \Rightarrow \alpha]]_\approx \leq_L [[\alpha]]_\approx$, i.e., $(\alpha \Rightarrow \alpha) \Rightarrow \alpha$ is a theorem. As $\alpha \Rightarrow \alpha$ is a theorem (an instance of T4), by detachment it follows that α is a theorem. $\square$

This result opens up a way for a proof of completeness of propositional calculus.

Proposition 3.6. *Propositional calculus is complete, i.e, every tautology is a theorem.*

Proof. The proof rests on the property of Boolean algebras of separating elements by maximal ideals, see Ch.1, sect. 11. Assume, to the contrary, that a formula α is not any theorem. Then, in the Lindenbaum–Tarski algebra, $[[\alpha]]_\approx \neq 1$, hence, there exists a maximal ideal I such that $[[\alpha]]_\approx \in I$. The quotient algebra $LT(Prop)/I = \{0,1\}$, see Ch. 1, sect. 11, and, as $1 \notin I$, $[[\alpha]]_\approx$ maps to 0, i.e., in the algebra $LT(Prop)/I$, α is valued as 0, hence, α is not any tautology ☐

The completeness theorem allows now for checking in an algorithmic manner whether a given meaningful expression is a theorem of our system: it is sufficient to this end to determine the truth value of this expression for each possible selection of literals and to check whether in each case the truth value of the expression is 1. If so, the expression, being a tautology, is a theorem; otherwise, the answer is in the negative. As the checking requires only at most 2^k steps for the number k of propositional variables in the expression, this procedure is finite. One says that propositional calculus is *decidable*.

We finally include in our discussion – in order to give a fairly complete picture of propositional calculus – a classical *deduction theorem* due to Herbrand [19], to be applied later on in a discussion of modal logics.

In few coming paragraphs, we use the Polish notation Cpq for implication, for conciseness' sake

Proposition 3.7. *(Herbrand [19]) If* $\gamma \vdash \beta$, *then* $\vdash C\gamma\beta$.

Proof. Let us observe that this proposition extends easily to a more general one viz. if $\gamma_1, .., \gamma_k \vdash \beta$ then $\gamma_1, ..., \gamma_{k-1} \vdash C\gamma_k\beta$. Simply please check that the proof below holds with minor changes in this more general case. Our proof below rests on some theses of the system (L).

Consider a derivation $\alpha_1, ..., \alpha_k$ of β from γ. The proof is by induction on i of the statement $\vdash C\gamma\alpha_i$. For $i = 1$, either α_1 is an axiom or α_1 is γ.

In the former case as $CqCpq$ is a theorem, we have after an adequate substitution $\vdash C\alpha_1 C\gamma\alpha_1$ and from this we infer by means of detachment that $\vdash C\gamma\alpha_1$.

In the latter case, by the theorem Cpp after the substitution p/γ we have $\vdash C\gamma\gamma$, i.e., $\vdash C\gamma\alpha_1$. Assuming that $\vdash C\gamma\alpha_i$ is proved for $i < n \leq k$, we consider α_n in the three cases.

1. α_n *is an axiom*;

2. α_n *is equiform with* γ;

3. α_n *follows by means of detachment from some* α_r, α_s *with* $r, s < n$, *i.e.,* α_r *is equiform with* $C\alpha_s\alpha_n$.

Clearly, only Case 3 is new, since Cases 1 and 2 may be done with similarly to the case when $i = 1$. For Case 3, we have

4. $\vdash C\gamma\alpha_s$;

5. $\vdash C\gamma C\alpha_s\alpha_n$.

As

$$CCpCqrCCpqCpr$$

is a theorem, substituting in it $p/\gamma, q/\alpha_s, r/\alpha_n$ gives

$$\vdash CC\gamma C\alpha_s\alpha_n CC\gamma\alpha_s C\gamma\alpha_n$$

from which by applying twice detachment with 4 and 5, we arrive at

$$\vdash C\gamma\alpha_n$$

The inductive step is completed and the deduction theorem follows $\qquad$ $\square$

We now consider logical calculi with more than two states of truth.

3.3   Many–Valued Calculi: 3–Valued Logic of Łukasiewicz

It had been assumed by schools and individuals developing logical systems up to the beginning of the 20th century that propositions of logic should be restricted to be either true or false and the twain should exclude themselves (cf., in this respect, Łukasiewicz [31]. Those systems are therefore called *two–valued*. On the other hand, from the very beginning of logic, as witnessed by the legacy of Aristotle, there was conviction that there are statements which cannot be true or false, e.g., statements about future and that such statements may be interpreted as *true to a degree*.

Łukasiewicz was the first to consider a many–valued logic (actually, a 3–valued one) in Łukasiewicz [25], cf. also, Łukasiewicz [30], [28], in which, in addition to already familiar truth values 1 (true) and 0 (false), other truth values, representing various partial *degrees* of truth were present (in particular, the value $\frac{1}{2}$ in 3–valued logic). The idea of such a logic was based on the truth matrix of propositional calculus, cf., Fig. 3.1, in which functors C, N, of, respectively, implication and negation, are represented.

Looking at the part referring to C, i.e., at the sub–matrix composed of columns 1 and 2, we may observe that the logical value $v(Cpq)$ depends on logical values $v(p), v(q)$, e.g., according to the formula

C	0	1	N
0	1	1	1
1	0	1	0

Fig. 3.1 Truth table of propositional calculus

C	0	1	2	N
0	1	1	1	1
1	0	1	2	0
2	2	1	1	2

Fig. 3.2 Truth table of the 3–valued calculus

$$v(Cpq) = min\{1, 1 - v(p) + v(q)\} \qquad (3.5)$$

If we admit an intermediate logical value $\frac{1}{2}$ then in the case when $v(p), v(q)$ take values $0, 1, \frac{1}{2}$, also $v(Cpq)$ does take one of this values. Taking into account a formula defining negation in propositional calculus,

$$v(Np) = 1 - v(p) \qquad (3.6)$$

we may construct the matrix for this new 3–valued logic as the following one (in which for typing convenience the value of $\frac{1}{2}$ has been replaced by the value of 2), cf., Fig. 3.2.

We denote this 3–valued logic with the symbol L₃. Let us remark here that formulas for C, N presented above allow for n–valued logics (with truth values $0, 1, ..., \frac{k}{n-1}$ where $k = 1, 2, ..., n - 2$) as well as for *countably valued logic* (with values as rational numbers in the interval $[0, 1]$), or *real valued logic* with values in the interval $[0, 1]$, all of which were proposed by Łukasiewicz and investigated in the 20's and 30's, cf., Łukasiewicz and Tarski [32]. Independently, n–valued logics were proposed in Post [39] and Gödel [15].

As with propositional logic, we will call a *tautology* of the 3–valued logic any meaningful expression whose truth value is 1 for any choice of truth values of propositional variables that occur in it.

An axiomatization of the logic L₃ was proposed in Wajsberg [62] by means of the following axioms

W1 $q \Rightarrow (p \Rightarrow q)$

W2 $(p \Rightarrow q) \Rightarrow [(q \Rightarrow r) \Rightarrow (p \Rightarrow r)]$

W3 $[(p \Rightarrow Np) \Rightarrow p] \Rightarrow p$

W4 $(\neg q \Rightarrow \neg p) \Rightarrow (p \Rightarrow q)$

Derivation rules are already familiar rules of substitution and detachment. We denote this deductive system with the symbol W.

As proved by Wajsberg (op. cit.), the system W is complete, i.e., a meaningful expression of this system is a theorem of this system if and only if it is a tautology of the 3–valued logic L₃.

Let us observe that – as one should expect – some axioms of propositional calculus are no longer tautologies of the 3–valued logic of Łukasiewicz, witness for this is, e.g., $(\neg p \Rightarrow p) \Rightarrow p$, its truth value in case $v(p) = 2$ is 2.

The original proof by Wajsberg of the completeness of his system may be found in Wajsberg [62]. We include here for the sake of completeness of exposition a proof proposed by Goldberg, Leblanc and Weaver [13]. The proof exploits basic properties of syntactic consequence and in its ideology is close to the proof of completeness of propositional calculus in sect. 3.2.

We use the symbol $\Gamma \vdash \alpha$ to denote the fact that a formula α has a proof from a set Γ of formulas. This means that in addition to a definition of a proof in sect. 3.2, one can assume that an intermediate α_i is in Γ, and the starting α_0 can either be in Γ or be an axiom instance.

In case $\gamma_1, ..., \gamma_k \vdash \beta$, it follows by applying the deduction theorem consecutively k number of times that $\vdash C\gamma_1 C\gamma_2 ... C\gamma_k \beta$.

We say that a set Γ is *syntactically consistent* if and only if there exists no α with $\Gamma \vdash \alpha$ and $\Gamma \vdash N\alpha$. A set Γ is a *maximal syntactically consistent set* if and only if Γ is syntactically consistent and for each meaningful expression α, if $\Gamma \cup \{\alpha\}$ is syntactically consistent then $\Gamma \vdash \alpha$. When Γ is not syntactically consistent, we will say that it is *syntactically inconsistent*.

As with propositional logic, we need a good deal of information about syntactic properties of W in order to attempt a proof of completeness. We list here the basic facts needed in the proof of completeness.

W5 $p \vdash q \Rightarrow p$

W6 $\vdash \neg p \Rightarrow (p \Rightarrow q)$

W7 $Cpq \vdash CCqrCpr$

W8 $p \Rightarrow q, q \Rightarrow r \vdash p \Rightarrow r$

W9 $\vdash \neg\neg p \Rightarrow (q \Rightarrow p)$

W10 $\vdash CCC[(p \Rightarrow q) \Rightarrow q] \Rightarrow (\neg p \Rightarrow q)$

W11 $\vdash \neg\neg p \Rightarrow p$

W12 $\vdash p \Rightarrow \neg\neg p$

W13 $\vdash p \Rightarrow p$

W14 $\vdash (p \Rightarrow q) \Rightarrow (\neg q \Rightarrow \neg p)$

W15 $\vdash CC(p \Rightarrow q) \Rightarrow (\neg\neg p \Rightarrow q)$

W16 $\vdash (\neg\neg p \Rightarrow q) \Rightarrow (\neg\neg p \Rightarrow \neg\neg q)$

W17 $\vdash (p \Rightarrow q) \Rightarrow (\neg\neg p \Rightarrow \neg\neg q)$

W18 $\vdash [p \Rightarrow (p \Rightarrow \neg p)] \Rightarrow (p \Rightarrow \neg p)$

W19 $\vdash CCCpNpNCpNpp$

W20 $\vdash \neg(p \Rightarrow q) \Rightarrow p$

W21 $\vdash \neg(p \Rightarrow q) \Rightarrow \neg q$

W22 $\vdash [p \Rightarrow (\neg q \Rightarrow \neg p)] \Rightarrow q)$

W23 $\vdash (p \Rightarrow \neg p) \Rightarrow [(\neg q \Rightarrow \neg negq) \Rightarrow (p \Rightarrow q)]$

W24 $\vdash [(p \Rightarrow (p \Rightarrow (q \Rightarrow r)] \Rightarrow [(p \Rightarrow (p \Rightarrow q) \Rightarrow (p \Rightarrow (p \Rightarrow r)]$

W25 $\neg(p \Rightarrow p)$ *is false for each substitution of truth values for p*

The proof begins with the following list of basic properties of consistent sets

Proposition 3.8. *The following are basic properties of consistent sets of formulas*

1. $\Gamma \vdash \alpha$, $\Gamma \subseteq \Gamma'$ imply $\Gamma' \vdash \alpha$;

2. $\Gamma \vdash \alpha$ implies the existence of a finite $\Gamma' \subseteq \Gamma$ with the property that $\Gamma' \vdash \alpha$;

3. $\Gamma \vdash \alpha$, $\Gamma \vdash C\alpha\beta$ imply $\Gamma \vdash \beta$;

4. $\Gamma \cup \{\alpha\} \vdash \beta$ implies $\Gamma \vdash C\alpha C\alpha\beta$;

5. Γ is not syntactically consistent if and only if $\Gamma \vdash f$ where f denotes any proposition such that Nf is a theorem;

6. If $\Gamma \cup \{\alpha\}$ is syntactically inconsistent then $\Gamma \vdash C\alpha N\alpha$;

7. If $\Gamma \cup \{C\alpha N\alpha\}$ is syntactically inconsistent then $\Gamma \vdash \alpha$.

Indeed, Properties 1, 2, 3 follow by definition of the relation $\vdash$; to prove Property 4, assume that $\Gamma \cup \{\alpha\} \vdash \beta$ and choose a proof $\alpha_1, \alpha_2, ..., \alpha_n$ of β from $\Gamma \cup \{\alpha\}$. By induction on j, we prove the following claim

(c_j) $\Gamma \vdash C\alpha C\alpha\alpha_j$

We assume (c_j) is valid for $j < i$. There are some cases to be considered.

Case 1. α_i *is* α. In this case (c_i) follows by W1 with $p, q/\alpha$.

Case 2. α_i *is an instance of an axiom or* $\alpha_i \in \Gamma$. As $\Gamma \vdash \alpha_i$, applying W5 twice yields $\Gamma \vdash CaCa\alpha_i$.

Case 3. α_i *follows by detachment from* α_k *and* $Ca_k\alpha_i$ *with* $k < i$. In this case we have by the inductive assumption

(i) $\Gamma \vdash CaCa\alpha_k$

(ii) $\Gamma \vdash CaCaCa_k\alpha_i$

It now suffices to apply W24 along with the Property 3 and (i), (ii). This concludes the proof of Property 4.

To begin with the proof of Property 5, assume that Γ is not syntactically consistent, i.e., $\Gamma \vdash \alpha$ and $\Gamma \vdash N\alpha$ for some meaningful expression α. Then we have $\vdash CN\alpha Caf$ by W5 and detachment applied twice yields $\Gamma \vdash f$.

Conversely, assuming that $\Gamma \vdash f$ for any false f, we have in particular $\Gamma \vdash NCpp$ by W25; as it follows from W13 that $\Gamma \vdash Cpp$, we conclude that Γ is syntactically inconsistent.

Concerning Property 6, assume that $\Gamma \cup \{\alpha\}$ is not syntactically consistent. From W5, as in the proof of Property 5, we infer that $\Gamma \cup \{\alpha\} \vdash \beta$ for any meaningful expression β in particular $\Gamma \cup \{\alpha\} \vdash N\alpha$ hence $\Gamma \vdash CaCaN\alpha$ by Property 4 and it follows from W18 by detachment that $\Gamma \vdash CaN\alpha$.

The converse Property 7 follows from W19 and Property 6. By Property 6, from the assumption that $\Gamma \cup \{CaN\alpha\}$ is syntactically inconsistent it follows that $\Gamma \vdash CCaN\alpha NcaN\alpha$ so W19 along with Property 3 implies $\Gamma \vdash \alpha$.

The proof of basic properties of the relation $\vdash$ concluded, we may pass to the proof of the completeness theorem. We recall our setting: we have a (countable) set of *propositional variables* $p_0, p_1, ..., p_n, ...$ from which meaningful expressions (formulas) are formed in same way as in propositional calculus.

A *valuation* is an assignment of truth values to propositional variables; clearly, any valuation v induces a valuation on meaningful expressions according to the semantic rules governing functors C, N. We denote by the symbol v^* the induced valuation on meaningful expressions. A set Γ of meaningful expressions is said to be *semantically consistent* if there exists a valuation v with the property that $v^*(\alpha) = 1$ for each $\alpha \in \Gamma$; we will use the shortcut $v^*(\Gamma) = 1$ in this case.

The notion of *semantic consequence* follows: a meaningful expression α is a semantic consequence of a set Γ of meaningful expressions if and only if for each valuation v with $v * (\Gamma) = 1$ we have $v * (\alpha) = 1$. The fact that α is a semantic consequence of Γ is denoted with the symbol $\Gamma \models \alpha$.

Following Goldberg–Leblanc–Weaver [13], we state

Proposition 3.9. *(Strong Completeness Theorem for L_3) For each meaningful expression α and each set Γ of meaningful expressions, $\Gamma \vdash \alpha$ if and only if $\Gamma \models \alpha$.*

Proof. One way the proof goes simply: if $\Gamma \vdash \alpha$ then we may choose a proof $\alpha_1, ..., \alpha_n$ of α from Γ and a valuation v with $v^*(\Gamma) = 1$ and check directly by induction that $v^*(\alpha_i) = 1$ for $i = 1, 2, ..., n$.

The proof in the reverse direction requires some work. It follows from the following Claim.

Claim. *If Γ is syntactically consistent that Γ is semantically consistent.*

The proof of Claim relies on the notion of a maximal consistent set. In this proof we use some results of Set Theory, consult Ch. 1. First, we enumerate all propositional variables into a countable set $V = \{p_0, p_1, ...\}$ and then we form the set M of meaningful expressions, enumerating its elements: $M = \{\alpha_0, \alpha_1, ...\}$; this is possible as the set M is countable since each meaningful expression uses only a finite number of propositional variables.

This done, we look at Γ and we construct a sequence $\Gamma_0 = \Gamma, \Gamma_1, ...$ of sets as follows. Given Γ_i we let

$$\Gamma_{i+1} = \Gamma_i \cup \{\alpha_{i+1}\}$$

in case $\Gamma_i \cup \{\alpha_{i+1}\}$ is syntactically consistent and

$$\Gamma_{i+1} = \Gamma_i$$

otherwise. Having the sequence $(\Gamma_i)_i$, we let

$$\Gamma^* = \bigcup_i \Gamma_i$$

Properties of Γ^* of importance to us are collected in the following sub–Claims.

Sub–Claim 1. *Γ^* is syntactically consistent.*

Indeed, assume to the contrary that $\Gamma^* \vdash \alpha$ and $\Gamma^* \vdash \neg\alpha$ for some α. Then by property 2, there exists a finite $\Delta \subseteq \Gamma^*$ with the property that $\Delta \vdash \alpha$, $\Delta \vdash \neg\alpha$. By construction of Γ^*, there exists Γ_i with $\Delta \subseteq \Gamma_i$ and thus property 1 implies that $\Gamma_i \vdash \alpha$, $\Gamma_i \vdash \neg\alpha$, a contradiction. It follows that Γ^* is syntactically consistent.

Sub–Claim 2. *Γ^* is a maximal syntactically consistent set.*

Assume that it is not true that $\Gamma^* \vdash \alpha$ for some meaningful expression α; as α is α_i for some i, we have that $\Gamma_{i-1} \cup \{\alpha\}$ is not syntactically consistent hence by Property 5, $\Gamma_{i-1} \cup \{\alpha\} \vdash f$ and Property 1 implies that $\Gamma^* \cup \{\alpha\} \vdash f$ so by Property 5 applied in the reverse direction we have that $\Gamma^* \cup \{\alpha\}$ is not syntactically consistent. It thus follows that if $\Gamma^* \cup \{\alpha\}$ is syntactically consistent then $\Gamma^* \vdash \alpha$ which means that Γ^* is a maximal consistent set.

We now exploit Γ^* to introduce a certain canonical valuation on the set V of propositional variables. To this end, we define a valuation v_0 as follows

$$
v_0(p) = \begin{cases} 1 \ \text{in case } \Gamma^* \vdash p \\ 0 \ \text{in case } \Gamma^* \vdash \neg p \\ \frac{1}{2} \ \text{otherwise} \end{cases}
\tag{3.7}
$$

Now, we establish the vital property of the induced valuation v_0^* on the set M expressed in

Sub–Claim 3.

$$
v_0^*(\alpha) = \begin{cases} 1 \ \text{in case } \Gamma^* \vdash \alpha \\ 0 \ \text{in case } \Gamma^* \vdash \neg\alpha \\ \frac{1}{2} \ \text{otherwise} \end{cases}
\tag{3.8}
$$

The proof of Sub–claim 3 goes by structural induction.

Case 1. α is of the form $\neg\beta$ and Sub–Claim 3 holds for β.

When $\Gamma^* \vdash \neg\beta$ we have $v_0^*(\beta) = 0$ hence $v_0^*(\neg\beta) = 1$ so the claim holds for α.

When $\Gamma^* \vdash \neg\neg\beta$ we have $v_0^*(\neg\beta) = 0$ so the claim again holds with α.

When finally neither of two previous cases hold we conclude by W11 that it is not true that $\Gamma^* \vdash \beta$ so $v_0^*(\beta) = \frac{1}{2}$ hence $v_0^*(\alpha) = \frac{1}{2}$. The claim holds with α.

Case 2. α is of the form $\beta \Rightarrow \gamma$ and Sub–Claim 3 holds for β, γ.

We have to examine the three possible cases again.

When $\Gamma^* \vdash \beta \Rightarrow \gamma$, we need to check that

$$
v_0^*(\beta \Rightarrow \gamma) = 1
$$

which depends on valuations of β, γ and we have some sub–cases to consider. When $v_0^*(\gamma) = 1$ which happens when $\Gamma^* \vdash \gamma$, we have $v_0^*(\alpha) = 1$; when $v_0^*(\beta) = 0$ we have again $v_0^*(\alpha) = 1$.

The sub–case $v_0^*(\gamma) = 0$, i.e., $\Gamma^* \vdash \neg\gamma$ remains but then, by W14, we have $\Gamma^* \vdash \neg\beta$ already settled in the positive.

So, finally, we have the sub–case, when neither of $\beta, \neg\beta, \gamma, \neg\gamma$ is derivable from Γ^* and then $v_0^*(\beta) = \frac{1}{2} = v_0^*(\gamma)$ a fortiori $v_0^*(\alpha) = 1$.

The next case is when

$$\Gamma^* \vdash \neg\beta \Rightarrow \gamma$$

By W20 and W21, we obtain $\Gamma^* \vdash \beta$ and $\Gamma^* \vdash \neg\gamma$ so $v_0^*(\beta) = 1$, $v_0^*(\gamma) = 0$ and it follows that $v_0^*(\alpha) = 0$.

Finally, we are left with the case when neither $\Gamma^* \vdash \beta \Rightarrow \gamma$ nor $\Gamma^* \vdash \neg\beta \Rightarrow \gamma$. It follows from W6 that it is not true that $\Gamma^* \vdash \neg\beta$ and similarly W1 implies that it is not true that $\Gamma^* \vdash \gamma$. Thus $v^*(\beta) \neq 0$ and $v^*(\gamma) \neq 1$. We are left with the following sub–cases.

We have $v^*(\beta) = 1$, i.e., $\Gamma^* \vdash \beta$ and then by W22, we have that it is not true that $\Gamma^* \vdash \neg\gamma$ so $v^*(\gamma) \neq 0$ and hence $v^*(\gamma) = \frac{1}{2}$ and in consequence $v^*(\alpha) = \frac{1}{2}$.

We have $v^*(\beta) = \frac{1}{2}$, i.e., it is not true that $\Gamma^* \vdash \beta$. It is finally here that we exploit maximality of Γ^*.

It follows by maximality of Γ^* that $\Gamma^* \cup \{\beta\}$ is not syntactically consistent so by Property 6, we have $\Gamma^* \vdash \beta \Rightarrow \neg\beta$; was $v^*(\gamma) = \frac{1}{2}$, we would have by the same argument that

$$\Gamma^* \vdash \neg\gamma \Rightarrow \neg\neg\gamma$$

and then we would have by W23 that $\Gamma^* \vdash \beta \Rightarrow \gamma$, a contradiction.

Thus $v^*(\gamma) = 0$ and $v^*(\alpha) = \frac{1}{2}$. It follows that Sub–Claim 3 holds for α which completes the inductive step and concludes the proof of Sub–Claim 3.

As $\Gamma \subseteq \Gamma^*$, we have $v^*(\alpha) = 1$ whenever $\alpha \in \Gamma$, i.e., Γ is semantically consistent. This concludes the proof of Claim.

Now, we return to the Strong Completeness Theorem; we assume that $\Gamma \models \alpha$. Thus $v^*(\Gamma) = 1$ implies $v^*(\alpha) = 1$ and hence $v^*(\alpha \Rightarrow \neg\alpha) = 0$, i.e.,

$$\Gamma \cup \{\alpha \Rightarrow \neg\alpha\}$$

is semantically inconsistent, hence, by Claim this set is syntactically inconsistent and by Property 7, we have $\Gamma \vdash \alpha$. This concludes the proof of the Strong Completeness Theorem $\qquad\qquad\qquad\qquad\qquad\qquad\qquad\qquad\quad\Box$

A specialization of the Strong Completeness Theorem (Proposition 3.9) to the case when Γ is the set of axioms W1–W4 yields,

Proposition 3.10. *(Completeness theorem for the system (W), Wajsberg [62], Goldberg–Leblanc–Weaver [13]) A meaningful expression α is a theorem of the system W if and only if it is a tautology of the logic L_3.*

Now, we proceed from 3- to n–valued logics.

3.4   Many–Valued Calculi: n–Valued Logic

We have seen above the proposition of Łukasiewicz for n–valued logic. In his logic, functors $\Rightarrow, \neg$ are defined via functions on truth values. In general, we may observe that statements of many–valued logic are of the form $F_i(p_1, ..., p_k)$ where F_i is a functor (e.g., $\Rightarrow, \neg$) and $p_1, .., p_k$ are propositional variables. Assuming that any of p_i may be assigned any of truth values $1, 2, ..., n$, the functor F_i may be described by a matrix of values $f_i(v_1, ..., v_k)$ where each v_i runs over truth values $1, 2, .., n$ which may be assigned to p_i. Any meaningful expression $\alpha(p_1, .., p_m)$ is then assigned a matrix of truth values $g(v_1, .., v_m)$ where $v_1, .., v_m$ run over truth values $1, .., n$ of propositional variables $p_1, ..., p_m$ occurring in α and g is the truth function of α composed of truth functions f_j of functors F_j occurring in α.

The idea of a true or false expression has to be generalized to the idea of acceptance or rejection of a meaningful expression which can be realized in an n–valued logic by a selection of a threshold value $1 \leq s < n$ and declaring a meaningful expression $\alpha(p_1, .., p_m)$ *accepted* when $g(v_1, ..., v_m) \leq s$ for any choice of truth values $1, 2, ..., m$ for $v_1, .., v_m$ where g is the truth function of α. Otherwise, α is rejected. We may observe that formulas with truth function taking the value 1 only are accepted regardless of choice of s (we may say that they are accepted with certainty) while formulas whose truth function takes on solely the value n are rejected regardless of choice of s (we may say that they are rejected with certainty). Other formulas may be accepted or rejected depending on the chosen value of s so their status is less precise and they may be termed to be *accepted to a degree*.

We have seen in case of 3–valued logic of Łukasiewicz that truth matrices have been described in a compressed form by functions: $f_1(v_1, v_2) = min\{1, v_2 - v_1 + 1\}$ for the functor $\Rightarrow$ and $f_2(v) = 1 - v$ for $\neg$. In the setting of an n–valued logic as proposed above, these formulas should be dualized to ensure that the greater truth value does correspond to a lesser acceptance degree. Thus the Łukasiewicz formulas in the case of n–valued logic are modified to the form,

$$f_1(v_1, v_2) = max\{1, v_2 - v_1 + 1\}, f_2(v) = 1 - v + n \qquad (3.9)$$

as proposed in Rosser and Turquette [45].

We now follow the scheme proposed by Rosser and Turquette (op. cit.). The functors in this scheme are $\Rightarrow, \neg$ as well as unary functors J_k for $k = 1, 2, ..., n$ construed as to satisfy the condition that the truth function j_k does satisfy the requirement that $j_k(v) = 1$ in case $v = k$ and $j_k(v) = n$ otherwise, for $k = 1, 2, ..., n$. Let us outline construction of $J_k's$. First, we let

$$\begin{cases} H_1(p) \ is \ \neg(p) \\ H_{i+1} \ is \ p \Rightarrow H_i(p) \end{cases} \qquad (3.10)$$

We denote by the symbol h_i the truth function of H_i. Then we have

Proposition 3.11. $h_i(v) = max\{1, n - (v-1) \cdot i\}$ *for each i.*

Proof. Indeed, inducting on i, we have $h_1(v) = f_2(v) = n - v + 1 = n - (v-1) = max\{1, n - (v-1)\}$ and $h_{i+1}(v) = f_1(v, h_i(v)) = max\{1, max\{1, n - (v-1) \cdot i - (v-1)\}\} = max\{1, max\{1, n - (v-1) \cdot (i+1)\}\} = max\{1, n - (v-1) \cdot (i+1)\}$ □

Now, it is possible to introduce functors $J_1, ..., J_n$; we resort to recursion. To begin with, we let

$$J_1(p) \; is \; N(H_{n-1}(p)) \tag{3.11}$$

Then, the truth function $j_1(v)$ has the required property, viz., $j_1(1) = 1, j_1(v > 1) = n$. Indeed, as $h_{n-1}(1) = n$ and $h_{n-1}(v) = 1$ for $v > 1$, it follows that $j_1(1) = n - n + 1 = 1$ and $j_1(v) = n - 1 + 1 = n$ for $v > 1$.

To define remaining functors J_k, one has to introduce auxiliary parameter κ; given $1 < k \le m$, we let $\kappa = \lceil \frac{m-k}{k-1} \rceil$. Clearly, κ is a decreasing function of k. Other properties of κ necessary in what follows are summed up below; the proofs are elementary and hence omitted:

$$\kappa < \frac{m-1}{k-1} \tag{3.12}$$

$$h_\kappa(k) = n - (k-1) \cdot \kappa \tag{3.13}$$

$$1 < h_\kappa(k) \le k \tag{3.14}$$

By property (3.14), either $k = h_\kappa(k)$ or $k > h_\kappa(k)$. Definition of J_k splits into two distinct formulas depending on which is the case.

For $1 < k < n - 1$, one defines J_k via

$$J_k(p) \; is \; J_1(C(H_\kappa(p) \vee p)(H_\kappa(p) \wedge p)) \tag{3.15}$$

in case $k = h_\kappa(k)$, and

$$J_k(p) \; is \; J_{\kappa(k)}(H_{\kappa(k)}(p)) \tag{3.16}$$

in case $k > h_\kappa(k)$.

Then, the truth function j_k of J_k does satisfy the requirement that $j_k(k) = 1$ and $j_k(v \ne k) = n$.

In case $k = h_\kappa(k)$, we have

$$j_k(k) = j_1(max\{1, max\{h_{\kappa(k)}, k\} - min\{h_{\kappa(k)}, k\} + 1\}) = j_1(1) = 1 \tag{3.17}$$

For $v < k$, we have,

$$h_\kappa(v) > h_\kappa(k) = k > v \tag{3.18}$$

hence,

$$j_k(v) = j_1(h_\kappa(v) - v + 1) = n \qquad (3.19)$$

In case $v > k$, we have $h_\kappa(v) < k < v$, hence,

$$j_k(v) = j_1(v - h_\kappa(v) + 1) = n \qquad (3.20)$$

In case $k > h_\kappa(k)$, we have

$$\begin{cases} j_k(k) = j_{h_\kappa(k)}(h_\kappa(k)) = 1 \\ j_k(v \neq k) = n \text{ when } h_\kappa(v) \neq h_\kappa(k) \end{cases} \qquad (3.21)$$

The remaining functor J_n will be defined via the formula

$$J_n(p) \ is \ J_1(Np)$$

We have $j_n(n) = j_1(n - n + 1) = 1$ and for $v < n$ it follows that $j_n(v) = j_1(n - v + 1) = n$.

In this way, functors $J_1, ..., J_n$ are defined and their truth functions $j_1, ..., j_n$ satisfy the requirement that $j_k(k) = 1$ and $j_k(v) = n$ whenever $v \neq k$.

Helped by these functors, one may define new functors which would play some of the roles played by two–valued functors $\Rightarrow, \neg$ in two–valued calculus of propositions. Notably, if one would like to preserve in an n–valued logic the property of two–valued negation that it does convert accepted statements into rejected ones and conversely, then one should modify the functor $\neg$: as defined in the Łukasiewicz n–valued calculus by means of the truth function $f_2(v) = n - v + 1$, it does convert statements accepted with certainty into statements rejected with certainty while statements accepted or rejected in a degree are not necessarily converted into the opposite kind.

To restore this property, one may define a functor $--$ via

$$-- (p) \ is \ J_{s+1}(p) \vee ... \vee J_n(p)$$

where $\vee$ (denoted also A) is the functor of alternation. Then α is accepted if and only if $--\alpha$ is rejected and conversely. Let us mention that the truth function $n(v)$ of this negation is given by the formula

$$n(v) = max\{j_{s+1}(v), ..., j_n(v)\}$$

as here the truth functions of alternation A and conjunction K are adopted from the Łukasiewicz logic.

Similarly, we may observe that the functor $\Rightarrow$ defined above does not ensure the validity of detachment in the form:

if statements α and $C\alpha\beta$ are accepted, then the statement β is accepted

To remedy this, we may invoke theses $(p \Rightarrow q) \Rightarrow (\neg p \vee q)$ and $(\neg p \vee q) \Rightarrow (p \Rightarrow q)$ of two–valued propositional logic and we may define a new functor $\sqsubseteq$ via

$$\sqsubseteq pq \text{ is } --p \vee q$$

Then detachment rule is valid in the form:

if statements α and $\sqsubseteq \alpha\beta$ are accepted then the statement β is accepted

n–valued logics described above are completely axiomatizable; axiom schemes and a proof of completeness will be found in Rosser and Turquette [45].

3.5 Many–Valued Calculi: [0,1]–Valued Logics

The Łukasiewicz semantics does allow for infinite sets of truth values, as indicated in Łukasiewicz. In this case, one may admit as a set of truth values any infinite set $T \subseteq [0,1]$ which is closed under functions

$$c : min(1, 1 - x + y), n : 1 - x$$

in the sense that $c(x,y), n(x) \in T$ whenever $x, y \in T$; we recall that the function c is the meaning of implication $\Rightarrow$ and the function n is the meaning of negation $\neg$ in the Łukasiewicz semantics. As shown in McNaughton [37], any such T needs to be dense in $[0,1]$; it is manifest that $0, 1 \in T$ for each admissible T. Thus, sets $Q_0 = Q \cap [0,1]$ of rational numbers in $[0,1]$ as well as the whole unit interval $[0,1]$ may be taken as sets of truth values leading to respective infinite valued logical calculi. The additional merit of these two sets of truth values is that both $Q, \mathbf{R}^1$ are from algebraic point of view, *algebraic fields* so they may be regarded as vector spaces over themselves, and this fact was exploited in completeness proofs, cf., Rose and Rosser [44].

It is well–known that [0,1]–valued logics are fundamentally important to fuzzy calculi, see sect. 3.6, below. With this in mind, we discuss in this Chapter infinite valued in particular [0,1]–valued logics with the Łukasiewicz semantics. Their completeness in the Łukasiewicz axiomatics, Łukasiewicz and Tarski [32] was conjectured by Łukasiewicz, loc.cit., under an additional axiom L5, cf., Hájek [17] for the *prelinearity axiom*, which was shown redundant, i.e., following from the four remaining axioms L1–L4, below, in Meredith [33] and Chang [5]

L1 $p \Rightarrow (q \Rightarrow p)$

L2 $(p \Rightarrow q) \Rightarrow ((q \Rightarrow r) \Rightarrow (p \Rightarrow r))$

L3 $(\neg p \Rightarrow \neg q) \Rightarrow (q \Rightarrow p)$

L4 $((p \Rightarrow q) \Rightarrow q) \Rightarrow ((q \Rightarrow p) \Rightarrow p)$

L5 $[(p \Rightarrow q) \Rightarrow (q \Rightarrow p)] \Rightarrow (q \Rightarrow p)$

had been proved in Rose and Rosser [44] by a syntactic argument.

It is convenient, cf., Rose and Rosser , op. cit., , to introduce auxiliary functors, kind of shortcuts

1. Apq is $(p \Rightarrow q) \Rightarrow q$;

2. Kpq is $\neg[(\neg p \Rightarrow \neg q) \Rightarrow \neg q]$;

3. Bpq is $\neg p \Rightarrow q$;

4. Lpq is $\neg(p \Rightarrow \neg q)$;

5. Epq is $LCpqCqp$.

In Łukasiewicz and Tarski [32], a conjecture by Łukasiewicz is stated that the set L1–L5 of axioms gives a complete axiomatization of infinite valued logical calculus with the Łukasiewicz semantics in case when the only designated value is 1.

Completeness of this axiomatization was announced in Wajsberg [49] without proof; the completeness proof was given in Rose and Rosser [44] and here we follow their development.

According to Łukasiewicz's semantics, truth functions c of $\Rightarrow$ and n of $\neg$ are given by the formulae quoted above, i.e., $c(x,y) = min\{1, 1 - x + y\}, n(x) = 1 - x$. Then truth functions a, k, b, l, e of derived functors, respectively, A, K, B, L, E are given by the following formulas

6. $a(x,y) = max\{x,y\}$.

7. $k(x,y) = min\{x,y\}$.

8. $b(x,y) = min\{1, x + y\}$.

9. $l(x,y) = max\{0, x + y - 1\}$.

10. $e(x,y) = min\{1 - x + y, 1 - y + x\}$.

We may check easily that

Proposition 3.12. *The following statements hold*

1. $a(x, y) = 1$ if and only if either $x = 1$ or $y = 1$;

z

2. $k(x, y) = 1$ if and only if $x = 1 = y$;

3. $b(x, y) = 1$ if and only if $x + y \geq 1$;

4. $l(x, y) = 1$ if and only if $x = 1 = y$;

5. $e(x, y) = 1$ if and only if $x = y$.

It follows that A, B are many–valued counterparts to classical disjunction, K, L are many–valued counterparts to classical conjunction, and E is a many–valued counterpart to classical equivalence.

As with any logical calculus, also in case of infinite valued logical calculus, its syntax, its semantics, and their interplay are subject of study. We will present results of this study here. We will simply assume that the set of truth values T is either $Q \cap [0, 1]$ or $[0, 1]$ and 1 is the only designated value. The resulting calculus is denoted L_∞.

Inference rules are detachment and substitution. We will use as usual the acceptance ("yields") symbol $\vdash$ and for a formula α as in earlier Chapters, the symbol $\vdash \alpha$ will mean that α is a *theorem* of the calculus, i.e., there exists a proof of α from axioms via detachment and substitution in the usual sense of sect. 2 ff. Similarly, the symbol $\Gamma \vdash \alpha$ will mean that α is provable from a set Γ of formulas.

We follow closely Rose and Rosser [44], and we give now a number of theorems and syntactic features of L_∞.

R1 $\Gamma \vdash p \Rightarrow q, \Gamma \vdash q \Rightarrow r$ *imply* $\Gamma \vdash p \Rightarrow r$

Indeed, this is a direct consequence of L2 and detachment. We let $p \Leftrightarrow q$ to denote that $p \Rightarrow q$ and $q \Rightarrow p$ hold.

R2 $\Gamma \vdash p \Leftrightarrow q$ *and* $\Gamma \vdash q \Leftrightarrow r$ *imply* $\Gamma \vdash p \Leftrightarrow r$
R2 follows by R1 and definition of $\Leftrightarrow$.

R3 $p \Rightarrow q, q \Rightarrow r \vdash p \Rightarrow r$
R3 follows by R1.

R4 $\vdash Apq \Leftrightarrow Aqp$
R4 follows by L3.

R5 *From* $\Gamma \vdash p \Leftrightarrow q$ *it follows that* $\Gamma \vdash p \Rightarrow r \Leftrightarrow q \Rightarrow r$
Indeed, by L2 and detachment, we have $p \Rightarrow q \vdash (q \Rightarrow r) \Rightarrow (p \Rightarrow r)$ and $q \Rightarrow p \vdash (p \Rightarrow r) \Rightarrow (q \Rightarrow r)$.

R6 $\vdash p \Rightarrow Aqp$

R6 is obtained by substitution $q/q \Rightarrow p$ in L1.

R7 $\vdash p \Rightarrow Apq$

R7 follows from R6 and L3.

R8 $\vdash [p \Rightarrow (q \Rightarrow r)] \Rightarrow [q \Rightarrow (p \Rightarrow r)]$

Indeed, $\vdash CqAqr$ by R7 and L2 implies

$$(i) \ \ \vdash [Aqr \Rightarrow (p \Rightarrow r)] \Rightarrow [q \Rightarrow (p \Rightarrow r)]$$

hence, substitution $q/q \Rightarrow r$ along with L2 yields

$$(ii) \ \ \vdash [p \Rightarrow (q \Rightarrow r)] \Rightarrow [(Aqr) \Rightarrow (p \Rightarrow r)]$$

From (i), (ii) one gets R8 via R1.

R9 $(q \Rightarrow r) \Rightarrow [(p \Rightarrow q) \Rightarrow (p \Rightarrow r)]$

R9 follows from R8 and L2.

R10 $\Gamma \vdash p \Leftrightarrow q$ *implies* $\Gamma \vdash (r \Rightarrow p) \Leftrightarrow (r \Rightarrow q)$.

R10 follows by substituting $p/q, q/p$ in R8 which yields

$$(i) \ \ \vdash (p \Rightarrow (q \Rightarrow))r \Leftrightarrow (q \Rightarrow (p \Rightarrow r)),$$

and using L2 in which substitutions $p/q, q/r, r/p$ have been made.

R11 $\vdash (p \Rightarrow p)$ Indeed, substituting q/p in R8 and applying L1 one obtains (i) $\vdash q \Rightarrow (p \Rightarrow p)$. Letting q to be any already proved theorem and applying detachment yields R11.

R12 $\vdash p \Leftrightarrow p$

Immediate from R11.

R13 $p \vdash q \Leftrightarrow (p \Rightarrow q)$

This follows from L1 where substitution $p/q, q/p$ has been made and from R7 in which A has been replaced by the right–hand side of 1.

R14 $\vdash q \Leftrightarrow Aqq$

Indeed, substitute $p/q \Rightarrow q$ in R13 and apply R11.

R15 $p \Rightarrow r, q \Rightarrow r \vdash Apq \Rightarrow r$

R16 *From* $p_1, .., p_k, p \vdash r$ *and* $q_1, ..., q_n, q \vdash r$ *it follows that* $p_1, ..., p_k, q_1, ..., q_n, Apq \vdash r$

R17 $\neg\neg p \Rightarrow p$

R18 $\vdash p \Rightarrow \neg\neg p$

R19 $\vdash p \Leftrightarrow \neg\neg p$
R19 follows by R17 and R18.

R20 $\vdash (p \Rightarrow q) \Leftrightarrow \neg q \Rightarrow \neg p$

R21 $\Gamma \vdash p \Leftrightarrow q$ *implies* $\Gamma \vdash \neg p \Leftrightarrow \neg q$
R21 follows from R20.

R19 and R20 imply

R22 $\vdash Lpq \Leftrightarrow Lqp$

R23 $\vdash Bpq \Leftrightarrow Bqp$

R24 $\vdash p \Rightarrow Bpq$
Indeed, substituting q/Nq in L1 obtains (i) $\vdash p \Rightarrow Bqp$ whence R24 follows
by R23.

R25 $\vdash Lqp \Rightarrow p$
R25 follows from R22.

R26 $\vdash Lpq \Rightarrow p$

R27 $\vdash (p \Rightarrow (q \Rightarrow r) \Leftrightarrow Lpq \Rightarrow r$
From R20 it follows that (i) $\vdash (p \Rightarrow (q \Rightarrow r)) \Leftrightarrow (p \Rightarrow (\neg r \Rightarrow \neg q))$ and R8
implies (ii) $\vdash (p \Rightarrow (q \Rightarrow r)) \Leftrightarrow (\neg r \Rightarrow (p \Rightarrow \neg q))$ so R27 follows by R23.

R28 $\vdash (p \Rightarrow (q \Rightarrow Lpq))$
It is enough to substitute in R27 r/Lpq and invoke the theorem $\vdash Lpq \Rightarrow Lpq$.

R29 $p \vdash q \Leftrightarrow Lpq$
R29 follows from R25 and R28.

R30 $\Gamma \vdash p \Leftrightarrow q$ *is equivalent to* $\Gamma \vdash Epq$
From R28 one infers (i) $p \Rightarrow q, q \Rightarrow p \vdash Epq$, and R25 with R26 imply (ii)
$Epq \vdash p \Leftrightarrow q$.

R31 $\vdash Epp$
It suffices to substitute q/p in R30.

R32 $Epq, Eqr \vdash Epr$
R32 follows from 30.

R31 $Epq \vdash E\neg p\neg q$
R31 follows from R30 and R21.

R32 $Epq, Ert \vdash E(p \Rightarrow r)(q \Rightarrow t)$
It follows by R5 and R10.

R33 $p, q \vdash p \Leftrightarrow q$
By R11 and L4, we obtain

$$(i) \ \vdash p \Rightarrow (r \Rightarrow r)$$

so by R28 we have

$$(ii) \ \vdash [(r \Rightarrow r) \Rightarrow p] \Rightarrow Ep(r \Rightarrow r)$$

and it follows from L1 that

$$(iii) \ p \Rightarrow [(r \Rightarrow r) \Rightarrow p]$$

(substitute $q/r \Rightarrow r$) so by R30,

$$(iv) \ p \vdash p \Leftrightarrow (r \Rightarrow r)$$

follows. Substitution p/q in (iv)) yields

$$(v) \ q \vdash q \Leftrightarrow (r \Rightarrow r)$$

whence R33 follows by R32.

R34 $\neg p, \neg q \vdash p \Leftrightarrow q$
By $p \vdash p \Leftrightarrow (r \Rightarrow r)$ ((iv) in R33), R19 and R21, it follows that

$$(i) \ \neg p \vdash p \Leftrightarrow \neg (r \Rightarrow r)$$

and from it R34 follows.

R35 $A\neg pq \Rightarrow (p \Rightarrow q)$
By R24 and L2,
$$(i) \ (Bpq \Rightarrow q) \Rightarrow (p \Rightarrow q)$$

follows which may be also written in the form of R35.

R36 $\vdash \neg p \Rightarrow (p \Rightarrow q)$
Indeed, L1 implies

$$(i) \vdash \neg p \Rightarrow (\neg q \Rightarrow \neg p)$$

whence R36 follows by R20.

We now close our discussion of syntax of L_∞ with a proof of the Deduction Theorem for this calculus.

We need some preliminary facts. To state them, we introduce a symbol $\Gamma_{i=1}^m r_i s$, cf., Rosser and Turquette [45], Rose and Rosser [44]. The definition is inductive on m and goes as follows,

$$\Gamma_{i=1}^m r_i s = \begin{cases} s \ in \ case \ m < 1 \\ Cr_m \Gamma_{i=1}^{m-1} r_i s \ otherwise \end{cases} \tag{3.22}$$

In case all r_i are a formula r, we write $(Cr)^m s$ instead of $\Gamma_{i=1}^m r_i s$.

Proposition 3.13. (Rosser and Turquette [45]) If $r_1, ..., r_m$ are formulas and $q_1, .., q_k$ with $k \leq m$ are among $r_1, ..., r_m$, then

$$\vdash \Gamma_{i=1}^k q_i s \Rightarrow \Gamma_{i=1}^m r_i s$$

Proof. By induction on m. In case $m = 0$ also $k = 0$ and the thesis reduces to $\vdash s \Rightarrow s$, i.e., to R11.

Assuming the thesis for m, we consider $m + 1$.

First, let $k = 0$; then by inductive assumption,

$$(i) \vdash s \Rightarrow \Gamma_{i=1}^m r_i s$$

and L1 gives us

$$(ii) \ \vdash \Gamma_{i=1}^m r_i s \Rightarrow \Gamma_{i=1}^{m+1} r_i s$$

The thesis follows by L2.

Now, let $k > 0$; then q_k is r_u. By the inductive assumption,

$$(iii) \ \vdash \Gamma_{i=1}^{k-1} q_i s \Rightarrow \Gamma_{i=u+1}^{m+1} r_i \Gamma_{i=1}^{u-1} r_i s$$

so R9 implies

$$(iv) \ \vdash \Gamma_{i=1}^k q_i s C q_u \Rightarrow \Gamma_{i=u+1}^{m+1} r_i \Gamma_{i=1}^{u-1} r_i s$$

There are two cases. For economy of space, we return to the symbol C for implication.

Case 1. $u = k + 1$. The thesis follows from (iv).

Case 2. $u < k + 1$. By R8,

$$(v) \ \vdash C(C q_u \Gamma_{i=u+1}^{m+1} r_i \Gamma_{i=1}^{u-1} r_i s C r_{m+1} C r_u \Gamma_{i=u+1}^{m+1} r_i) \Gamma_{i=1}^{u-1} r_i s$$

The inductive assumption implies

$$(vi) \quad \vdash C(Cr_u\Gamma_{i=u+1}^m r_i \Gamma_{i=1}^{u-1} r_i s)\Gamma_{i=1}^m r_i s$$

whence R9 yields

$$(vii) \quad \vdash C(Cr_{m+1}Cr_u\Gamma_{i=u+1}^m r_i s)\Gamma_{i=1}^{u-1} r_i s$$

Now, L2 applied to (iv), (v), (vii), gives the thesis   □

In a similar way the following proposition may be proved.

Proposition 3.14. *(Rosser and Turquette [45])* $\vdash CCqrC\Gamma_{i=1}^m p_i q\Gamma_{i=1}^m p_i r.$

Proof. Again, we induct on m. In case $m = 0$ the theorem becomes $\vdash r \Rightarrow r$, i.e., R11.

The inductive step from m to $m + 1$ involves R9, which yields

$$(i) \quad \vdash C(C\Gamma_{i=1}^m p_i q\Gamma_{i=1}^m p_i r)(C(\Gamma_{i=1}^{m+1} p_i q)\Gamma_{i=1}^{m+1} p_i r)$$

It suffices now to apply the inductive assumption and L2   □

Proposition 3.15. *(Rose and Rosser [44])*
$\Gamma_{i=1}^m p_i q, \Gamma_{i=1}^n s_i(q \Rightarrow r) \vdash \Gamma_{i=1}^m p_i \Gamma_{i=1}^n s_i r.$

Proof. Assume (i) $\Gamma_{i=1}^n s_i Cqr$ and apply to it R8 to obtain (ii) $q \Rightarrow \Gamma_{i=1}^n s_i r$. Proposition 3.14 yields

$$(iii) \quad C\Gamma_{i=1}^m p_i q\Gamma_{i=1}^m p_i \Gamma_{i=1}^n s_i r$$

□

Letting p_i, q_i to be p we get a corollary, cf., Novák [38], Hájek [17].

Proposition 3.16. $\vdash C(C((Cp)^m q)((Cp)^n Cqr))((Cp)^{n+m} r).$

We now are in position to prove the Deduction theorem for L_∞.

Proposition 3.17. *(Deduction Theorem, Rose and Rosser [44], Novák [38], Hájek [16])* If $\Gamma, p \vdash q$, then $\Gamma \vdash (Cp)^n q$ for a natural number $n \geq 0$.

Proof. We refer to the proof of Deduction Theorem in the case of classical propositional calculi, sect. 3.2. We carry out that proof; observe, that the proof goes through in our new case here except in the case when α_n is gotten by detachment from $\alpha_r \Rightarrow \alpha_n$ with some $r < n$, in which case we apply Proposition 3.16   □

Here the survey ends of basic syntactic properties of the infinite valued calculus L_∞. In the part concerned with semantics, some further syntactic properties emerge when necessary. Now, we pass to semantics of L_∞ with the intention of giving an outline of the completeness proof by Rose and Rosser [44].

The Rose and Rosser idea of the completeness proof rests on the separation properties in finite–dimensional vector spaces. We simplify the arguments there by replacing matrix dichotomies used by Rose and Rosser with Propositions 18–20 below. Our first lemma is a well–known statement of Linear Algebra. We include a proof.

Proposition 3.18. *Consider an algebraic infinite field $F \subseteq \mathbf{R}^1$. We are concerned with either $F = Q$, the field of rational numbers or $F = \mathbf{R}^1$. Consider linear functionals $f_1, ..., f_k, g$ on F^n with values in F. If*

$$(i)\ f_1(x) = 0, ..., f_k(x) = 0\ imply\ g(x) = 0\ for\ every\ x \in F^n$$

then there exist $\lambda_1, .., \lambda_k \in F$ with the property that $g = \sum_{i=1}^{k} \lambda_i f_i$.

Proof. Consider $W = \{(f_1(x), ..., f_k(x)) : x \in F^n\}$. Then $W \subseteq F^k$ is a vector subspace of F^k. We define a functional h on W by letting $h((f_1(x), ..., f_k(x)))$ to be $g(x)$. Clearly, h is well–defined by our assumption (i) and then h extends to a functional $H : F^k \to F$. There exists the matrix of H say $[\lambda_1, ..., \lambda_k]$, i.e.,

$$H(x) = [\lambda_1, ..., \lambda_k]x^T = \sum_{i=1}^{k} \lambda_i x_i$$

By restricting H to W, we obtain a particular result that $g = \sum_{i=1}^{k} \lambda_i f_i$ $\quad\square$

A simple corollary follows.

Proposition 3.19. *In the notation of Proposition 3.18, if*

$$(ii)\ f_i(x) \geq 0\ for\ i = 1, 2, ..., k\ imply\ g(x) \geq 0\ for\ each\ x \in F^n$$

then $g = \sum_{i=1}^{k} \lambda_i f_i$ for some $\lambda_1 \geq 0, .., \lambda_k \geq 0 \in F$.

Indeed, the assumption (i) follows easily from (ii). That λ's be non–negative is also forced by (ii).

We now consider a more general case when our functionals are *affine functions*, i.e.,

$$f_i(x) = c_i + \sum_{i=1}^{n} a_{ij}x_j$$

and

$$g(x) = d + \sum_{i=1}^{n} b_i x_i$$

where $c_i, a_{ij}, d, b_j \in F$. As clearly, $f_i - c_i, g - d$ are linear functionals, it follows from already proved cases that,

Proposition 3.20. *If affine functions $f_1, .., f_k, g$ satisfy the condition that*

$$f_1(x) \geq 0, .., f_k(x) \geq 0 \ imply \ g(x) \geq 0 \ for \ every \ x \in F^n$$

then $g = \mu + \sum_{i=1}^{k} \lambda_i f_i$ with $\mu \geq 0, \lambda_1 \geq 0, ..., \lambda_k \geq 0 \in F$.

The crux of the completeness proof is in approximating general formulas with formulas of a special type, called *polynomial formulas*.

Here, by a *polynomial* a function of the form $c + \sum_{i=1}^{k} b_i x_i$ is meant with integral coefficients a, b_i. For each such polynomial f, a set PF(f) of formulas will be defined. To this end, we define the *norm* $|f|$ of f by letting $|f| = \sum_{i=1}^{k} |b_i|$. The definition of $PF(f)$ splits into few cases.

PF1 *In case $|f| = 0$, we let: when $c \geq 1$, then $PF(f) = \{Cx_ix_j\}$, where x_i, x_j are variables, and when $c \leq 0$, then $PF(f) = \{NCx_ix_j\}$, where x_i, x_j are variables.*

In case $|f| > 0$, we assume that PF(g) has been already defined for every g with $|g| < |f|$. Again, there are some subcases.

PF2 *When there is $b_i > 0$, we let $PF(f) = \{LBqx_jr\}$, where $q \in PF(f - x_j), r \in PF(f + 1 - x_j)$.*

PF3 *When there is $b_i < 0$, we let $PF(f) = \{LBqNx_jr\}$, where $q \in PF(f + x_j - 1), r \in PF(f + x_j)$.*

For a polynomial formula p, we denote by $v(p)$ the truth value of p stemming from truth values of variables x_i. Actually then, $v(p)$ is a truth function. Let also *sgn* be the *sign function* defined via $sgn(x) = 1$ in case $x > 1, sgn(x) = 0$ in case $x < 0$ and $sgn(x) = x$, otherwise. Then we have

Proposition 3.21. *For each polynomial formula $p \in PF(f)$ and every set $x_i \in [0, 1]$ of variable values, we have $v(p) = sgn(f)$.*

Proof. Proof is by inducting on $|f|$. In case $|f| = 0$, both sides are 1. Assuming truth of Proposition in case of all g with $|g| < |f|$, we consider $p \in PF(f)$.

Again, there are some cases to consider, according to definitions of polynomial formulas given above.

1. *Some $b_i > 0$ so p is $LBqx_ir$ with some $q \in PF(f - x_i), r \in PF(f + 1 - x_i)$.*

   Then we have four sub–cases.
   (a) $f > 1 + x_i$ in which sub–case $sgn(f - x_i), sgn(f + 1 - x_i) = 1$ hence $v(q) = v(r) = 1$ by the inductive assumption and thus by semantics of B, L it follows that $v(p) = 1$. Clearly, $sgn(f) = 1$.
   Other sub–cases (b) $x_i \leq f < 1 + x_i$ (c) $-1 + x_i \leq f < x_i$ (d) $f < -1 + x_i$ are discussed in analogous way,
2. *Some $b_i < 0$ so p is $LBqNx_ir$ with $q \in PF(f + x_i - 1), r \in PF(f + x_i)$.*
   We have four sub–cases, viz., (a) $f + x_i > 2$ (b) $1 \leq f + x_i \leq 2$ (c) $0 \leq f + x_i < 1$ (d) $f + x_i < 0$ in which we argue as in Case 1.   □

This ends the basic introduction to polynomial formulas. Now, we list, following Rose and Rosser [44], some deeper properties of polynomial formulas necessary in the proof of completeness. We would omit the tedious technical lemmas intervening in these proofs which otherwise would obscure the main ideas of the proof and we refer the interested reader to Rose and Rosser, op. cit., Thms. 10.1–10.7, 13.1 – 13.18, for proofs of these properties.

Proposition 3.22. *(Rose and Rosser [44]) The following are among properties of polynomial formulas*

1. $\vdash p \Leftrightarrow q$ *whenever* $p, q \in PF(f)$;

2. $\vdash ANpq$ *whenever* $p \in PF(f), q \in PF(f + 1)$;

3. $\vdash p \Leftrightarrow Nq$ *whenever* $p \in PF(f), q \in PF(1 - f)$;

4. $\vdash Apq$ *whenever* $p \in PF(f), q \in PF(2 - f)$;

5. $t \vdash CpCCqrs$ *whenever* $q \in PF(f), r \in PF(g), s \in PF(1 - f + g), t \in PF(2 - f), p \in PF(g + 1)$;

6. $\vdash CpCqr$ *whenever* $q \in PF(f), r \in PF(g), p \in PF(1 - f + g)$;

7. $p_1, .., p_k \vdash q$ *whenever* $p_i \in PF(1 + f_i), q \in PF(1 + \sum_{i=1}^{k} f_i)$ *for some* $k > 0$;

8. $\vdash p$ *whenever* $p \in PF(1 + x_i)$;

9. $\vdash p$ *whenever* $p \in PF(2 - x_i)$;

10. $p_1, ..., p_m \vdash q$ *whenever* $p_i \in PF(1 + f_i), q \in PF(1 + g), f_1, ..., f_m, g$ *have integral coefficients, and* $f_1(x) \geq 0, ..., f_m(x) \geq 0$ *imply* $g(x) \geq 0$ *for every* $x \in [0, 1]^n$ *whose each coordinate is rational.*

Assuming truth of statements in Proposition 3.22, we now may proceed with concluding steps in completeness proof.

For a formula p and a polynomial f, a new formula $\alpha(p, f)$ is defined as follows. For an arbitrarily chosen $q \in PF(f), r \in PF(2-f), t \in PF(1+f)$, we let $\alpha(p, f)$ to be $LLEpqrs$. Let us observe that any other choice of q, r, s subject to conditions yields an equivalent formula $\alpha(p, f)$ due to Proposition 3.22(1).

With help of this new class of formulas, one may prove the following principal statement.

Proposition 3.23. *(Rose and Rosser [44]) For each formula* p, *there exist* $m \geq 0$, *polynomial formulas* $q_1, ..., q_m, r_1, ..., r_m$ *and polynomials* $f_1, ..., f_{2^m}$ *with the properties*

1. $\vdash Aq_i r_i$ for $i = 1, 2, ..., m$.
2. *for each subset* $\{j_1, .., j_k\} \subseteq \{1, 2, ..., m\}$ *with the complementing subset* $\{i_1, .., i_{m-k}\}$ *there exists* $1 \leq k \leq 2^m$ *with the property that*

$$q_{j_1}, ..., q_{j_k}, r_{i_1}, ..., r_{i_{m-k}} \vdash \alpha(p, f).$$

Proof. We induct on complexity of p.

In case p consists of a single symbol, it is of the form x_i; consider $m = 0$ so only f_1 needs to be defined and choose $f_1(x) = x_i$. Then (i) $\vdash r$ (ii) $\vdash t$ by Proposition 3.22 (8, 9). By Proposition 3.22 (1), we have

$$\vdash q \Leftrightarrow LB\neg ssx_i \Rightarrow Css$$

which is

$$\vdash q \Leftrightarrow LBNCsspCss$$

By R11 and R33, we get that

$$\vdash q \Leftrightarrow BNCssp$$

whence

$$\vdash q \Leftrightarrow p$$

follows by R11, R19, and R13. Then $\vdash Epq$ follows by R3 and (i), (ii), R28, and Proposition 3.22 (6), imply $\vdash \alpha(p, f)$.

To get through the inductive step, assume that the proposition is true for all t with complexity less than l, and consider p with l symbols. There are two cases in need of consideration: 1. p is Nt 2. p is Ctr.

Case 1. There exist an integer n, polynomial formulas $q_1, ..., q_n, r_1, ..., r_n$ and polynomials $g_1, ..., g_{2^n}$ satisfying 1,2, in Proposition 3.23 with t. Keeping $m = n$, we let $f_i = 1 - g_i$ for $1 \leq i \leq 2^n$.

Assuming q^*, r^*, s^* were chosen from, respectively, $PF(g_i)$, $PF(2 - g_i)$, $PF(1 + g_i)$ and choosing q, r, s from respectively $PF(f_i), PF(2 - f_i), PF(1 + f_i)$ we have by Proposition 3.22 (1, 3), that

(i) $\vdash r \Leftrightarrow s^*$

(ii) $\vdash s \Leftrightarrow r^*$

(iii) $\vdash q \Leftrightarrow Nq^*$

Now, we have $Etq^* \vdash ENtNq^*$ by R31, i.e.,

(iv) $Etq^* \vdash Epq$

We now consider a subset $\{j_1, ..., j_k\}$; we have by inductive assumption that

$$q_{j_1}, ..., q_{j_k}, r_{i_1}, ..., r_{i_{m-k}} \vdash \alpha(t, g_i)$$

for some i. It follows from (i), (ii), (iii), (iv), R25, R26, and R28 that

$$\alpha(t, g) \vdash \alpha(p, f)$$

which concludes the proof in this case.

Case 2. As p is $t \Rightarrow r$, the inductive assumption yields us $m, q_1, ..., q_m$, $r_1, ..., r_m$, $g_1, ..., g_{2m}$ satisfying conditions 1,2 of Proposition 3.23 with t and $n, q_1^*, ..., q_n^*, r_1^*, ..., r_n^*, h_1, ..., h_{2n}$ satisfying those conditions with r.

It remains to glue these constructs together to produce the respective set for p.

We let $k = m + n + 2^{m+n}$. We make use of q_i's, q_i^*'s, r_i's, and r_i^*'s whose number is $m + n$ so we need 2^{m+n} new polynomial formulae. To construct them, for $1 \leq u \leq 2^m$ and $1 \leq v \leq 2^n$ we select $s_{uv} \in PF(1 - g_u + h_v), s_{uv}^* \in PF(1 + g_u - h_v)$. As $\vdash As_{uv}s_{uv}^*$ by Proposition 3.22 (4), sets

$$\{q_1, ..., q_m, r_1, ..., r_n, s_{11}, ..., s_{2^m2^n}\}$$

and

$$\{q_1^*, ..., q_m^*, r_1^*, ..., r_n^*, s_{11}^*, ..., s_{2^m2^n}^*\}$$

satisfy condition 1 of Proposition 3.23. We denote elements of the first set with symbols a_i and elements of the second set with symbols a_i^*.

Now, we need polynomials satisfying condition 2 with new sets of formulas. We select a subset $\{j_1, ..., j_w\}$ of the set $\{1, ..., k\}$, and we consider the set $M = \{a_{j_1}, ..., a_{j_w}, a_{i_1}^*, ..., a_{i_{k-w}}^*\}$.

Let Γ be the subset of M consisting of formulas of the form q_i, q_i^* and Δ be the subset of r_i, r_i^* of M. By inductive assumption, we have $u \leq m$, $v \leq n$ such that

(i) $\Gamma \vdash \alpha(t, g_u)$

(ii) $\Delta \vdash \alpha(r, h_v)$

Applying Proposition 3.22 (6), R25, and R26, we infer from (i), (ii) that

(iii) $\Gamma, \Delta \vdash Ett_1, t_2, t_3, Err_{1,r}, r_{2,r}, r_{3,r}$

where $\alpha(t, g_u)$ is $LLEtt_1t_2t_3$ and $\alpha(r, h_v)$ is $LLErr_{1,r}r_{2,r}r_{3,r}$.

By (iii) and R32, it follows that

(iv) $\Gamma, \Delta \vdash EpCt_1 r_{1,r}$

By Proposition 3.22 (6), one obtains

(v) $\vdash Cs_{uv} Ct_1 r_{1,r}$

and Proposition 3.22 (5), implies via (v) that

(vi) $t_2 \vdash Cr_{3,r} CCt_1 r_{1,r} s_{uv}$

Now, it follows from (iii), (iv) that

(vii) $\Gamma, \Delta \vdash Eps_{uv}$

It is time now to show the existence of the appropriate polynomial f_M for this case. There are two cases to discuss: either $s_{uv} \in M$ (Case 1) or $s_{uv} \notin M$ (Case 2).

Case 1. Let f_M be identically 1. By R11 and R33, we have,

(viii) $s_{uv} \vdash Es_{uv} Czz$ *for an arbitrarily chosen formula z*

whence by (vii) and R32,

(ix) $\Gamma, \Delta, s_{uv} \vdash EpCzz$

By Proposition 3.22 (1), and the definition of f_M, we have for $p_1 \in PF(f_M), p_2 \in PF(2 - f_M), p_3 \in PF(1 + f_M)$ that $p_1 \vdash Czz$, $p_2 \vdash Czz$, and $p_3 \vdash Czz$, and thus by R11 and R28 one arrives at,

$$\Gamma, \Delta, s_{uv} \vdash \alpha(p, f_M)$$

concluding the proof in this case.

Case 2. We have $s_{uv}^* \in M$. Let $f_M = 1 - g_u + h_v$ and select p_1, p_2, p_3 appropriately in respective PF's. By Proposition 3.22 (1), $\vdash p_1 \Leftrightarrow s_{uv}$ and $\vdash p_2 \Leftrightarrow s_{uv}^*$ which imply via (vii) and R28 that

(x) $\Gamma, \Delta, s_{uv}^* \vdash LEpp_1 p_2$

By Proposition 3.22 (7), it follows that

(xi) $t_2, r_{3,r} \vdash p_3$

Finally, by (iii) and (x), one infers that

$$\Gamma, \Delta, s_{uv}^* \vdash \alpha(p, f_M)$$

concluding the proof in Case 2 and hence the proof of the proposition   □

We only need one more auxiliary statement

Proposition 3.24. *For each formula p and each polynomial f the following hold*

1. $\Gamma \vdash \alpha(p, f)$ implies $\Gamma, p \vdash p_1$.
2. $\Gamma \vdash \alpha(p, f)$ implies $\Gamma \vdash Cp_1p$.

Proof. Concerning Property 1, we begin with $\Gamma \vdash LLEpp_1p_2p_3$ whence by R22 first $\Gamma \vdash Lp_3LEpp_1p_2$ and next $\Gamma \vdash Lp_3Lp_2Epp_1$ follow. Assuming p, we have $\Gamma, p \vdash Lp_3Lp_2p_1$ and applying twice R25, one obtains $\Gamma, p \vdash p_1$.

For Property 2, we begin as in case of Property 1, until $\Gamma \vdash Lp_3Lp_2Epp_1$ and at this point one applies R25 twice getting $\Gamma \vdash Epp_1$ from which $\Gamma \vdash Cp_1p$ follows   □

We now may state the completeness theorem. Clearly, $Ł_\infty$ is sound: whenever a formula p is a theorem, it is a tautology as well, i.e., $v(p)$ is identically 1. The converse is also true.

Proposition 3.25. *(The Completeness Theorem, announced in Wajsberg [49], Rose and Rosser [44]) For each formula p, if p is a tautology, i.e., the value $v(p)$ is identically 1, then p is a theorem.*

Proof. Assume that $v(p)$ is identically 1 for a formula p. By Proposition 3.23 there exist $m \geq 0$, polynomial formulas $q_1, ..., q_m, r_1, ..., r_m$ and polynomials $g_1, ..., g_m, h_1, ..., h_m$ as well as $f_1, ..., f_{2^m}$ with the properties following from Proposition 3.24

(i) $\vdash Aq_ir_i$ for $i = 1, 2, ..., m$

(ii) For each subset $\{j_1, .., j_k\} \subseteq \{1, 2, ..., m\}$ with the complementing subset $\{i_1, ..., i_{m-k}\}$, we have a polynomial f_k such that

(a) $q_{j_1}, ..., q_{j_k}, r_{i_1}, ..., r_{i_{m-k}}, p \vdash p_1$

(b) $q_{j_1}, ..., q_{j_k}, r_{i_1}, ..., r_{i_{m-k}} \vdash Cp_1p$

with some $p_1 \in PF(f_k)$ and $q_i \in PF(g_i), r_i \in PF(h_i)$ for $i = 1, 2, ..., m$

From (a), by Proposition 3.21 it follows that whenever $g_i(x), h_i(x) \geq 1$, then $f_k(x) \geq 1$ for each truth value vector x with coordinates in $[0, 1]$. Then by Proposition 3.22 (10), we have

(iii) $q_{j_1}, ..., q_{j_k}, r_{i_1}, ..., r_{i_{m-k}} \vdash p_1$

whence by (b) and detachment

(iv) $q_{j_1}, ..., q_{j_k}, r_{i_1}, ..., r_{i_{m-k}} \vdash p$

Now, we have eliminated p_1, hence f_k, a fortiori, no more do we need formulas $\alpha(p, f_i)$ and we are left only with the fact that (iv) holds for each choice of $\{j_1, ..., j_k\} \subseteq \{1, 2, ..., m\}$.

Using R16 along with Property 1 an appropriate number of times yields

(v) $\vdash p$

which concludes the proof   □

Hence, the calculus $\mathbf{L}_\infty$ is complete. Please observe that the proof above covers cases when the set of truth values is an infinite subset of $[0, 1]$ (with additional above mentioned requirements), e.g., the set of rational values, or simply the segment $[0, 1]$.

3.5.1   MV–Algebras

Another approach, along the lines of algebraic proofs of completeness, was proposed in Chang [5]. The idea pursued by Chang was to give a counterpart of the algebraic proof of completeness of propositional calculus due to Rasiowa and Sikorski, see sect. 3.2, by constructing an algebra in which formulas of infinite–valued calculus could be interpreted and the maximal ideal property would hold. The result was an MV–algebra (i.e., many–valued algebra).

An MV–algebra is a structure $(X, +, \cdot, -, 0, 1)$, in which $+, \cdot$ are binary operations which are symmetric and associative, $-$ is a unary operation of complement, $0, 1$ are constants, and moreover the properties hold

1. $x + (- - x) = 1, x \cdot (-x) = 0$;

2. $x + 1 = 1, x \cdot 1 = x$;

3. $x + 0 = x, x \cdot 0 = 0$;

4. $x = - - x, -0 = 1$;

5. $-(x + y) = (-x) \cdot (-y); -(x \cdot y) = (-x) + (-y)$.

The connectives $\vee, \wedge$ are introduced by means of the formulas

6. $x \vee y = x \cdot (-y) + y$.

7. $x \wedge y = [x + (-y)] \cdot y$.

Then, $\vee, \wedge$ are symmetric and associative, and the lattice $(X, \vee, \wedge)$ is distributive.

Moreover

8. $x \vee 1 = 1, x \wedge 1 = x$.

9. $x \vee 0 = x; x \wedge 0 = 0$.

10. $x \vee x = x = x \wedge x$.

11. $x \wedge (x \vee y) = x = x \vee (x \wedge y)$.

Let us observe that $\vee$ conforms semantically to A and $\wedge$ conforms semantically to K in the Rose and Rosser's scheme.

The ordering relation $\leq_L$ is defined by means of

$$x \leq_L \Leftrightarrow x \vee y = y \Leftrightarrow x \wedge y = x \qquad (3.23)$$

Then, Chang [5], par. 1.3–1.17, proves

Proposition 3.26. *The ordering $\leq_L$ is a congruence, i.e.,*

1. $x \leq_L y$ and $t \leq_L z$ imply $x \vee t \leq_L y \vee z$;

2. $x \leq_L y$ and $t \leq_L z$ imply $x \wedge t \leq_L y \wedge z$;

3. $x \leq_L y$ and $t \leq_L z$ imply $x + t \leq_L y + z$;

4. $x \leq_L y$ and $t \leq_L z$ imply $x \cdot t \leq_L y \cdot z$.

It follows that $\vee, \wedge$ are,. respectively, the join and the meet operations. Moreover, the algebra $B(X) = \{x \in X : x + x = x\}$ is a Boolean algebra in which $x + y = x \vee y$ and $x \wedge y = x \cdot y$.

The Lindenbaum–Tarski algebra $LT(\infty)$ of the calculus $Ł_\infty$ is defined as in case of propositional calculus, by considering the equivalence classes $[[\alpha]]_\approx$ of formulas, where $\alpha \approx \beta$ holds if and only if $\alpha \Rightarrow \beta$ and $\beta \Rightarrow \alpha$ are theorems of $Ł_\infty$, with induced operations,

1. $[[\alpha]]_{\approx} + [[\beta]]_{\approx} = [[B\alpha\beta]]_{\approx}$;

2. $[[\alpha]]_{\approx} \cdot [[\beta]]_{\approx} = [[L\alpha\beta]]_{\approx}$;

3. $[[\alpha]]_{\approx} \vee [[\beta]]_{\approx} = [[A\alpha\beta]]_{\approx}$;

4. $[[\alpha]]_{\approx} \wedge [[\beta]]_{\approx} = [[K\alpha\beta]]_{\approx}$;

5. $[[\alpha]]_{\approx} = 1 \Leftrightarrow \vdash \alpha$;

6. $[[\alpha]]_{\approx} = 0 \Leftrightarrow \vdash \neg\alpha$.

Hence, see Chang [5], sect. 2, $LT(\infty)$ is an MV–algebra.

Other example, see Chang [5], sect. 2, is submitted by the interval $[0,1]$ with operations

1. $x + y = min\{1, x + y\}$;

2. $x \cdot y = max\{0, x + y - 1\}$;

3. $-x = 1 - x$;

4. $x \vee y = max\{x, y\}$;

5. $x \wedge y = min\{x, y\}$;

6. $x \leq_L y \Leftrightarrow x \leq y$,

which represent the semantics of the Łukasiewicz logic.

For the completeness proof on these lines, see Chang [5], cf., also an exposition in Hájek [17].

3.6 Many–Valued Calculi: Logics of Residual Implications

These logics are also termed 'fuzzy logics' because of their association with fuzzy theory of approximate reasoning due to Zadeh [24], for which they serve as a theoretical foundation. Classical propositional logics are related to the classical notion of a set/collection/class/concept, understood as an object in which some other objects may or not belong, in either case with certainty; as with many–valued departure from the true/false dichotomy, the classical notion of a set did undergo a relaxation to the notion of a *fuzzy set/concept* in Zadeh [24]. The relaxation had made usage of the manner of representing a set by means of a *characteristic function*: given a set $X \subseteq U$, U being a universe of considered objects, the characteristic function χ_X of the set X is defined by means of the formula,

$$\chi_X(u) = \begin{cases} 1 \ u \in X \\ 0 \ u \notin X \end{cases} \tag{3.24}$$

The fuzzy set X is defined by means of a *fuzzy membership function* μ_X : $U \to [0,1]$; a subjective choice of μ_X determines the understanding of the nature of objects with respect to the concept X.

In sect. 3.5, above, we have witnessed functors A, B, K, L, whose semantic interpretations have come in pairs: $l(x,y) = max\{0, x + y - 1\}, b(x,y) = min\{1, x+y\}$, and, $k(x,y) = min\{x,y\}, a(x,y) = max\{x,y\}$. Denoting l, k with T and b, a with S, we observe the duality

$$S(x,y) = 1 - T(1 - x, 1 - y); \ T(x,y) = 1 - S(1 - x, 1 - y) \tag{3.25}$$

A more general setting is provided by the framework of $t - norms$ (triangular norms) due originally to Menger [33], cf., Schweitzer–Sklar [62]. A t–norm $T : [0,1]^2 \to [0,1]$ does satisfy the requirements

T1 (Symmetry): $T(x,y) = T(y,x)$

T2 (Associativity): $T(x, T(y,z)) = T(T(x,y), z)$

T3 (Monotonicity): if $x > x'$ then $T(x,y) \geq T(x',y)$

T4 (Boundary conditions): $T(x,0) = 0; T(x,1) = x$

It is a simple task to verify that l, k are t–norms; we may reveal their names,

$L(x,y) = max\{0, x + y - 1\}$ (*the Łukasiewicz t–norm*)

$M(x,y) = min\{x,y\}$ (*the t–norm minimum*)

We add one more classical t–norm,

$P(x,y) = x \cdot y$ (*the Product t–norm, or, the Menger t–norm*)

The Łukasiewicz formula for interpreting implication $\Rightarrow$, i.e., $c(x,y) = min\{1, 1 - x + y)\}$, can be recovered from the t–norm $L(x,y)$; the tool is the *residual implication* $\Rightarrow_T$ defined by means of the equivalence

$$x \Rightarrow_T y \geq z \iff T(x,z) \leq y \tag{3.26}$$

Equivalently,

$$x \Rightarrow_T y = sup\{z : T(x, z) \le y\} \tag{3.27}$$

In case the t–norm $T(x, y)$ is *continuous*, the operator *sup* becomes the operator *max*. Helped by (3.27), one can easily find the explicit formulas for residual implications induced, respectively, by L, P, M:

1. $x \Rightarrow_L y = min\{1, x - y + 1\}$;

2. $x \Rightarrow_P y = 1 \Leftrightarrow x \le y$ and $\frac{y}{x}$ *otherwise*;

3. $x \Rightarrow_M y = 1 \Leftrightarrow x \le y$ and y *otherwise*.

Residual implications $\Rightarrow_L$, $\Rightarrow_P$, $\Rightarrow_M$ are known as, respectively, *the Łukasiewicz implication, the Goguen implication, the Gödel implication*.

For a given continuous t–norm $T(x, y)$, the *T–many–valued logic*, for initial logical ideas in the realm of fuzziness see Goguen [12] and Zadeh [52], see in this respect Hájek [17], is built of a countable set of propositional variables $\{p_i : i = 1, 2, ...\}$ along with connectives (functors) & (strong conjunction), $\land$ (conjunction), $\lor$ (disjunction), $\Rightarrow$ (implication), $\neg$ (negation) and a constant Z.

Formulas of the T–logic are built from propositional variables by means of connectives. Semantics is defined by interpreting & as $T(x, y)$, $\Rightarrow$ as $\Rightarrow_T$, the constant Z as 0 and negation $\neg$ as $\Rightarrow_T 0$.

The Łukasiewicz logic has been already introduced and described in sect. 3.5: it is the logic of the Łukasiewicz t–norm and its residual implication. Other logics, among others, are the *product logic* in which the implication is evaluated as the residual implication $\Rightarrow_P$, the strong conjunction as the product t–norm $P(x, y)$, and negation $\neg$ as $\neg_G$, and the *Gödel logic* with, respectively, $\Rightarrow_M$, $Min(x, y)$, $\neg_G$, representing semantically, respectively, the implication, the strong conjunction, and negation. The reader may consult Hájek [17] for details concerning axiomatization and semantic models for those logics.

For the purpose of future reference, we quote the axiomatic system of the *basic logic* of Hájek [17], i.e., the fuzzy logic whose theorems are valid under any continuous t–norm induced semantics. The axioms are, see loc.cit.

1. $(p \Rightarrow q) \Rightarrow ((q \Rightarrow r) \Rightarrow (p \Rightarrow r))$;

2. $p\&q \Rightarrow p$;

3. $p\&q \Rightarrow q\&p$;

4. $[p\&(p \Rightarrow q)] \Rightarrow [q\&(q \Rightarrow p)]$;

5. $[(p \& q) \Rightarrow r] \Leftrightarrow [p \Rightarrow (q \Rightarrow r)]$;

6. $[(p \Rightarrow q) \Rightarrow r] \Rightarrow [((p \Rightarrow q) \Rightarrow r) \Rightarrow r]$;

7. $Z \Rightarrow p$.

An important tool for automated reasoning, the resolution technique is discussed next.

3.7 Automated Reasoning

The old idea of Leibniz, of a symbolic reasoning performed in an automatic way by a machine, realizes in the field of, e.g., propositional calculus (a corresponding scheme holds for predicate calculus, among others). Deductive reasoning employs devices known collectively as an *inference engine* which allow to deduce a consequence from a set of premises. The set of premises is often called a *knowledge base*, KB in symbols. A *proof* of a statement α from KB is a sequence of statements $(\alpha_i : i = 1, 2, ..., k)$ such that $\alpha_1 \in KB$, $\alpha_k = \alpha$, and each α_i is obtained from $\alpha_1, ..., \alpha_{i-1}$ by means of an application of MP or $\alpha_i \in KB$; a statement α provable from KB is *KB–deductible*, $KB \vdash \alpha$, in symbols. A *model* for KB is a valuation on sentential variables under which all statements in KB are true. As MP preserves truth, it follows that $KB \vdash \alpha$ entails that α is true in each model for KB; this fact is known as *semantic entailment*, $KB \models \alpha$, in symbols.

The *inference problem* consists in verifying whether a statement α follows semantically (is entailed by) from a knowledge base KB. Propositional calculus got to this end an automatic tool: the *resolution technique* of Robinson [43]. Resolution is applied to formulas in the clausal conjunctive normal form (CNF); to represent a formula α in CNF, one lists sentential variables in α as $p_1, ..., p_k$ and one considers all valuations on those k variables i.e., the set $2^k = \{0, 1\}^k$. For each $v \in 2^k$, one finds the truth value $[\alpha]_v$ of α under v and defines the set $T(\alpha) = \{v \in 2^k : [\alpha]_v = 1\}$. For each $t \in T([\alpha])$, one lets $p_i^t = p_i$ in case $t(p_i) = 1$ and $p_i^t = \neg p_i$ in case $t(p_i) = 0$. Then the formula $\bigvee_{t \in T([\alpha])} \bigwedge_i p_i^t$ is *disjunctive normal form* (DNF) of α.

Well–known de Morgan formulas $\neg(p \vee q) \Leftrightarrow (\neg p) \wedge (\neg q)$ and $\neg(p \wedge q) \Leftrightarrow (\neg p) \vee (\neg q)$ allow to produce CNF of α: first, produce DNF of $\neg \alpha$ and then convert it by the de Morgan laws to CNF of $\neg\neg\alpha$, i.e., to CNF of α. It follows that CNF of α is a conjunction $\bigwedge_m C_m$ of *clauses* $C_m : l_1^m \vee ... \vee l_{j_m}^m$, where each l_i^m is a *literal*, i.e., it is of the form of a variable or its negation.

The technique of resolution rests on the *resolution rule* applied to pairs of clauses $l_1 \vee ... \vee l_j$, $p_1 \vee .... \vee p_m$ which contain a pair l_i, p_j of contradictory literals, i.e., l_i is $\neg p_j$,

$$Resolution\ rule: \quad \frac{l_1 \vee \ldots \vee l_i \vee \ldots \vee l_j, p_1 \vee \ldots \vee p_j \vee \ldots \vee p_m}{l_1 \vee . \vee \overbrace{l_i} \vee .. \vee l_j \vee p_1 \vee \ldots \overbrace{p_j} \vee \ldots \vee p_m}, \qquad (3.28)$$

where the symbol $\overbrace{l_i}$ denotes the fact that the literal l_i is omitted in the disjunction. In plain wording, the resolution rule allows for cancelation of contradictory literals in two clauses, leading to one clause rid of the complementary pair. Successive application of the resolution rule to a set of clauses, can result in one of the two final outcomes: (1) the set of clauses does stabilize without producing the empty clause $\Box$ (2) the empty clause $\Box$ is produced at some step.

The empty clause $\Box$ is *unsatisfiable*, i.e., it is true under no valuation. As the resolution rule yield true clauses from true clauses, the appearance of the empty clause $\Box$ does witness that the original set of clauses was unsatisfiable. It remains to observe that

Proposition 3.27. $KB \models \alpha$ *if and only if the formula* $(\bigwedge KB) \wedge \neg\alpha$ *is unsatisfiable,*

in order to conclude that for verification whether $KB \models \alpha$ it suffices to apply the resolution rule recurrently to the set of clauses of the formula $(\bigwedge KB) \wedge \neg\alpha$ until the result (1) or (2) is obtained. In case (2), $KB \models \alpha$ is confirmed.

The satisfiability problem SAT is known to be NP–complete since Cook [6], hence the entailment problem is co–NP–complete. It is computationally feasible, however, in a most important case of *definite clauses*. A *Horn clause* is a clause $l_1 \vee \ldots \vee l_j$ in which at most one literal is positive, i.e., non–negated, and a *definite clause* is a Horn clause with exactly one positive literal. A tautology $(p \Rightarrow q) \Leftrightarrow \neg p \vee q$ allows to write down a definite clause $l_1 \vee \ldots \vee l_j$ where l_j is positive, as the implication $\bigwedge_{i=1}^{j-1} \neg l_i \Rightarrow l_j$, i.e., as a *decision rule* or a *fact* in case when $j = 1$, i.e, the premise of the rule is empty. Resolution with definite clauses and facts is linear in size of the clause set, hence of practically low complexity, and in most cases of applications, knowledge bases consist of Horn clauses.

These results transfer partially to T–logics. Let us denote with, respectively, SAT(2), SAT(L), SAT(M), SAT(P), the sets of satisfiable formulas of propositional calculus, the Łukasiewicz logic, the Gödel logic, and the product logic, i.e., for each $X = 2, L, M, P$ and $\alpha \in SAT(X)$, there exists a valuation v such that $v(\alpha) = 1$. Clearly, each valuation for a proposition is a valuation for fuzzy logics as well hence $SAT(2) \subseteq SAT(Y)$ for $Y = L, M, P$. The converse inclusions $SAT(P), SAT(M) \subseteq SAT(2)$ hold as well, see Hájek [17], sect. 6.2, hence $SAT(P) = SAT(M) = SAT(2)$ and in consequence $SAT(P), SAT(M)$ are NP–complete. For the Łukasiewicz logic, one has $SAT(2) \subseteq SAT(L)$, Hájek loc. cit.; actually, $SAT(L)$ is NP–complete, see Mundici [36].

3.8   Predicate Logic

Expressions that state a certain property of objects are called *predicates*. A symbol $P(x_1, ..., x_k)$ denotes a *k–ary* predicate which does express that a *k*–tuple $(x_1, ..., x_k)$ of objects has a property P. In addition to expressing predicates, this calculus allows for saying that all objects have the given property or some of them have it. To this end, *quantifiers* $\forall$ (meaning: for all) and $\exists$ (meaning: there exists) are used: the formula $\forall x.P(x)$ is read 'all objects substituted for x have the property P' and the formula $\exists x.P(x)$ reads 'there is an object substituted for x which has the property P'.

As with propositional calculus, we discern in the structure of predicate calculus various sets of symbols used to denote various types of variables, and various types of objects we reason about and then we have some rules to form meaningful expressions of the language of this calculus. Next, we have some axioms, i.e., expressions we accept as true on the basis of their syntax only and derivation rules used to produce in a systematic, as it were mechanical way, theses of the system.

We are also taking into account the semantic aspect by defining meanings of expressions in order to select true expressions of the system. Finally, we may compare semantic and syntactic aspects via, e.g., completeness property. The predicate calculus is also called the *first order theory, first order functional calculus*, and the term *(pure) predicate calculus* is reserved for a theory in which no functional symbols intervene. In the coming presentation of basic ideas and results about predicate calculus, we follow the exposition in Rasiowa and Sikorski [42].

We will make use of the following sets of symbols

1. *A countable set V of individual variables;*

2. *A countable set F of function symbols;*

3. *A countable set Pr of predicate symbols;*

4. *The set of propositional connectives* $\Rightarrow, \neg, \vee, \wedge$;

5. *The set of quantifier symbols* $Q = \{\forall, \exists\}$;

6. *A set of auxiliary symbols* $\{(,), [,], \{,\}, \}$.

From symbols, more complex expressions are constructed. We represent the set F of function symbols as the union $\bigcup \{F_n :\in N\}$ and we call elements of the set F_n *n–ary* function symbols for each $n \in N$. From function symbols, parentheses and individual variable symbols expressions called *terms* are constructed; formally, a term is an expression of the form $\phi(\tau_1, ..., \tau_m)$ where $\phi \in F_m$ and $\tau_1, ..., \tau_m$ have already been declared terms.

In order to make this idea into a formal definition, we resort to induction on complexity of formulas involved, and we define the set $Term$ of terms as the intersection of all sets X such that

1. *All individual variable symbols are in X;*

2. *All 0-ary function symbols are in X;*

3. *For each $n \geq 1$, if $\tau_1, ..., \tau_n$ are in X and ϕ is an $n-$ary function symbol, then $\phi(\tau_1, ..., \tau_n)$ is in X.*

The set $Term$ is the least set satisfying requirements 1–3. Having terms defined, we proceed on to form more complex expressions, *formulas*, and again we will perform this task in a few steps. So we define the set $Form$ of formulas as the intersection of all sets Y such that

1. *For each $m \geq 1$, if $\tau_1, ..., \tau_n$ are terms and ρ is an $n-$ary predicate symbol, then $\rho(\tau_1, ..., \tau_n)$ is in Y;*

2. *If $\alpha, \beta \in Y$, then $\neg\alpha, \alpha \Rightarrow \beta \in Y$;*

3. *If $\alpha \in Y$, then $\forall x\alpha, \exists x\alpha \in Y$.*

Now, as with propositional calculus, we have to assign meanings to formulas (meaningful expressions) of predicate calculus. Unlike in the former case, we cannot assume the existence of two states for variables, i.e., truth and falsity. To account for individual variables, we have to interpret them in a set. Thus, we introduce the notion of an *interpretation frame* as a non–empty set I. Given I, we define an *interpretation* in I as a function M_I with properties,

1. *For each $n \in N$, M_I maps each $n-$ary function symbol ϕ onto a function $f_\phi : I^n \to I$ (in case $n = 0$, f_ϕ is a constant, i.e., a fixed element in I);*

2. *For each $n \in N$, M_I maps each $n-$ary predicate symbol ρ onto a relation $R_\rho \subseteq I^n$.*

Let us observe that under the function M_I, formulas of predicate calculus become *propositional functions* when in addition individual variable symbols are interpreted as variables ranging over the set I. This means that to every term of the form $\phi(x_1, ..., x_n)$, where $x_1, ..., x_n$ are individual variable symbols, the function M_I assigns the function

$$f_\phi(x_1, ..., x_n) : I^n \to I$$

where $x_1, ..., x_n$ are variables ranging over I.

Clearly, in case of a term of the form $\phi(\tau_1, ..., \tau_n)$, where $\tau_1, ..., \tau_n$ have been assigned functions $g_1, ..., g_n$, with $g_j : I^{n_j} \to I$, the term $\phi(\tau_1, ..., \tau_n)$ is assigned the composition

$$f_\phi(g_1(x_1^1, ..., x_{n_1}^1), ..., g_n(x_1^n, ..., x_{n_n}^n))$$

Then, to each formula of the form $\rho(\tau_1, ..., \tau_n)$, the function M_I assigns the propositional function

$$R_\rho(g_1(x_1^1, ..., x_{n_1}^1), ..., g_n(x_1^n, ..., x_{n_n}^n))$$

which becomes a proposition after substituting the specific elements of I for variables x_i^j. We will call such assignments *elementary*.

Now, in order to check the truth of formulas, we have to interpret in the set I variables. To this end, we introduce the notion of a *valuation* being a function $v : V \to I$; thus, v assigns to each individual variable x_i an element $v(x_i) \in I$.

Under v, each elementary assignment becomes a proposition

$$R_\rho(g_1(v(x_1^1), ..., v(x_{n_1}^1)), ..., g_n(v(x_1^n), ..., v(x_{n_n}^n)))$$

for an elementary assignment α, we denote by the symbol $[[\alpha]]_M^v$ the logical value (true/false) of α under the valuation v. In the sequel, when no confusion is possible, we will denote this value with the symbol $[[\alpha]]^v$.

Having defined elementary assignments, we define truth values of more complex formulas under any valuation v, see Tarski [47]. To this end, we let

1. $[[\neg\alpha]]^v = 1$ *if and only if* $[[\alpha]]^v = 0$;

2. $[[\alpha \Rightarrow \beta]]^v = 1$ *if and only if either* $[[\alpha]]^v = 0$ *or* $[[\beta]]^v = 1$;

3. $[[\forall x\alpha]]^v = 1$ *if and only if* $[[\alpha]]^{v(x/c)} = 1$ *for every choice of* $c \in I$, *where* $v(x/c)$ *is the valuation* v *in which the value* $v(x)$ *is changed to* c;

4. $[[\exists x\alpha]]^v = 1$ *if and only if* $[[\alpha]]^{v(x/c)} = 1$ *for some choice of* $c \in I$.

Let us observe that valuation of a formula obtained by means of propositional connectives corresponds to the case of propositions and interpretation of quantifiers in symbolic formulae is done by a common–sense interpretation of phrases "every choice", "some choice".

We call a formula α *true under* M_I *in* I when $[[\alpha]]^v = 1$ for every valuation v. The formula α is said to be *true in* I when it is true under every function M_I. Finally, α is *true* if and only if it is true in every interpretation frame I.

We now apply this notion of truth towards the exhibition of basic derivation rules in formal syntax of predicate calculus.

We formulate the *detachment rule* as usual, i.e., if $\alpha, \alpha \Rightarrow \beta$ are theorems, then β is accepted as a theorem. The substitution rule (S) will be modified adequately to the predicate calculus case as follows.

For a formula $\forall x \alpha$ as well as for a formula $\exists x \alpha$, if x occurs in α then we say that x is *bound* in α; any variable y not bound by a quantifier is *free* in α. To denote free variables in α we will write $\alpha(x_1, .., x_k)$ in place of α and all other variables not listed in $\alpha(x_1, .., x_k)$ are either bound or do not occur in α.

(S) *(Substitution for free individual variables) If $\alpha(x_1, ..., x_k)$ is a theorem and $\tau_1, ..., \tau_k$ are terms then the formula $\alpha(\tau_1, ..., \tau_k)$ obtained from $\alpha(x_1, ..., x_k)$ by the substitution $x_1/\tau_1, ..., x_k/\tau_k$ is accepted as a theorem.*

We observe that

Proposition 3.28. *The detachment rule and the substitution rule give true formulas when applied to true formulas.*

Proof. For the case of detachment, if $\alpha, \alpha \Rightarrow \beta$ are true then for every I, M_I and every valuation v we have $[[\alpha]]^v = 1$ hence as $[[\alpha \Rightarrow \beta]]^v = 1$ we have $[[\beta]]^v = 1$, i.e., β is true.

For substitution, if $[[\alpha(x_1, ..., x_k)]]^v = 1$ for every v then clearly $[[\alpha(\tau_1, ..., \tau_k)]]^v = 1$ for every v as in the latter formula values taken by $\tau_1, ..., \tau_k$ form a subset of I^k which is the set of values taken by $(x_1, ..., x_k)$ in the former formula $\qquad \qquad \qquad \square$

We turn to rules governing use of quantifiers; these rules describe, respectively, admissible cases in which we introduce, respectively, omit, quantifier symbols.

IE *(Introduction of existential quantifier) For formulas $\alpha(x), \beta$ with y not bound in $\alpha(x)$ if $\alpha(x) \Rightarrow \beta$ is true then $\exists y \alpha(y) \Rightarrow \beta$ is true*

IU *(Introduction of universal quantifier) For formulas $\alpha, \beta(x)$ with no occurrence of x in α and y not bound in α if $\alpha \Rightarrow \beta(x)$ is true, then $\alpha \Rightarrow \forall x \beta(x)$ is true*

EE *(Elimination of existential quantifier) For formulas $\alpha(x), \beta$ with no occurrence of bound variable y in $\alpha(x)$ if $\exists y \alpha(y) \Rightarrow \beta$ is true, then $\alpha(x) \beta$ is true*

EU *(Elimination of universal quantifier) For formulas $\alpha, \beta(x)$ with no occurrence of bound variable y in $\beta(x)$ if $\alpha \Rightarrow \forall y \beta(y)$ is true then $\alpha \Rightarrow \beta(x)$ is true*

We will define formally syntax of predicate calculus. We will take as axioms the following schemes

T1 $(\alpha \Rightarrow \beta) \Rightarrow [(\alpha \Rightarrow \gamma) \Rightarrow (\beta \Rightarrow \gamma)]$

T2 $(\neg\alpha\alpha) \Rightarrow \alpha$

T3 $\alpha \Rightarrow (\neg\alpha \Rightarrow \beta)$

and we use detachment MP, substitution for free variables S, IE, IU, EE, and EU as derivation rules.

The symbol **2** denotes the Boolean algebra consisting of truth values $0, 1$ with Boolean operations $x \cup y = max\{x, y\}$, $x \cap y = min\{x, y\}$, $-x = 1 - x$, $x \Rightarrow y = -x \cup y$ and distinguished elements $0 < 1$.

As with propositional calculus, we construct the Lindenbaum–Tarski algebra of predicate calculus $LT(Pred)$, by considering the equivalence relation $\approx$ defined as before by the condition that $\alpha \approx \beta$ if and only if $\alpha \Rightarrow \beta$ and $\beta \Rightarrow \alpha$ are theorems of predicate calculus.

We denote by the symbol $[[\alpha]]_\approx$ the equivalence class of α with respect to $\approx$. The Lindenbaum–Tarski algebra of predicate calculus is a Boolean algebra (cf., a historic note in Rasiowa and Sikorski [42], pp 209, 245–6). The ordering $\leq_L$ is defined as before by $[[\alpha]]_\approx \leq_L [[\beta]]_\approx$ if and only if $\alpha \Rightarrow \beta$ is a theorem.

Proposition 3.29. *(Rasiowa and Sikorski [42]) The algebra Form/ $\approx$ is a Boolean algebra with operations*

1. $[[\alpha]]_\approx \cup [[\beta]]_\approx = [[\alpha \vee \beta]]_\approx$;

2. $[[\alpha]]_\approx \cap [[\beta]]_\approx = [[\alpha \wedge \beta]]_\approx$;

3. $[[\alpha]]_\approx \Rightarrow [[\beta]]_\approx = [[\alpha \Rightarrow \beta]]_\approx$;

4. $-[[\alpha]]_\approx = [[\neg\alpha]]_\approx$.

Moreover, for each formula of the form $\alpha(x)$, we have

5. $[[\exists y\alpha(y)]]_\approx = sup\{[[\alpha(\tau)]]_\approx : \tau \in Term\}$, *where $\alpha(\tau)$ comes from $\alpha(x)$ as a result of substitution of a term τ for the free variable x.*

6. $[[\forall y\alpha(y)]]_\approx = inf\{[[\alpha(\tau)]]_\approx : \tau \in Term\}$.

Proof. We first show that $[[\alpha \vee \beta]]_\approx$ is the join of classes $[[\alpha]]_\approx, [[\beta]]_\approx$; the theorem

$$\alpha \Rightarrow (\alpha \vee \beta)$$

implies that

$$[[\alpha]]_\approx \leq_L [[\alpha \vee \beta]]_\approx \tag{3.29}$$

and, similarly,

$$[[\beta]]_\approx \leq_L [[\alpha \vee \beta]]_\approx \tag{3.30}$$

hence,

$$[[\alpha]]_\approx \cup [[\beta]]_\approx \leq_L [[\alpha \vee \beta]]_\approx \qquad (3.31)$$

On the other hand, if $[[\alpha]]_\approx \leq_L [[\gamma]]_\approx$ and $[[\beta]]_\approx \leq_L [[\gamma]]_\approx$, i.e., $\alpha \Rightarrow \gamma$ and $\beta \Rightarrow \gamma$ are theorems, then applying detachment twice to the theorem

$$(\alpha \Rightarrow \gamma) \Rightarrow [(\beta \Rightarrow \gamma) \Rightarrow (\alpha \vee \beta) \Rightarrow \gamma]$$

we get $(\alpha \vee \beta) \Rightarrow \gamma$, i.e.,

$$[[\alpha \vee \beta]]_\approx \leq_L [[\gamma]]_\approx \qquad (3.32)$$

It follows that

$$[[\alpha]]_\approx \cup [[\beta]]_\approx = [[\alpha \vee \beta]]_\approx \qquad (3.33)$$

A similar proof shows that

$$[[\alpha]]_\approx \cap [[\beta]]_\approx = [[\alpha \wedge \beta]]_\approx \qquad (3.34)$$

To show distributivity, it suffices to prove the existence of a relative pseudo–complementation, cf., Ch. 1, sect. 14.

For classes $[[\alpha]]_\approx, [[\beta]]_\approx$, we prove that

$$[[\alpha]]_\approx \Rightarrow [[\beta]]_\approx = [[\alpha \Rightarrow \beta]]_\approx \qquad (3.35)$$

To this end, consider any $[[\gamma]]_\approx$ with the property that

$$[[\alpha]]_\approx \cap [[\gamma]]_\approx \leq_L [[\beta]]_\approx$$

i.e., $(\alpha \wedge \gamma) \Rightarrow \beta$ is a theorem. It follows from a theorem

$$[(\alpha \wedge \gamma) \Rightarrow \beta] \Rightarrow (\gamma \Rightarrow (\alpha \Rightarrow \beta))$$

via detachment that $\gamma \Rightarrow (\alpha \Rightarrow \beta)$ is a theorem, i.e., $[[\gamma]]_\approx \leq_L [[\alpha \Rightarrow \beta]]_\approx$. To conclude this part of the proof it suffices to consider the theorem $[\alpha \Rightarrow (\beta \wedge \alpha)] \Rightarrow \beta$, which implies that

$$[[\alpha \Rightarrow (\beta \wedge \alpha)]]_\approx = [[\alpha \Rightarrow \beta]]_\approx \cap [[\alpha]]_\approx \leq_L [[\beta]]_\approx \qquad (3.36)$$

It follows that $[[\alpha \Rightarrow \beta]]_\approx$ is the relative pseudo–complement $[[\alpha]]_\approx \Rightarrow [[\beta]]_\approx$. In consequence of the last proved fact, there exists the unit element in this lattice, 1 and

$$1 = [[\alpha \Rightarrow \alpha]]_\approx = [[\alpha]]_\approx \Rightarrow [[\alpha]]_\approx \qquad (3.37)$$

for each α. As $(\alpha \wedge \neg\alpha) \Rightarrow \beta$ is a theorem, we have,

$$[[\alpha \wedge \neg\alpha]]_\approx = [[\alpha]]_\approx \cap [[\neg\alpha]]_\approx \leq_L [[\beta]]_\approx$$

for any formula β, i.e., $[[\alpha \wedge \neg\alpha]]_\approx$ is the null element 0. Thus

$$[[\alpha]]_\approx \cap [[\neg\alpha]]_\approx = 0 \tag{3.38}$$

i.e., $[[\neg\alpha]]_\approx \leq_L -[[\alpha]]_\approx$. The theorem $[\alpha \Rightarrow (\alpha \wedge \neg\alpha)] \Rightarrow \neg\alpha$ implies,

$$[[\alpha \Rightarrow (\alpha \wedge \neg\alpha)]]_\approx = [[\alpha]]_\approx \Rightarrow 0 = -[[\alpha \leq_L [[\neg\alpha]]_\approx \tag{3.39}$$

hence, $[[\neg\alpha]]_\approx = -[[\alpha]]_\approx$. Finally, the theorem $\alpha \vee \neg\alpha$ renders the equality

$$[[\alpha]]_\approx \cup -[[\alpha]]_\approx = 1 \tag{3.40}$$

for each formula α, i.e., $-[[\alpha]]_\approx$ is the complement of $[[\alpha]]_\approx$ and $Form/\approx$ is a Boolean algebra.

There remain statements 5 and 6 to verify. Let us look at statement 5, and consider a formula $\alpha(x)$. As the set of individual variable symbols is infinite, there exists a variable symbol y which does not occur in $\alpha(x)$ and no quantifier in $\alpha(x)$ binds y; as $\gamma \Rightarrow \gamma$ is a theorem of predicate calculus for every formula γ, also the formula $\exists y\alpha(y) \Rightarrow \exists y\alpha(y)$ is a theorem and the derivation rule EE (of elimination of existential quantifier) yields the theorem $\alpha(x) \Rightarrow \exists y\alpha(y)$.

Substitution of a term τ for the free variable x in the last theorem via the substitution rule gives us the theorem $\alpha(\tau) \Rightarrow \exists y\alpha(y)$. Thus $[[\alpha(\tau)]]_\approx \leq_L [[\exists y\alpha(y)]]_\approx$ for every term $\tau \in Term$.

It follows that

$$sup\{[[\alpha(\tau)]]_\approx : \tau \in Term\} \leq_L [[\exists y\alpha(y)]]_\approx \tag{3.41}$$

For the converse, let us assume that for some formula γ

$$[[\alpha(\tau)]]_\approx \leq_L [[\gamma]]_\approx \tag{3.42}$$

for every term $\tau \in Term$. We may find a variable z which does not occur in γ; as $[[\alpha(z)]]_\approx \leq_L [[\gamma]]_\approx$, i.e., $\alpha(y) \Rightarrow \gamma$, it follows by the rule IE (of introduction of existential quantifier) that $\exists y\alpha(y) \Rightarrow \gamma$ is a theorem, i.e.,

$$[[\exists y\alpha(y)]]_\approx \leq_L [[\gamma]]_\approx \tag{3.43}$$

Hence,

$$[[\exists y\alpha(y)]]_\approx = sup\{[[\alpha(\tau)]]_\approx : \tau \in Term\} \tag{3.44}$$

The statement 6 may be proved on similar lines, with rules EU, IU in place of rules EE, IE (these results are due to Rasiowa [41], [40] and Henkin [18], cf., Rasiowa and Sikorski [42], p. 251) □

The equalities

1. $[[\exists y \alpha(y)]]_\approx = sup\{[[\alpha(\tau)]]_\approx : \tau \in Term\}$.
2. $[[\forall y \alpha(y)]]_\approx = inf\{[[\alpha(\tau)]]_\approx : \tau \in Term\}$,

are called (Q)–*joins* and *meets*, respectively, see Rasiowa and Sikorski. [42]

We now may produce a proof of the main result bridging syntax and semantics of predicate calculus.

Proposition 3.30. *A formula α of predicate calculus is a theorem if and only if $[[\alpha]]_\approx = 1$.*

Proof. (after Rasiowa and Sikorski [42]) Assume that α is a theorem. By the theorem $\alpha \Rightarrow CC\alpha\alpha$ of predicate calculus, it follows via detachment that $(\alpha \Rightarrow \alpha) \Rightarrow \alpha$ is a theorem, hence, $[[\alpha \Rightarrow \alpha]]_\approx \leq_L [[\alpha]]_\approx$, i.e., $1 \leq_L [[\alpha]]_\approx$ and thus $1 = [[\alpha]]_\approx$.

Conversely, in case $[[\alpha]]_\approx = 1$ we have $[[\beta]]_\approx \leq_L [[\alpha]]_\approx$ for each β, hence, in particular, $[[\alpha \Rightarrow \alpha]]_\approx \leq_L [[\beta]]_\approx$, i.e., $(\alpha \Rightarrow \alpha) \Rightarrow \alpha$ is a theorem and detachment yields that α is a theorem   □

We should now extend the last result to any interpretation. Consider an interpretation frame I along with an interpretation M_I and a valuation v. Then we have the following result.

Proposition 3.31. *For any formulas α, β of predicate calculus, if $[[\alpha]]_\approx = [[\beta]]_\approx$, then $[[\alpha]]_M^v = [[\beta]]_M^v$.*

Proof. Assume that $[[\alpha]]_\approx| = [[\beta]]_\approx$ i.e. the formulas $\alpha \Rightarrow \beta$ and $\beta \Rightarrow \alpha$ are theorems of predicate calculus. Then the conditions hold,

1. *Either* $[[\alpha]]_M^v = 0$ *or* $[[\beta]]_M^v = 1$.
2. *Either* $[[\beta]]_M^v = 0$ *or* $[[\alpha]]_M^v = 1$.

The conjunction of conditions 1 and 2 results in the condition,

$$[[\alpha]]_M^v = 0 = [[\beta]]_M^v \vee [[\alpha]]_M^v = 1 = [[\beta]]_M^v$$

i.e., $[[\alpha]]_M^v = [[\beta]]_M^v$ □

This result shows that the function $[[\cdot]]_M^v$ defined by the valuation v from the set $Form$ of formulas of predicate calculus into the Boolean algebra **2** can be factored through the Boolean algebra $Form/\approx$:

$$[[\cdot]]_M^v = h \circ q(\cdot)$$

where $q : Form \rightarrow Form/\approx$ is the quotient function and $h : Form/\approx \rightarrow$ **2** is defined by the formula

$$h([[\alpha]]_\approx) = [[\alpha]]_M^v$$

Actually, more is true,

Proposition 3.32. *(Rasiowa and Sikorski, op.cit.) The function* $h : Form/$ $\approx\to 2$ *is a homomorphism preserving (Q) –joins and (Q)-meets.*

Proof. Given formulas α, β of predicate calculus, we have

$$h([[\alpha]]_\approx \cup [[\beta]]_\approx) = h([[\alpha \vee \beta]]_\approx) = [[\alpha \vee \beta]]_M^v = [[\alpha]]_M^v \cup [[\beta]]_M^v =$$

$$h([[\alpha]]_\approx \cup h([[\beta]]_\approx)$$

Similarly,

$$h([[\alpha]]_\approx \cap [[\beta]]_\approx) = h([[\alpha]]_\approx) \cap h([[\beta]]_\approx)$$

and

$$h(-[[\alpha]]_\approx) = -h([[\alpha]]_\approx)$$

Moreover,

$$h(1) = h([[\alpha \vee \neg\alpha]]_\approx) = [[\alpha]]_M^v \cup [[\neg\alpha]]_M^v = 1$$

For (Q)–joins, we have

$$h([[\exists y\alpha(y)]]_\approx) = [[\exists y\alpha(y)]]_M^v = sup_x[[\alpha(y)]]_M^{v(y/x)} = sup_\tau[[\alpha(\tau)]]_M^v =$$

$$sup_\tau h([[\alpha(\tau)]]_\approx)$$

Similarly, the result for meets follows   □

We may now sum up latest developments in a version of completeness theorem. Completeness of predicate calculus was first demonstrated in Gödel [14].

Proposition 3.33. *(after Rasiowa and Sikorski [42]) (A completeness theorem for predicate calculus) The following statements are equivalent for each formula* α *of predicate calculus.*

1. α is a theorem of predicate calculus;

2. $h([[\alpha]]_\approx) = 1_{\mathbf{B}}$ where $1_{\mathbf{B}}$ is the unit element in the algebra $\mathbf{B}$ for every interpretation frame $I, h, \mathbf{B}$ which preserves (Q)–meets and (Q)–joins;

3. $[[\alpha]]_\approx = 1$.

Proof. Property 1 implies Property 2 by Propositions 3.30, 3.31 and Property 2 implies Property 3 obviously. That Property 3 implies Property 1 is one part of Proposition 3.30   □

As with propositional calculus, completeness allows us for verifying whether a formula is a theorem on semantic grounds. For instance the following list consists of 'standard' often in use theorems of predicate calculus.

Proposition 3.34. *The following are theorems of predicate calculus*

1. $\forall x \alpha(x) \vee \beta \Rightarrow \forall x(\alpha(x) \vee \beta);$

2. $\forall x(\alpha(x) \vee \beta) \Rightarrow \forall x \alpha(x) \vee \beta;$

3. $\exists x \alpha(x) \vee \beta \Rightarrow \exists x(\alpha(x) \vee \beta));$

4. $\exists x(\alpha(x) \vee \beta) \Rightarrow \exists x \alpha(x) \vee \beta;$

5. $\forall x \alpha(x) \wedge \beta \Rightarrow \forall x(\alpha(x) \wedge \beta);$

6. $\forall x(\alpha(x) \wedge \beta) \Rightarrow \forall x \alpha(x) \wedge \beta;$

7. $\exists x \alpha(x) \wedge \beta \Rightarrow \exists x(\alpha(x) \wedge \beta);$

8. $\exists x(\alpha(x) \wedge \beta) \Rightarrow \exists x \alpha(x) \wedge \beta;$

9. $\forall x \alpha(x) \Rightarrow \exists x \alpha(x);$

10. $\forall x(\alpha(x) \Rightarrow \beta) \Rightarrow (\exists \alpha(x) \Rightarrow \beta);$

11. $(\exists \alpha(x) \Rightarrow \beta) \Rightarrow \forall x(\alpha(x) \Rightarrow \beta);$

12. $\forall x(\alpha(x) \Rightarrow \beta) \Rightarrow (\exists x \alpha(x) \Rightarrow \beta);$

13. $(\exists x \alpha(x) \Rightarrow \beta) \Rightarrow \forall x(\alpha(x) \Rightarrow \beta);$

14. $\forall x(\alpha \Rightarrow \beta(x)) \Rightarrow (\alpha \Rightarrow \forall x \beta(x));$

15. $\forall x(\alpha(x) \Rightarrow \beta(x)) \Rightarrow (\forall x \alpha(x) \Rightarrow \forall x \beta(x));$

16. $\forall x(\alpha(x) \Rightarrow \beta(x)) \Rightarrow (\exists x \alpha(x) \Rightarrow \exists x \beta(x));$

17. $\forall x \alpha(x) \wedge \beta(x) \Rightarrow \forall x \alpha(x) \wedge \forall x \beta(x);$

18. $\forall x \alpha(x) \wedge \forall x \beta(x) \Rightarrow \forall x \alpha(x) \wedge \beta(x);$

19. $(\exists x \alpha(x) \vee \beta(x)) \Rightarrow \exists x \alpha(x) \vee \exists x \beta(x);$

20. $\exists x \alpha(x) \vee \exists x \beta(x) \Rightarrow \exists x(\alpha(x) \vee \beta(x));$

21. $\forall x \alpha(x) \vee \forall x \beta(x) \Rightarrow \forall x(\alpha(x) \vee \beta(x));$

22. $\exists x \alpha(x) \wedge \beta(x) \Rightarrow \exists x \alpha(x) \wedge \exists x \beta(x).$

From the completeness theorem some consequences follow of which we would like to remark on the following.

1. *Predicate calculus is consistent (meaning that there is no formula α such that both α and $\neg\alpha$ are theorems); indeed as $[[\alpha]]_\approx = 1$ it implies that $[[\neg\alpha]]_\approx = 0$;*

2. *For a formula $\alpha(x_1, .., x_k)$ with free variable symbols $x_1, ..., x_k$, we call syntactic closure of $\alpha(x_1, .., x_k)$ the formula*

$$\forall y_1 \forall y_2 ... \forall y_k \alpha(x_1/y_1, .., x_k/x_k)$$

   It follows from the semantic interpretation that $\alpha(x_1, .., x_k)$ is a tautology of predicate calculus if and only if the syntactic closure

$$\forall y_1 \forall y_2 ... \forall y_k \alpha(x_1/y_1, .., x_k/y_k)$$

   is a tautology of predicate calculus;

3. *A formula α is prenex if all quantifier operations in α precede predicate operations, i.e., α is of the form*

$$Q x_1 ... Q x_m \beta(x_1, .., x_m),$$

   where $\beta(x_1, ..., x_m)$ is quantifier–free and the symbol Q stands for $\forall, \exists$.

   A prenex formula α which is equivalent to a formula γ, i.e., both $\gamma \Rightarrow \alpha$ and $\alpha \Rightarrow \gamma$ are theorems is said to be the prenex form of the formula γ.

We have

Proposition 3.35. *For every formula γ there exists a prenex form α of γ.*

Proof. We can construct a prenex form of α by structural induction. Elementary formulas of the form $\rho(\tau_1, ..., \tau_m)$ are already prenex. If formulas α, β are in prenex forms, respectively,

$$Q x_1 ... Q x_m \gamma(x_1, ..., x_m), Q y_1 ... Q y_k \delta(y_1, ..., y_k)$$

where variables x_i do not occur in δ and variables $y_1, .., y_k$ do not occur in γ, then $\alpha \vee \beta$, i.e.,

$$Q x_1 ... Q x_m \gamma(x_1, ..., x_m) \vee Q y_1 ... Q y_k \delta(y_1, ..., y_k) \qquad (3.45)$$

may be represented equivalently by virtue of Proposition 3.34 1–4 as

$$Q x_1 [Q x_2 ... Q x_m \gamma(x_1, ..., x_m) \vee Q y_1 ... Q y_k \delta(y_1, ..., y_k)] \qquad (3.46)$$

and by successive repeating of 1–4 we arrive finally at

$$Qx_1...Qx_mQy_1...Qy_k[\gamma(x_1,...,x_m) \vee \delta(y_1,...,y_k)] \qquad (3.47)$$

which is a prenex form of $\alpha \vee \beta$.

A similar argument using theses 5–8 of Proposition 3.34 in place of theses 1–4 will give a prenex form of $\alpha \wedge \beta$. Letting $Q* = \forall$ in case $Q = \exists$ and $Q* = \exists$ in case $Q = \forall$, we may write down the formula $\neg\alpha$ in the prenex form $Q*x_1...Q*x_m\neg\gamma(x_1,..,x_m)$. Finally, if $\alpha*$ is a prenex form of α then $Qx\alpha*$ is a prenex form of $Qx\alpha$ □

A formula is *open*, when there are no occurrences of quantifier symbols in it. Denoting by the symbol $Form^0$ the set of all open formulas of predicate calculus, we may define a theorems of this calculus as an expression which may be derived from axioms by means of detachment and substitution rules only. Thus, a theorem of the calculus of open expressions is a theorem of predicate calculus as well. Denoting by $[[\alpha]]_\approx^0$ the element of the Lindenbaum–Tarski algebra of the open formulas calculus, we have the following

Proposition 3.36. *(Rasiowa and Sikorski [42], Thm. 15.6) The function h assigning to the class $[[\alpha]]_\approx^0$ the class $[[\alpha]]_\approx$ is an isomorphism of the Lindenbaum–Tarski algebra of calculus of open expressions with the Lindenbaum–Tarski algebra of predicate calculus.*

We now consider a particular case when all predicates are *unary*, i.e., they require a single argument, so they are of the form $P(x)$. It turns out that in this case when verifying truth of formulas, we may restrict ourselves to interpretation frames of small cardinality,

Proposition 3.37. *(Rasiowa and Sikorski [42]) For a formula $\alpha(P_1,...,P_k; x_1,...,x_n)$ where $P_1,...,P_k$ are all (unary) predicate symbols occurring in α and $x_1,...,x_n$ are all variable symbols in α, if $\alpha(P_1,...,P_k; x_1,...,x_n)$ is true in an interpretation frame M, then it is also true in an interpretation frame $M^* \subseteq M$ such that cardinality $|M^*|$ is at most 2^k.*

Proof. We may assume that α is in prenex form

$$Qx_1...Qx_m\beta(P_1,...,P_k; x_1,...,x_n),$$

and for each sequence $I = i_1,...,i_k$ of $0's$ and $1's$, we may form the set $M(I) = \{x \in M : [[P_m(x)]] = i_m; m = 1,...,k\}$. Then the family $\mathcal{P} = \{M(I) : I \in \{0,1\}^k\}$ is a partition of the set M.

We may select then for each $x \in M$ the partition element $\mathcal{P}(x)$ containing x and for each non–empty set $\mathcal{P}(x)$ we may select an element $s(x) \in \mathcal{P}(x)$. Then the set $M^* = \{s(x) : x \in M\}$ has cardinality at most that of the set $\{0,1\}^k$, i.e., 2^k □

Let us observe that formulas

$$\begin{cases} (i)\ \beta(P_1, ..., P_k; x_1, ..., x_n) \\ (ii)\ \beta(P_1, ..., P_k; s(x_1), ..., s(x_n)) \end{cases} \tag{3.48}$$

are equivalent, hence,

$$[[(Qx_1...Qx_m\beta(P_1, ..., P_k; x_1, ..., x_n)]]_M^v =$$

$$[[Qx_1...Qx_m\beta(P_1, ..., P_k; s(x_1), ..., s(x_n)]]_M^{s(v)} =$$

$$[[Qx_1...Qx_m\beta(P_1, ..., P_k; x_1, ..., x_n)]]_{M^*}^{s(v)}$$

It follows that α is true in M if and only if α is true in M^*.

In consequence of the above result, in order to verify the truth of α, it is sufficient to check its truth on a set M with $|M| \leq 2^k$. In this case α reduces to the proposition, e.g., in case α is $\forall x P(x)$ the corresponding proposition is $P(x_1) \wedge ... \wedge P(x_{2^k})$.

Thus to verify the truth of this proposition, we may employ the method of truth tables known in propositional calculus. This method requires a finite number of steps, in our exemplary case at most 2^{2^k}. It follows that calculus of unary predicates is *decidable*. Let us mention that in general predicate calculus is undecidable.

3.9 Modal Logics

Somewhat at crossroads between deductive and inductive reasoning, modal logics are endowed with a formal deductive mechanisms, yet they qualify their statements as *necessarily true* or *possibly true*, giving in this way a subjective touch to their formulas. Modal expressions appear in Aristotle, see Łukasiewicz [31], e.g., a statement 'tomorrow there is a sea–battle'. Also in Aristotle there appear basic postulates about the semantic nature of necessity L and possibility M, i.e., the duality postulate

$$L\alpha \Leftrightarrow \neg M \neg \alpha \tag{3.49}$$

and the postulate (K), viz.,

$$(K)\ L(\alpha \Rightarrow \beta) \Rightarrow (L\alpha \Rightarrow L\beta) \tag{3.50}$$

3.9.1 Modal Logic K

Syntax of K in addition to propositional variables and propositional connectives is endowed with the necessity functor L, and the set of formulas is enlarged by addition of the rule that if α is a formula, then $L\alpha$ is a formula.

Axiom schemes for K are are axiom schemes T1–T3 for propositional calculus, and the axiom scheme (K). Derivation rules are substitution, detachment and a new *necessitation rule*

$$\frac{\alpha}{L\alpha} \tag{3.51}$$

Semantics for modal logics grew from analysis of intensional utterances going back to Frege's 'Sinn und Bedeutung' [10], in which *sense* (Sinn) and *reference* (Bedeutung) are two facets of a statement, the first responsible for the modus of utterance and the second for the object referred by the utterance to.

A classical example is provided by the utterances 'The morning star' and 'The evening star' which both refer to the planet Venus by means of distinct senses. This idea got a happy form in Carnap [4] who introduced the notion of a state and interpreted intensional statements semantically as functions on the set of states into logical values. A final form modal semantics acquired in Kripke [21] where *possible world semantics* was introduced. Montague [35], Gallin [11], Van Benthem [1] introduce into intensional logic, and Hughes and Creswell [20] is a tract on modal logic.

Kripke semantics uses a set W of *possible worlds* along with an *accessibility relation* $R \subseteq W \times W$. The pair (W, R) is a *frame*. A *valuation* v assigns to each propositional variable p and each world w in the frame (W, R) a value $v(p, w) \in \{T, F\}$ in the set of truth (T) and falsity (F). By standard truth tables, this assignment determines the value $v(\alpha, w)$ of each propositional formula α at each w, i.e.,

1. $v(\neg\alpha, w) = T \Leftrightarrow v(\alpha, w) = F$;

2. $v(\alpha \vee \beta, w) = T \Leftrightarrow v(\alpha, w) = T$ or $v(\beta, w) = T$;

3. $v(\alpha \wedge \beta, w) = T \Leftrightarrow v(\alpha, w) = T$ and $v(\beta, w) = T$;

4. $v(\alpha \Rightarrow \beta, w) = F \Leftrightarrow v(\alpha, w) = T$ and $v(\beta, w) = F$.

The value $v(L\alpha, w)$ is determined by means of the accessibility relation R; $v(L\alpha, w) = T$ if and only if $v(\alpha, w_1) = T$ for each world $w_1 \in W$ such that $(w, w_1) \in R$, i.e., w_1 is accessible from w.

A modal formula α is true in the frame (W, R) when $v(\alpha, w) = T$ for each valuation v and each world $w \in W$. A formula α is a *tautology* of the modal logic K if and only if α is true in each frame (W, R).

One can for instance check the validity of (K); for an arbitrary frame (W, R) and a valuation v, for an arbitrary world $w \in W$, assume that $v(L(\alpha \Rightarrow \beta), w) = T$ and $v(L\alpha, w) = T$. Consider an arbitrary world w_1 accessible from w, i.e., $(w, w_1) \in R$. By assumptions, $v(\alpha \Rightarrow \beta, w_1) = T$, $v(\alpha, w_1) = T$, hence, $v(\beta, w_1) = T$ by 4 above. It follows that $v(L\beta, w) = T$ proving that (K) is a tautology.

This argument proves as well that detachment preserves tautologies and that axiom schemes T1–T3 have as their instances tautologies. Obviously, necessitation leads from tautologies to tautologies. Hence,

Proposition 3.38. *Each formula which has a proof from axioms in modal logic K is a tautology of this logic.*

Modal logic K is complete. The proof employs the technique due to Lemmon and Scott [22] of *canonical models*, cf., Hughes and Creswell [20]. Informally speaking, these are frames in which worlds are construed as maximal consistent sets of formulas.

Let us recall that a set Γ of formulas is *consistent* if and only if there is no formula α with $\Gamma \vdash \alpha$ and $\Gamma \vdash N\alpha$. In this case we use a meta–predicate Con and $Con(\Gamma)$ would mean that Γ is consistent.

We will say that Γ is *maximal consistent* if and only if $Con(\Gamma)$ and for every consistent set Δ if $\Gamma \subseteq \Delta$ then $\Gamma = \Delta$. We use a meta–predicate Con_{max} to denote this fact so $Con_{max}(\Gamma)$ would mean that Γ is maximal consistent. We list some basic properties of maximal consistent sets.

Proposition 3.39. *For Γ with $Con_{max}(\Gamma)$, the following hold*

1. for every α: $\alpha \in \Gamma$ if and only if $N\alpha \notin \Gamma$;

2. $\alpha \vee \beta \in \Gamma$ if and only if $\alpha \in \Gamma$ or $\beta \in \Gamma$;

3. $\alpha \wedge \beta \in \Gamma$ if and only if $\alpha \in \Gamma$ and $\beta \in \Gamma$;

4. if $\vdash \alpha$ then $\alpha \in \Gamma$;

5. if $\alpha \in \Gamma$ and $\alpha \Rightarrow \beta \in \Gamma$ then $\beta \in \Gamma$.

Proof. In case of Property 1, assume that $\alpha \notin \Gamma$ for a meaningful expression α. This means that $\Gamma \cup \{\alpha\}$ is not consistent, a fortiori for a finite subset $\Lambda \subseteq \Gamma$ we have $\Lambda, \alpha \vdash \delta$ and $\Lambda, \alpha \vdash N\delta$.

By the deduction theorem, we have $\Lambda \vdash \alpha \Rightarrow \delta$, $\Lambda \vdash \alpha \Rightarrow \neg\delta$. By the theorem $(p \Rightarrow q) \Rightarrow (p \Rightarrow \neg q) \Rightarrow p$ in which we substitute $p/\alpha, q/\delta$ we have applying detachment twice that $\Lambda \vdash \neg\alpha$.

It follows that $\Gamma \cup \neg\alpha$ is consistent hence $\neg\alpha \in \Gamma$.

For Property 2, assume to the contrary $\alpha \vee \beta \in \Gamma$, $\alpha \notin \Gamma$ and $\beta \notin \Gamma$. Then, by already proven Property 1, $\neg\alpha, \neg\beta \in \Gamma$ and by the theorem $p \Rightarrow (q \Rightarrow (p \wedge q)))$ with substitutions $p/\neg\alpha, q/\neg\beta$ we obtain applying detachment twice that $\Gamma \vdash \neg\alpha \wedge \neg\beta$ i.e. $\Gamma \vdash \neg(\alpha \vee \beta)$ hence Γ is not consistent, a contradiction. The proof of Property 3 is on the same lines. For Property 4, it is manifest that $\vdash \alpha$ implies $\Gamma \cup \{\alpha\}$ consistent hence by maximality $\alpha \in \Gamma$. Property 5 follows from Property 1 as $\alpha \Rightarrow \beta$ is equivalent to $\neg\alpha \vee \beta$ so $\alpha \in \Gamma$ implies $\beta \in \Gamma$ □

Let us also observe that a set Γ of meaningful expressions is consistent if and only if for no finite subset $\Lambda \subseteq \Gamma$ we have $\vdash \neg \bigwedge \Lambda$. Indeed, to see necessity of this condition, assume Γ consistent and $\vdash \neg \bigwedge \Lambda$ for some finite subset $\Lambda \subseteq \Gamma$. By the theorem $p \Rightarrow (q \Rightarrow (p \wedge q))$ applied $|\Lambda| - 1$ times, we get $\Gamma \vdash \bigwedge \Lambda$ hence $\Gamma \vdash \bigwedge \Lambda, \neg \bigwedge \Lambda$ contradicting the consistency of Γ.

To see sufficiency, assume that Γ is not consistent, i.e., $\Gamma \vdash \beta, \neg\beta$ for some meaningful expression β. By compactness, $\Lambda \vdash \beta, \neg\beta$ for some finite $\Lambda \subseteq \Gamma$.

Let $\Lambda = \{\gamma_1, ..., \gamma_k\}$; by the theorem $p_1 \wedge ... \wedge p_k \Rightarrow p_i$ for $i = 1, 2, ..., k$, it follows that $\bigwedge \Lambda \vdash \beta, \neg\beta$ and then the theorem $(p \Rightarrow q) \wedge (p \Rightarrow \neg q) \Rightarrow \neg p$ implies that $\vdash \neg \bigwedge \Lambda$. Thus Γ is not consistent.

We now define a canonical interpretation frame $M^c = (W^c, R^c, v^c)$ by letting

1. $W^c = \{\Gamma : Con_{max}(\Gamma)\}$, *i.e., worlds are maximal consistent sets of meaningful expressions;*

2. $\Gamma R^c \Gamma'$ *if and only if* $L\alpha \in \Gamma$ *implies* $\alpha \in \Gamma'$ *for every meaningful expression* α;

3. $v^c(\Gamma, p) = 1$ *if and only if* $p \in \Gamma$.

We then have,

Proposition 3.40. *For each meaningful expression α of the system K, we have $v^c(\Gamma, \alpha) = 1$ if and only if $\alpha \in \Gamma$.*

Proof. By induction on structural complexity of α. We will assume that α, β already satisfy our claim. For $\neg\alpha$, we have $v^c(\Gamma, \neg\alpha) = 1$ if and only if $v^c(\Gamma, \alpha) = 0$, i.e., $\alpha \notin \Gamma$ which by Proposition 3.39 (1) is equivalent to $\neg\alpha \in \Gamma$.

For $\alpha \Rightarrow \beta$, we have $v^c(\Gamma, \alpha \Rightarrow \beta) = 1$ if and only if $v^c(\Gamma, \alpha) = 0$ or $v^c(\Gamma, \beta) = 1$, i.e., $\alpha \notin \Gamma$ or $\beta \in \Gamma$, i.e., $\neg\alpha \vee \beta \in \Gamma$ by Proposition 3.39 (3), i.e., $\alpha \Rightarrow \beta \in \Gamma$.

Finally, we consider $L\alpha$; there are two cases
(i) $L\alpha \in \Gamma$ (ii) $L\alpha \notin \Gamma$
In case (i), for every Γ' with $\Gamma R^c \Gamma'$ we have $\alpha \in \Gamma'$, i.e., $v^c(\Gamma', \alpha) = 1$ from which $v^c(\Gamma, L\alpha) = 1$ follows. In case (ii), when $L\alpha \notin \Gamma$, we have $\neg L\alpha \in \Gamma$. We check that the following holds.

Claim $Con(\{\delta : L\delta \in \Gamma\} \cup \{\neg\alpha\})$

Indeed, was the contrary true, we would have a finite set $\{\gamma_1, ..., \gamma_k\}$ with $\{L\gamma_1, .., L\gamma_k\} \subseteq \Gamma$ such that $\vdash \neg(\gamma_1 \wedge ... \wedge \gamma_k \wedge \neg\alpha)$, i.e., $\vdash C\gamma_1 \wedge ... \wedge \gamma_k \alpha$ hence $\vdash LC\gamma_1 \wedge ... \wedge \gamma_k \alpha$ by necessitation and thus it follows by the axiom (K) that $\vdash CL\gamma_1 \wedge ... \wedge L\gamma_k L\alpha$ implying that $\vdash \neg(L\gamma_1 \wedge ... \wedge L\gamma_k \wedge \neg L\alpha)$ contrary to the facts that $Con(\Gamma)$ and $\{L\gamma_1, ..., L\gamma_k, \neg L\alpha\} \subseteq \Gamma$. Our claim is proved.

Now, as $\{\delta : L\delta \in \Gamma\} \cup \{\neg\alpha\}$ is consistent it extends to a maximal consistent set Γ'; clearly, $\Gamma R^c \Gamma'$; as $N\alpha \in \Gamma'$ we have $v^c(\Gamma', \alpha) = 0$ and finally $v^c(\Gamma, L\alpha) = 0$ □

We may now state the main result of this discussion.

Proposition 3.41. *(The Completeness Theorem)* $\models \alpha$ *implies* $\vdash \alpha$ *for every formula* α.

Proof. Implication: if $\models \alpha$, then $\vdash \alpha$ is equivalent to the implication (i) if non $\vdash \alpha$, then non $\models \alpha$ or, equivalently, if non $\vdash \alpha$ then there exists $M = (W, R, v)$ with $v(\alpha, w) = 0$ for some $w \in W$.

As the condition non $\vdash \neg\alpha$ is equivalent to the condition $Con(\{\alpha\})$ one may formulate a condition equivalent to (i), viz., (ii) if $Con(\alpha)$, then there exists $M = (W, R, v)$ with $v(\alpha, w) = 1$ for some $w \in W$.

But, $Con(\{\alpha\})$ implies the existence of Γ with $Con_{max}(\Gamma)$ and $\alpha \in \Gamma$ hence in the canonical interpretation frame M^c we have $v^c(\Gamma, \alpha) = 1$ proving (ii) hence (i) and the completeness theorem as well   □

We apply the completeness theorem to verify that the formula $Lp \Rightarrow p$ is not a tautology of the system K.

It is sufficient to consider an interpretation frame $M = (W, R, v)$ in which there are worlds $w \neq w^* \in W$ with $W = \{w, w^*\}$, wRw^*, non wRw, $v(p, w) = 0$, and $v(p, w^*) = 1$. Then $\models_M Lp$ and non $\models_M p$ thus non $\models_M Lp \Rightarrow p$.

3.9.2 Modal Logic T

Adding to the axiom schemes of the system K a new axiom scheme, Feys [7]

$$(T)\ L\alpha \Rightarrow \alpha \tag{3.52}$$

would lead to a new system denoted T. The collapse method shows that T is consistent.

Let us observe that the axiom (T) is true in every interpretation frame $M = (W, R, v)$ in which the relation R is *reflexive*, i.e., wRw for each $w \in W$. If $v(L\alpha, w) = 1$ for a $w \in W$, then $v(\alpha, w_1) = 1$ for each w_1 with $(w, w_1) \in R$, hence, $v(\alpha, w) = 1$ and $v(L\alpha \Rightarrow \alpha, w) = 1$ for each $w \in W$. The converse is also true: if (T) is true in an interpretation frame $M = (W, R, v)$ then R is reflexive.

We may state a completeness theorem for the system T. We denote by the symbol $\mathcal{M}_r$ the set of all interpretation frames with the relation R reflexive.

Proposition 3.42. *(The Completeness Theorem for T) For each formula α, α is derivable in T if and only if α is true in every frame in $\mathcal{M}_r$.*

3.9.3   Modal Logic S4

We may look at the formula

$$(S4) \ L\alpha \Rightarrow LL\alpha \tag{3.53}$$

see, Lewis and Langford [23]

It is easy to see that (S4) is true in an interpretation frame $M = (W, R, v)$ if and only if R is *transitive*, i.e., $(w, w_1) \in R$ and $(w_1, w_2) \in R$ imply $(w, w_2) \in R$ for each triple $w, w_1, w_2 \in W$.

Adding (S4) to T gives a new consistent system $S4$. Let us observe that we may introduce graded modalities: for $n \in N$, we define the modality L^n as follows: $L^0 p = p$, $L^{n+1} p = LL^n p$. Then (S4) yields by subsequent substitutions $\vdash L^n \alpha \Rightarrow L^{n+1} \alpha$ for each $n \in N$. As T allows for the converse, $\vdash L^{n+1} \alpha \Rightarrow L^n \alpha$ for each $n \in N$, we finally get in S4 equivalences $L\alpha \Leftrightarrow L^n \alpha$ for $n = 2, 3, \dots$.

As truth of (T) forces R to be reflexive and truth of (S4) forces R to be transitive, we get a completeness theorem for the system $S4$ by considering the set $\mathcal{M}_{rt}$ of frames in which the relation R is reflexive and transitive.

Proposition 3.43. *(The Completeness Theorem for S4) For each formula α, α is derivable in S4 if and only if α is true in every frame in $\mathcal{M}_{rt}$.*

3.9.4   Modal Logic S5

Let us now consider the formula

$$(S5) \ M\alpha \Rightarrow LM\alpha \tag{3.54}$$

see Lewis and Langford [23]. It is distinct from the previously considered in that it does express a certain interplay between the two modalities L and M. We observe,

Proposition 3.44. *In the system obtained by adding (S5) to the system T we obtain (S4).*

Proof. First, the axiom (T): $L\alpha \Rightarrow \alpha$ may be dualized to

$$N\alpha \Rightarrow NL\alpha$$

and by the tautology

$$\alpha \Rightarrow NN\alpha$$

it may be restated as

$$N\alpha \Rightarrow NLNN\alpha$$

i.e.,

$$N\alpha \Rightarrow MN\alpha$$

hence, substituting $\alpha/N\alpha$ and using the equivalence

$$NN\alpha \Leftrightarrow \alpha$$

we get

$$(T^*)\ \alpha \Rightarrow M\alpha \qquad\qquad (3.55)$$

as an equivalent formulation of (T). Substitution $\alpha/L\alpha$ in (T^*) yields $L\alpha \Rightarrow ML\alpha$.

An analogical dualization of (S5) yields the expression

$$ML\alpha \Rightarrow L\alpha$$

hence, by (T^*) we have

$$(i)\ ML\alpha \Leftrightarrow L\alpha$$

or, dually,

$$(ii)\ M\alpha \Leftrightarrow LM\alpha$$

Now, it follows from (i) that

$$L\alpha \Rightarrow ML\alpha$$

and by (ii), we have

$$L\alpha \Rightarrow LML\alpha$$

which again by (i), yields

$$L\alpha \Rightarrow LL\alpha$$

i.e., (S4)   □

The consequence of this result is that the system $S4$ is contained in the system $S5$. We also demonstrate that in the system $S5$, the following meaningful expression

$$(B)\ N\alpha \Rightarrow LNL\alpha$$

is derivable. Indeed,

$$\vdash_{S5} ML\alpha \Rightarrow L\alpha$$

as the dual form of (S5), thus, by (T),

$$\vdash_{S5} ML\alpha \Rightarrow \alpha$$

follows, which is the dual form of (B). Let us observe that (B) is true in any model $M = (W, R, v)$ in which the relation R is *symmetric*, i.e., $(w, w_1) \in R$ implies $(w_1, w) \in R$. Indeed, for a model $M = (W, R, v)$, and for a $w \in W$, assume that $v(ML\alpha, w) = 1$, hence, there is $w_1 \in W$ with $(w, w_1) \in R$ and $v(L\alpha, w_1) = 1$, so $v(\alpha, w_2) = 1$ for every $w_2 \in W$ with $(w_1, w_2) \in R$, hence, in particular, $v(\alpha, w) = 1$, as by symmetry of R we have $(w_1, w) \in R$. It follows that $v(ML\alpha \Rightarrow \alpha, w) = 1$. Hence, formulas derivable in the system $S5$ are true in any interpretation frame $M = (W, R)$ in which the relation R is an *equivalence* relation. The completeness theorem for $S5$ acquires the following form. We denote by $\mathcal{M}_{eq}$ the set of all interpretation frames in which the relation R is an equivalence relation.

Proposition 3.45. *For any formula α, α is derivable in the system $S5$ if and only if α is true in every frame in $\mathcal{M}_{eq}$.*

We actually pointed above in Proposition 3.42, Proposition 3.43, Proposition 3.45 to sufficiency of the condition on interpretation frames in each case. To show the necessity of these conditions in respective cases, we check that relations in each of canonical interpretation frames for respective systems are such as indicated in completeness theorems.

Proposition 3.46. *The relation R in canonical interpretation frames for T, $S4$, $S5$ is, respectively, reflexive, reflexive and transitive, and an equivalence.*

Proof. In case of the system T, by (T) $L\alpha \Rightarrow \alpha$, for every Γ with $Con_{max}(\Gamma)$ we have that $L\alpha \in \Gamma$ does imply $\alpha \in \Gamma$ hence $\Gamma R^c \Gamma$, i.e., the relation R^c is reflexive. In case of the system $S4$, R^c is reflexive as (T) is a theorem of this system. Assume $\Gamma R^c \Gamma^*$ and $\Gamma^* R^c \Gamma^{**}$; from $L\alpha \in \Gamma$ it follows by the axiom $(S4)$ $L\alpha \Rightarrow LL\alpha$ that $LL\alpha \in \Gamma$ hence $L\alpha \in \Gamma^*$ and $\alpha \in \Gamma^{**}$. Thus $L\alpha \in \Gamma$ implies $\alpha \in \Gamma^{**}$ i.e. $\Gamma R^c \Gamma^{**}$ proving that R^c is transitive. It remains to prove for the system $S5$ that R^c is symmetric, as reflexivity and transitivity in this case follow by the fact that the system $S5$ contains (T) and $(S4)$. We recall the theorem (B) $N\alpha \Rightarrow LNL\alpha$ of this system. Assume that $\Gamma R^c \Gamma^*$, i.e., $L\alpha \in \Gamma$ implies $\alpha \in \Gamma^*$.

If it was not the case that $\Gamma^* R^c \Gamma$, we would have an expression β with $L\beta \in \Gamma^*$ and it would not be true that $\beta \in \Gamma$. Thus $N\beta \in \Gamma$ and by (B) we have that $LNL\beta \in \Gamma$ so $NL\beta \in \Gamma^*$. It follows that Γ^* has both $L\beta, NL\beta$ as elements, contradicting consistency of Γ^*. Thus, $\Gamma^* R^c \Gamma$ and R^c is symmetric   □

We add that systems T, S4, S5 were independently introduced under notation, respectively, M, M', M'' in Von Wright [50]. An exposition of first–order intensional logic can be found in Fitting [8].

References

1. Van Benthem, J.: A Manual of Intensional Logic. CSLI Stanford University, Stanford (1988)
2. Bocheński, I.M.: Die Zeitgönossischen Denkmethoden. A. Francke AG, Bern (1954)
3. Bocheński, I.M.: A History of Formal Logic. Notre Dame University Press, Notre Dame (1971)
4. Carnap, R.: Necessity and Meaning. Chicago University Press, Chicago (1947)
5. Chang, C.C.: A new proof of the completeness of the Lukasiewicz's axioms. Trans. Amer. Math. Soc. 93, 74–80 (1959)
6. Cook, S.A.: The complexity of theorem–proving procedures. In: Proceedings of the 3rd Annual ACM Symposium on Theory of Computing, pp. 151–158. ACM Press, New York (1971)
7. Feys, R.: Les logiques nouvelles des modalités. Revue Néoscholastique de Philosophie 40, 41, 217–252, 517–553 (1937)
8. Fitting, M.C.: First–order intensional logic. Annals of Pure and Applied Logic 127, 171–193 (2004)
9. Frege, G.: Begriffsschritt. Eine der Arithmetischen Nachgebildete Formelsprache des Reinen Denkens. Verlag von Louis Nebert, Halle a/S (1879)
10. Frege, G.: Über Sinn und Bedeutung. Zeitschrift für Philosophie und Philosophische Kritik NF 100, 25–50 (1892)
11. Gallin, D.: Intensional and higher–order modal logic. North Holland, Amsterdam (1975)
12. Goguen, J.: −9) The logic of inexact concepts. Synthese 19, 325–373 (1968)
13. Goldberg, H., Leblanc, H., Weaver, G.: A strong completeness theorem for 3–valued logic. Notre Dame J. Formal Logic 15, 325–332 (1974)
14. Gödel, K.: Die Vollstandigkeit der Axiome des Logischen Funktionenkalküls. Monats. Math. Phys. 37, 349–360 (1930)
15. Gödel, K.: Zum intuitionistichen Aussagenkalküls. Anzeiger Akademie der Wissenschaften Wien. Math-naturwisschen. Klasse 69, 65–66 (1932)
16. Hájek, P.: Fuzzy logic and arithmetical hierarchy II. Studia Logica 58, 129–141 (1997)
17. Hájek, P.: Metamathematics of Fuzzy Logic. Kluwer, Dordrecht (1998)
18. Henkin, L.: The completeness of the first–order functional calculus. J. Symb. Logic 14, 159–166 (1949)
19. Herbrand, J.: Recherches sur la théorie de la déemonstration. Travaux de la Soc. Sci. Lettr. de Varsovie Cl. III 33, 33–160 (1930)

20. Hughes, G.E., Creswell, M.J.: An Introduction to Modal Logic. Methuen, London (1972)
21. Kripke, S.: Semantical considerations on modal logics. Acta Philosophica Fennica. Modal and Many–Valued Logics, pp. 83–94 (1963)
22. Lemmon, E.J., Scott, D.S., Segerberg, K. (eds.): The 'Lemmon Notes': An Introduction to Modal Logic. Blackwell, Oxford (1963)
23. Lewis, C.I., Langford, C.H.: Symbolic Logic, 2nd edn. Dover, New York (1959)
24. Łukasiewicz, J.: W sprawie odwracalności stosunku racji i następstwa (Concerning the invertibility of the relation between the premise and the conclusion (in Polish)). Przegląd Filozoficzny 16 (1913)
25. Łukasiewicz, J.: Farewell lecture at Warsaw University (1918); Borkowski L. (ed.) Jan Łukasiewicz. Selected Works. North Holland, Amsterdam (1970)
26. Łukasiewicz, J.: O logice trójwartościowej (On three–valued logic, in Polish). Ruch Filozoficzny 5, 170–171 (1920)
27. Łukasiewicz, J.: Zagadnienia prawdy (On problems of truth, in Polish). Księga Pamiątkowa XI Zjazdu Lekarzy i Przyrodników Polskich (Commemorating Book of the XI–th Meeting of the Polish Medics and Naturalists), pp. 84 ff (1922)
28. Łukasiewicz, J.: Philosophische bemerkungen zu mehrwertige Systemen des Aussagenkalkuls. C. R. Soc. Sci. Lettr. Varsovie 23, 51–77 (1930)
29. Łukasiewicz, J.: Aristotle's Syllogistic from the Standpoint of Modern Formal Logic, 2nd edn. Oxford University Press, Oxford (1957)
30. Łukasiewicz, J.: Elements of Mathematical Logic, 1st edn. Pergamon Press – Polish Scientific Publishers, Oxford – Warsaw (1963)
31. Łukasiewicz, J.: On the history of the logic of propositions (1970); Borkowski L. (ed.) Jan Łukasiewicz. Selected Works. North Holland, Amsterdam (1970)
32. Łukasiewicz, J., Tarski, A.: Untersuchungen über den Aussagenkalkuls. C. R. Soc. Sci. Lettr. Varsovie 23, 39–50 (1930)
33. Menger, K.: Statistical metrics. Proc. Nat. Acad. Sci. USA 8, 535–537 (1942)
34. Meredith, C.A.: The dependence of an axiom of Łukasiewicz. Trans. Amer. Math. Soc. 87, 54 (1958)
35. Montague, R.: Pragmatics and intensional logic. Synthese 22, 68–94 (1970)
36. Mundici, D.: Satisfiability in many–valued sentential logic is NP–complete. Theoretical Computer Science 52, 145–153 (1987)
37. McNaughton, R.: A theorem about infinite–valued sentential logic. J. Symbolic Logic 16, 1–13 (1951)
38. Novák, V.: On the syntactico–semantical completeness of first–order fuzzy logic. Kybernetika 2, 47–62, 134–152 (1990)
39. Post, E.L.: Introduction to a general theory of elementary propositions. Amer. J. Math. 43, 163–185 (1921)
40. Rasiowa, H.: A proof of the Skoiem-Löwenheim theorem. Fundamenta Mathematics 38, 230–232 (1951)
41. Rasiowa, H.: On satisfiability and deducibility in non–classical functional calculi. Bull. Pol. Ac.: Math (Cl.III) 1, 229–231 (1953)
42. Rasiowa, H., Sikorski, R.: The Mathematics of Metamathematics. PWN-Polish Scientific Publishers, Warszawa (1963)
43. Robinson, J.A.: A machine–oriented logic based on the resolution principle. Journal ACM 12, 23–41 (1965)
44. Rose, A., Rosser, J.B.: Fragments of many–valued statement calculi. Trans. Amer. Math. Soc. 87, 1–53 (1958)

45. Rosser, J.B., Turquette, A.R.: Many–valued Logics. North Holland, Amsterdam (1958)
46. Schweizer, B., Sklar, A.: Probabilistic Metric Spaces. North Holland, Amsterdam (1983)
47. Tarski, A.: Pojęcie prawdy w językach nauk dedukcyjnych (On the notion of truth in languages of deductive sciences, in Polish). Travaux de la Soc. des Sci. et des Lettres de Varsovie Cl. III 34, VII+116 (1933)
48. Wajsberg, M.: Axiomatization of the 3–valued sentential calculus (in Polish, a summary in German). C. R. Soc. Sci. Lettr. Varsovie 24, 126–148 (1931)
49. Wajsberg, M.: Beiträge zum Metaaussagenkalkül I. Monat. Math. Phys. 42, 221–242 (1935)
50. Von Wright, G.H.: An Essay in Modal Logic. North Holland, Amsterdam (1951)
51. Zadeh, L.A.: Fuzzy sets. Information and Control 8, 338–353 (1965)
52. Zadeh, L.A.: Fuzzy logic and approximate reasoning. Synthese 30, 407–428 (1975)

Chapter 4
Reductive Reasoning Rough and Fuzzy Sets as Frameworks for Reductive Reasoning

Reductive reasoning, in particular inductive reasoning, Bocheński [9], Łukasiewicz [30], is concerned with finding a proper p satisfying a premise $p \Rightarrow q$ for a given conclusion q. With some imprecision of language, one can say that its concern lies in finding a right cause for a given consequence. As such, inductive reasoning does encompass many areas of research like Machine Learning, see Mitchell [37], Pattern Recognition and Classification, see Duda et al. [16], Data Mining and Knowledge Discovery, see Kloesgen and Zytkow [26], all of which are concerned with a right interpretation of data and a generalization of findings from them. The matter of induction opens up an abyss of speculative theories, concerned with hypotheses making, verification and confirmation of them, means for establishing optimality criteria, consequence relations, non–monotonic reasoning etc. etc., see, e.g., Carnap [12], Popper [55], Hempel [22], Bochman [10].

Our purpose in this chapter is humble; we wish to give an insight into two paradigms intended for inductive reasoning and producing decision rules from data: rough set theory and fuzzy set theory.

We pay attention to structure and basic tools of these paradigms; rough sets are interesting for us, as forthcoming exposition of rough mereology borders on rough sets and uses knowledge representation in the form of information and decision systems as studied in rough set theory. Fuzzy set theory, as already observed in Introduction, is to rough mereology as set theory is to mereology, a guiding motive; in addition, main tools of fuzzy set theory: t–norms and residual implications are also of fundamental importance to rough mereology, as demonstrated in following chapters.

4.1 Rough Set Approach Main Lines

Introduced in Pawlak [44], [45] rough set theory is based on ideas that go back to Gottlob Frege, Gottfried Wilhelm Leibniz, Jan Łukasiewicz, Stanislaw Leśniewski, to mention a few names of importance. Its characteristics is that they divide notions (concepts) into two classes: exact as well as inexact.

L. Polkowski: Approximate Reasoning by Parts, ISRL 20, pp. 145–190.
springerlink.com   © Springer-Verlag Berlin Heidelberg 2011

The idea for a dividing line between the two comes from Frege [19]: an inexact concept should possess a boundary region into which objects fall which can be classified with certainty neither to the concept nor to its complement. This boundary to a concept is constructed in the rough set theory from indiscernibility relations induced by attributes (features) of objects.

Knowledge is assumed in rough set theory to be a *classification* of objects into categories (decision classes). As a frame for representation of knowledge we adopt following Pawlak [44], *information systems*.

One of languages for knowledge representation is the *attribute–value* language in which notions representing things are described by means of *attributes* (features) and their *values*; information systems are pairs of the form (U, A) where U is a set of objects – representing things – and A is a set of attributes; each attribute a is modeled as a mapping $a : U \to V_a$ from the set of objects into the *value set* V_a. For an attribute a and its value v, the *descriptor*, see Pawlak [45], Baader et al. [2], $(a = v)$ is a formula interpreted in the set of objects U as $[[(a = v)]] = \{u \in U : a(u) = v\}$.

Descriptor formulas are the smallest set containing all descriptors and closed under sentential connectives $\vee, \wedge, \neg, \Rightarrow$. Meanings of complex formulas are defined recursively

1. $[[\alpha \vee \beta]] = [[\alpha]] \cup [[\beta]]$;

2. $[[\alpha \wedge \beta]] = [[\alpha]] \cap [[\beta]]$;

3. $[[\neg \alpha]] = U \setminus [[\alpha]]$;

4. $[[\alpha \Rightarrow \beta]] = [[\neg \alpha \vee \beta]]$.

In descriptor language each object $u \in U$ can be encoded over a set B of attributes as its information vector $Inf_B(u) = \{(a = a(u)) : a \in B\}$. A formula α is *true* (is a tautology) if and only if its meaning $[[\alpha]]$ equals the set U.

The Leibniz Law (the Principle of Identity of Indiscernibles and Indiscernibility of Identicals), Leibniz [28], affirms that two things are identical if and only if they are indiscernible, i.e., no available operator acting on both of them yields distinct values

$$\forall F.F(x) = F(y) \Leftrightarrow x = y \qquad (4.1)$$

In the context of information systems, indiscernibility relations are introduced and interpreted in accordance with the Leibniz Law from sets of attributes: given a set $B \subseteq A$, the indiscernibility relation relative to B is defined as

$$Ind(B) = \{(u, u') : a(u) = a(u') \text{ for each } a \in B\} \qquad (4.2)$$

Objects u, u' in relation $Ind(B)$ are said to be B–*indiscernible* and are regarded as identical with respect to knowledge represented by the information system (U, B). The class $[u]_B = \{u' : (u, u') \in Ind(B)\}$ collects all objects identical to u with respect to B.

Each class $[u]_B$ is B–*definable*, i.e., the decision problem whether $v \in [u]_B$ is decidable: one says that $[u]_B$ is *exact*. More generally, an *exact concept* is a set of objects in the considered universe which can be represented as the union of a collection of indiscernibility classes; otherwise, the set is *inexact* or *rough*. In this case, there exist a boundary about the notion consisting of objects which can be with certainty classified neither into the notion nor into its complement. To express the B–*boundary* of a concept X, induced by the set B of attributes, *approximations over B*, see Pawlak, op. cit., op. cit., are introduced, i.e., the B–*lower approximation*

$$\underline{B}X = \bigcup\{[u]_B : [u]_B \subseteq X\} \tag{4.3}$$

and the B–*upper approximation*

$$\overline{B}X = \bigcup\{[u]_B : [u]_B \cap X \neq \emptyset\} \tag{4.4}$$

The difference

$$Bd_B X = \overline{B}X \setminus \underline{B}X \tag{4.5}$$

is the B–*boundary* of X; when non-empty, it does witness that X is B–inexact. The reader has certainly noticed the topological character of the approximations as, respectively, the closure and the interior of a concept in the topology generated by indiscernibility relations, cf., Ch. 2, sect. 12.

The inductive character of rough set analysis of uncertainty stems from the fact that the information system (U, A) neither exhausts the world of objects W nor the set A^* of possible attribute of objects. On the basis of induced classification $\Delta : U \to Categories$, one attempts to classify each possible unknown object $x \in W \setminus U$; this classification is subject to uncertainty.

In order to reduce the complexity of classification task, a notion of a *reduct* was introduced, see Pawlak [45]; a reduct B of the set A of attributes is a minimal subset of A with the property that $Ind(B) = Ind(A)$. An algorithm for finding reducts based on Boolean reasoning, see Brown [11], was proposed in Skowron and Rauszer [66]; given input (U, A) with $U = \{u_1, ..., u_n\}$ it starts with the *discernibility matrix*

$$M_{U,A} = [c_{i,j} = \{a \in A : a(u_i) \neq a(u_j)\}]_{1 \geq i, j \leq n} \tag{4.6}$$

and the Boolean function

$$f_{U,A} = \bigwedge_{c_{i,j} \neq \emptyset, i < j} \bigvee_{a \in c_{i,j}} \overline{a}, \tag{4.7}$$

where $\overline{a}$ is the Boolean variable associated with the attribute $a \in A$.

The function $f_{U,A}$ is converted to its DNF form

$$f_{U,A}^* : \bigvee_{j \in J} \bigwedge_{k \in K_j} \overline{a_{j,k}} \qquad (4.8)$$

Then, sets of the form $R_j = \{a_{j,k} : k \in K_j\}$ for $j \in J$, corresponding to *prime implicants* $\bigwedge_{k \in K_j} \overline{a_{j,k}}$ of $f_{U,A}^*$ are all reducts of A.

To verify this thesis, it suffices to consider for a subset $B \subseteq A$, the valuation v_B: $v_B(\overline{a}) = 1$ if and only if $a \in B$, and observe that $v_B(f_{U,A}) = 1$ if and only if $Ind(B) = Ind(A)$. On the other hand, as $f_{U,A}$ and $f_{U,A}^*$ are semantically equivalent, we have $v_B(f_{U,A}^*) = 1$ and this happens if and only if $\{\overline{a_{j,k}} : k \in K_j\} \subseteq B$ for some prime implicant $\bigwedge_{k \in K_j} \overline{a_{j,k}}$, so minimal B with $Ind(B) = Ind(A)$ are of the form $B = \{a_{j,k} : k \in K_j\}$ for some prime implicant $\bigwedge_{k \in K_j} \overline{a_{j,k}}$.

For a reduct B, the reduced information system (U, B), secures the same classification as the full system (U, A), hence, it does preserve knowledge.

Moreover, as $Ind(B) \subseteq Ind(C)$ for any subset $C \subseteq A$, one can establish a functional dependence of C on B: as for each object $u \in U$, $[u]_B \subseteq [u]_C$, the assignment $f_{B,C} : Inf_B(u) \to Inf_C(u)$ is functional, i.e., values of attributes in B determine functionally values of attributes in C on U, object-wise. Thus, any reduct determines functionally the whole system.

4.2 Decision Systems

A *decision system* is a triple (U, A, d) in which d is the *decision attribute, decision*, the attribute not in A, that does express the evaluation of objects by an external oracle, an expert. Attributes in A are called in this case *conditional*, in order to discern them from the decision d. Values v_d of decision d can be regarded as codes for categories into which the decision classifies objects.

Inductive reasoning by means of rough sets aims at finding as faithful as possible description of the concept d in terms of conditional attributes in A in the language of descriptors. This description is effected by means of *decision rules*, i.e., formulas of descriptor logic of the form of an implication

$$\bigwedge_{a \in B} (a = v_a) \Rightarrow (d = v) \qquad (4.9)$$

The formula (4.9) is *true* in case

$$[[\bigwedge_{a \in B} (a = v_a)]] = \bigcap_{a \in B} [[(a = v_a)]] \subseteq [[(d = v)]]$$

Otherwise, the formula is *partially true*. An object o which is matching the rule, i.e., $a(o) = v_a$ for $a \in B$ can be classified to the class $[[(d = v)]]$; often a partial match based on a chosen distance measure has to be performed.

The simplest case is when the decision system is *deterministic*, i.e., when $Ind(A) \subseteq Ind(d)$. In this case, the relation between A and d is functional, given by the unique assignment $f_{A,d}$ or in the decision rule form (4.9) as the set of rules: $\bigwedge_{a \in A}(a = a(u)) \Rightarrow (d = f_{A,d}(u))$ for $u \in U$. In place of A any reduct R of A can be substituted leading to shorter rules.

In the contrary case of a non–deterministic system, some classes $[u]_A$ are split into more than one decision classes $[v]_d$ leading to ambiguity in classification. In that case, decision rules are divided into *true* (or, exact, certain) and *possible*; to induce the certain rule set, the notion of a δ–*reduct* was proposed in Skowron and Rauszer [66].

To define δ–reducts, the *generalized decision* δ_B is defined: for $u \in U$,

$$\delta_B(u) = \{v \in V_d : \exists u'.d(u') = v \wedge (u, u') \in Ind(B)\} \qquad (4.10)$$

A subset B of A is a δ–reduct, when it is a minimal subset od A with respect to the property that $\delta_B = \delta_A$. δ–reducts can be obtained from the modified Skowron and Rauszer algorithm as pointed in there; it suffices to modify the entries $c_{i,j}$ of the discernibility matrix to new entries $c'_{i,j}$, by letting

$$c^d_{i,j} = \{a \in A \cup \{d\} : a(u_i) \neq a(u_j)\} \qquad (4.11)$$

and

$$c'_{i,j} = \begin{cases} c^d_{i,j} \setminus \{d\} \text{ in case } d(u_i) \neq d(u_j) \\ \emptyset \text{ in case } d(u_i) = d(u_j) \end{cases} \qquad (4.12)$$

The algorithm described above input with entries $c'_{i,j}$ forming the matrix $M^\delta_{U,A}$ outputs all δ–reducts to d encoded as prime implicants of the associated Boolean function $f^\delta_{U,A}$.

The problem of finding a reduct of minimal length is NP-hard [66], therefore, one may foresee that no polynomial algorithm is available for computing reducts. Thus, the algorithm based on discernibility matrix has been proposed with stop rules that permit to stop the algorithm and obtain a partial set of reducts, see Bazan [4]. In finding approximations to reducts methods of artificial intelligence like swarm algorithms are applied, see Hong Yu et al. [23]; in Moshkov et al. [38], an idea of a descriptive set of attributes, computed by a polynomial algorithm is discussed.

In order to precisely discriminate between certain and possible rules, the notion of a *positive region* along with the notion of a *relative reduct*, see Pawlak [45], was studied in Skowron and Rauszer [66].

Positive region $Pos_B(d)$ is the set

$$\{u \in U : [u]_B \subseteq [u]_d\} \qquad (4.13)$$

which is equivalent to

$$\bigcup_{v \in V_d} \underline{B}[[(d = v)]] \tag{4.14}$$

$Pos_B(d)$ is the greatest subset X of U such that (X, B, d) is deterministic; it generates certain rules.

Objects in $U \setminus Pos_B(d)$ are subjected to ambiguity; given such u, and the collection $v_1, .., v_k$ of decision values on the class $[u]_B$, the decision rule describing u can be written down as

$$\bigwedge_{a \in B} (a = a(u)) \Rightarrow \bigvee_{i=1,...,k} (d = v_i) \tag{4.15}$$

Each of the rules $\bigwedge_{a \in B}(a = a(u)) \Rightarrow (d = v_i)$ is possible but not certain, as only for a fraction of objects in the class $[u]_B$ the decision takes the value v_i on.

Relative reducts are minimal sets B of attributes with the property that $Pos_B(d) = Pos_A(d)$; they can also be found by means of discernibility matrix $M_{U,A}^*$ in as pointed to in [66] with entries

$$c_{i,j}^* = \begin{cases} c_{i,j}^d \setminus \{d\} \text{ in case either } d(u_i) \neq d(u_j) \text{ and } u_i, u_j \in Pos_A(d) \\ \text{or } pos(u_i) \neq pos(u_j) \\ \emptyset \text{ otherwise} \end{cases}$$

$$\tag{4.16}$$

For a relative reduct B, certain rules are induced from the deterministic system $(Pos_B(d), A, d)$, possible rules are induced from the non-deterministic system $(U \setminus Pos_B(d), A, d)$. In the last case, one can find δ–reducts and turn the system into a deterministic one $(U \setminus Pos_B(d), A, \delta)$ inducing certain rules of the form $\bigwedge_{a \in B}(a = a(u)) \Rightarrow \bigvee_{v \in \delta(u)}(d = v)$.

4.3 Decision Rules

Forming a decision rule consists in searching in the pool of available semantically non–vacuous descriptors for their combination that describes closely a chosen decision class. The very basic idea of inducing rules consists in considering a set B of attributes: the lower approximation $Pos_B(d)$ allows for rules which are certain, the upper approximation $\bigcup_{v \in V_d} \overline{B}[(d = v)]$ adds rules which are possible.

We write down concisely a decision rule in the form $\phi/B, u \Rightarrow (d = v)$ where $\phi/B, u$ is a descriptor formula $\bigwedge_{a \in B}(a = a(u))$ over B, see Pawlak and Skowron [47]. A method for inducing decision rules in a systematic way of Pawlak and Skowron [47] and Skowron [65]consists in finding the set of all δ–reducts $\mathbf{R}=\{R_1, ..., R_m\}$, and defining for each reduct R_j and each object $u \in U$, the rule $\phi/R_j, u \Rightarrow (d = d(u))$. Rules obtained by this method are not minimal usually in the sense of the number of descriptors in the premise ϕ.

A method for obtaining decision rules with minimal number of descriptors, in Pawlak and Skowron [47] and Skowron [65], consists in reducing a given rule $r : \phi/B, u \Rightarrow (d = v)$ by finding a set $R_r \subseteq B$ consisting of irreducible attributes in B only, in the sense that removing any $a \in R_r$ causes the inequality

$$[\phi/R_r, u \Rightarrow (d = v)] \neq [\phi/R_r \setminus \{a\}, u \Rightarrow (d = v)] \qquad (4.17)$$

to hold. In case $B = A$, reduced rules $\phi/R_r, u \Rightarrow (d = v)$ are called *optimal basic rules (with minimal number of descriptors)*. The method for finding of all irreducible subsets of the set A in Skowron [65], consists in considering another modification of discernibility matrix: for each object $u_k \in U$, the entry $c'_{i,j}$ into the matrix $M^\delta_{U,A}$ for δ–reducts is modified into

$$c^k_{i,j} = \begin{cases} c'_{i,j} \text{ in case } d(u_i) \neq d(u_j) \text{ and } i = k \vee j = k \\ \emptyset \text{ otherwise} \end{cases} \qquad (4.18)$$

Matrices $M^k_{U,A}$ and associated Boolean functions $f^k_{U,A}$ for all $u_k \in U$ allow for finding all irreducible subsets of the set A and in consequence all basic optimal rules (with minimal number of descriptors).

Decision rules can also be found in an *exhaustive way* by selecting a maximal consistent set of decision rules, incrementally adding descriptors object–wise and eliminating contradicting cases, see RSES system [59]. A method for inducting a minimal set of rules is proposed in LERS system of Grzymala–Busse [20]

Sequential covering idea of inducing decision rules, see Mitchell [37], leading to a minimal in a sense set of rules covering all objects, has been realized as covering algorithms in RSES system [59].

Association rules in the sense of Agrawal et al. [1], induced by means of Apriori algorithm, loc.cit., i.e., rules with high confidence coefficient can also be induced in information system context by regarding descriptors as boolean features in order to induce certain and possible rules, Kryszkiewicz and Rybiński [27].

Decision rules are judged by their quality on the basis of the training set, and by quality in classifying new unseen as yet objects, i.e., by their performance on the test set. Quality evaluation is done on the basis of some measures; for a rule $r : \phi \Rightarrow (d = v)$, and an object $u \in U$, one says that u *matches* r in case $u \in [[\phi]]$. $match(r)$ is the number of objects matching r.

Support, $supp(r)$, of r is the number of objects in $[[\phi]] \cap [[(d = v)]]$; the fraction $cons(r) = \frac{supp(r)}{match(r)}$ is the *consistency degree* of r: $cons(r) = 1$ means that the rule is certain.

Strength, *strength*(r), of the rule r is defined, Michalski et al. [36], Bazan [5], Grzymala–Busse and Ming Hu [21], as the number of objects correctly classified by the rule in the training phase; *relative strength* is defined as the fraction,

$$rel - strength(r) = \frac{supp(r)}{|[[(d = v)]]|} \qquad (4.19)$$

Specificity of the rule r, $spec(r)$, is the number of descriptors in the premise ϕ of the rule r, cf., op. cit., op. cit.

In the testing phase, rules vie among themselves for object classification when they point to distinct decision classes; in such case, negotiations among rules or their sets are necessary. In these negotiations rules with better characteristics are privileged.

For a given decision class $c : d = v$, and an object u in the test set, the set $Rule(c, u)$ of all rules matched by u and pointing to the decision v, is characterized globally by,

$$Support(Rule(c, u)) = \sum_{r \in Rule(c,u)} strength(r) \cdot spec(r) \qquad (4.20)$$

The class c for which $Support(Rule(c, u))$ is the largest wins the competition and the object u is classified into the class $c : d = v$.

It may happen that no rule in the available set of rules is matched by the test object u and partial matching is necessary, i.e., for a rule r, the matching factor $match - fact(r, u)$ is defined as the fraction of descriptors in the premise ϕ of r matched by u to the number $spec(r)$ of descriptors in ϕ. The rule for which the partial support, *Part-Support(Rule(c,u))*, given as

$$\sum_{r \in Rule(c,u)} match - fact(r, u) \cdot strength(r) \cdot spec(r) \qquad (4.21)$$

is the largest, wins the competition and it does assign the value of decision to u.

In a similar way, notions based on relative strength can be defined for sets of rules and applied in negotiations among them [21].

4.4 Dependencies

Decision rules are particular cases of dependencies among attributes or their sets; certain rules of the form $\phi/B \Rightarrow (d = v)$ establish functional dependency of decision d on the set B of conditional attributes. Functional dependence of the set B of attributes on the set C, $C \mapsto B$, in an information system (U, A) means that $Ind(C) \subseteq Ind(B)$.

Minimal sets $D \subseteq C$ of attributes such that $D \mapsto B$ can be found from a modified discernibility matrix $M_{U,A}$, see Skowron and Rauszer [66]: letting

$\langle B \rangle$ to denote the global attribute representing B: $\langle B \rangle(u) = \langle b_1(u), ..., b_m(u) \rangle$ where $B = \{b_1, ..., b_m\}$, for objects u_i, u_j, one sets $c_{i,j} = \{a \in C \cup \{\langle B \rangle\}$: $a(u_i) \neq a(u_j)\}$ and then $c_{i,j}^B = c_{i,j} \setminus \{\langle B \rangle\}$ in case $\langle B \rangle$ is in $c_{i,j}$; otherwise $c_{i,j}^B$ is empty.

The associated Boolean function $f_{U,A}^B$ gives all minimal subsets of C on which B depends functionally; in particular, when $B = \{b\}$, one obtains in this way all subsets of the attribute set A on which b depends functionally, see Skowron and Rauszer [66]. A number of contributions are devoted to this topic in an abstract setting of semi–lattices, see Pawlak [45] and Novotny and Pawlak [41].

Partial dependence of the set B on the set C of attributes takes place when there is no functional dependence $C \mapsto B$; in that case, some measures of a degree to which B depends on C were proposed in Novotny and Pawlak [41]: the degree can be defined, e.g., as the fraction $\gamma_{B,C} = \frac{|Pos_C B|}{|U|}$, where the C-positive region of B is defined in analogy to already discussed positive region for decision, i.e.,

$$Pos_C(B) = \{u \in U : [u]_C \subseteq [u]_B\}$$

In this case, B depends on C partially to the degree $\gamma_{B,C}$: $C \mapsto_{\gamma_{B,C}} B$.

The relation $C \mapsto_r B$ of partial dependency is transitive in the sense: if $C \mapsto_r B$ and $D \mapsto_s C$, then $D \mapsto_{max\{0,r+s-1\}} B$, where $t(r,s) = max\{0, r + s - 1\}$ is the Łukasiewicz t-norm, see, e.g., Polkowski [54], sect. 10. 4.

4.5 Topology of Rough Sets

We spend some time on topological issues in rough sets; consult Ch. 2 for notions of topology. Clearly, in a finite context of an information system (U, A), any topology can only be finite. The simplest case is that when we consider indiscernibility relation $Ind(B)$ for some attribute set B. As it does induce a partition on the set U, it induces also the partition topology π_B on U; as we know, cf., Ch. 2, sect. 10, each indiscernibility class is clopen, and any intersection of clopen sets is again clopen. Therefore the quotient space $U/Ind(B)$ is discrete. In order to obtain an interesting topology, one is bound to enter the realm of infinity and consider at least the case when the attribute set and the object set are countably infinite. We can assume that attributes form a descending sequence $\{a_n : n = 1, 2, ...\}$, i.e., $Ind(a_{n+1}) \subseteq Ind(a_n)$, each n, and let π_n be the partition topology induced by $Ind(a_n)$ for each n.

We denote by π_0 the topology on the set U, whose open basis is the union $\bigcup\{\pi_n : n = 1, 2, ...\}$. We define π_0–exact sets and π_0–rough sets as follows, where Cl_{π_0}, Int_{π_0} is closure, respectively, interior operator with respect to π_0.

1. A set $Z \subseteq U$ is π_0–exact if and only if $Int_{\pi_0} Z = Cl_{\pi_0} Z$.
2. Otherwise, Z is π_0–rough, i.e., in this case, $Int_{\pi_0} Z \neq Cl_{\pi_0} Z$.

A proposition follows, Polkowski [48], [49], [50], [52], on representation of π_0–rough sets

Proposition 4.1. *A pair (Q, T) of π_0–closed subsets in U satisfies conditions $Q = Cl_{\pi_0} X, T = U \setminus Int_{\pi_0} X$ with a rough subset $X \subseteq U$ if and only if Q, T satisfy the following conditions*

1. $U = Q \cup T$;
2. $Q \cap T \neq \emptyset$;
3. $Q \cap T$ does not contain any point x such that the singleton $\{x\}$ is Π_0–open.

This proposition allows for a representation of each π_0–rough set as a pair (Q, T) of π_0–closed sets satisfying 1–4.

It turns out that the space of π_0–rough sets is metrizable, see Polkowski [49]–[52].

Proposition 4.2. *The topology π_0 is metrizable by means of a metric $d(x, y) = \sum_n 10^{-n} \cdot d_n(x, y)$, where the metric d_n is defined by means of*

$$d_n(x, y) = \begin{cases} 1 \ in \ case \ [x]_n \neq [y]_n \\ 0 \ otherwise \end{cases} \tag{4.22}$$

Now, we introduce a new metric d_H, modeled on the Hausdorff–Pompéïu metric, see Ch. 2, sect. 11, into the family $\mathcal{C}(U)$ of closed subsets of U in the topology π_0.

For closed subsets K, H of U, we let

$$d_H(K, H) = max\{max_{x \in K} dist(x, H), max_{y \in H} dist(y, K)\}$$

(recall that $dist(x, H) = min\{d(x, z) : z \in H\}$ is the *distance* of x to the set H). The standard proof shows that d_H is a metric on $\mathcal{C}$.

We may propose an algorithm for computing $d_H(K, H)$ based on the comparison of closures $Cl_n K, Cl_n H$ for $n = 1, 2, ...$ where Cl_n is the closure operator in the topology π_n.

When $Cl_n K \neq Cl_n H$, there exists, e.g., $x \in Cl_n K \setminus Cl_n H$, and thus, $[x]_n \subseteq Cl_n K, [x]_n \cap Cl_n H = \emptyset$ from which it follows that,

$$d(x, z) \geq \frac{1}{9} \cdot 10^{-n+1} \tag{4.23}$$

for every $z \in Cl_n H$ implying that

$$d_H(K, H) \geq \frac{1}{9} \cdot 10^{-n+1} \tag{4.24}$$

We have the following,

Proposition 4.3. *For any pair K, H of closed sets, we have*

1. *If $Cl_n K = Cl_n H$ for every n, then $d_H(K, H) = 0$;*

2. *If n is the first among indices j such that $Cl_j K \neq Cl_j H$, then $d_H(K, H) = \frac{1}{9} \cdot 10^{-n+1}$.*

For any pair $(Q_1, T_1), (Q_2, T_2)$ of rough sets, we let

$$D((Q_1, T_1), (Q_2, T_2)) = max\{d_H(Q_1, Q_2), d_H(T_1, T_2)\} \qquad (4.25)$$

and

$$D^*((Q_1, T_1), (Q_2, T_2)) = max\{D((Q_1, T_1), (Q_2, T_2)), d_H(Q_1 \cap Q_2, T_1 \cap T_2)\} \qquad (4.26)$$

As d_H is a metric on closed sets, D, D^* are metrics on rough sets. From Properties 1, 2, above, we infer that

Proposition 4.4. *For any pair $(Q_1, T_1), (Q_2, T_2)$ of rough sets*

1. *If n is the first among indices j with the property that either $Cl_j Q_1 \neq Cl_j Q_2$ or $Cl_j T_1 \neq Cl_j T_2$, then $D((Q_1, T_1), (Q_2, T_2)) = \frac{1}{9} \cdot 10^{-n+1}$, otherwise, $D((Q_1, T_1), (Q_2, T_2)) = 0$;*

2. *If n is the first among indices j with the property that either $Cl_j Q_1 \neq Cl_j Q_2$ or $Cl_j T_1 \neq Cl_j T_2$, or $Cl_j(Q_1 \cap T_1) \neq Cl_j(Q_2 \cap T_2)$, then $D^*((Q_1, T_1), (Q_2, T_2)) = \frac{1}{9} \cdot 10^{-n+1}$, otherwise, $D^*((Q_1, T_1), (Q_2, T_2)) = 0$.*

We assume that

C *For each descending sequence $([x_n]_n)_n$ of equivalence classes, we have*

$$\bigcap_n [x_n]_n \neq \emptyset \qquad (4.27)$$

Condition C does express our positive assumption that the universe of our information system has no "holes"; on the other hand, C does imply that in the case where infinitely many of relations R_n are non–trivial, the universe U would be uncountable, so information systems satisfying C should be searched for among those constructed on sets of real numbers or real vectors.

Condition C allows us to demonstrate the basic completeness property of metric spaces of rough sets, see Polkowski op. cit., op. cit. The completeness property here is more complex, then in the general case of a metric space as it requires that both metrics D, D^* cooperate in the following sense

Proposition 4.5. *Under (C), each D^*–fundamental sequence $((Q_n, T_n))_n$ of rough sets converges in the metric D to a rough set.*

We can also consider a finiteness condition

F *Each relation R_n induces a finite number of equivalence classes*

It turns out that, see [49]

Proposition 4.6. *Under C+F, the space of π_0–rough sets is compact in metric D, i.e., any sequence $((Q_j, T_j))_j$ of rough sets contains a sub–sequence $((Q_{j_k}, T_{j_k}))_k$ convergent in the metric D to a limit rough set (Q^*, T^*).*

An interesting application rough set topology finds in fractal geometry.

4.6 A Rough Set Reasoning Scheme: The Approximate Collage Theorem

We give an example of approximate reasoning with rough sets in topological context, see Polkowski [54].

The Collage Theorem has been proposed to be applied in fractal compression; from technical point of view it is a version of the Banach fixed–point theorem adopted to the case of compact sets (fractals) endowed with the Hausdorff-Pompéïu metric induced by the Euclidean metric in a Euclidean space, cf., Barnsley [3]. We know that those spaces are complete, cf., Ch. 2, sect. 11.

Here, we consider the following case, see Polkowski [54], [54]. We assume that a sequence $(F_n)_n$ of compact sets (fractals) is given in a Euclidean space E (say 2– or 3–dimensional). We denote by the symbol D_E the Hausdorff –Pompéïu metric on the space of compact sets in E induced by a standard metric ρ on E (e.g., to fix our attention, we may adopt the metric

$$\rho(\mathbf{x}, \mathbf{y}) = \sum_i ((x_i - y_i)^2)^{\frac{1}{2}}$$

We consider a natural information system (R^k, A), where A is a countable set of attributes such that the n–th attribute a_n does induce the partition of R^k into squares of side length $1/2^n$.

We already know that every fractal in our sense (i.e., having a fractional dimension) is a π_0–rough set with respect to the information system $\mathcal{A}$ on E. We may thus consider the sequence $(F_n)_n$ as a sequence in the complete space of π_0–rough sets and thus we may apply the Banach fixed – point theorem (or, a rough set counterpart of it).

We denote by the symbol $a_m^+ F$ the upper approximation of the compact F with respect to the attribute a_m of the information system $\mathcal{A}$; we may formulate our question related to the sequence $(F_n)_n$. We assume that the sequence $(F_n)_n$ does converge in the metric d to the limit set F. Then a question does arise

Q *Estimate the least natural number n with the property that $a_m^+ F_n = a_m^+ F$ for a given natural number m*

Let us comment on the meaning of this problem. Given F_n, F, we replace those sets with their upper approximations $a_m^+ F_n, a_m^+ F$. Let us observe that description of $a_m^+ F$ requires only a finite number of names of m–cubes in the partition P_m into indiscernibility classes of the relation Ind_{a_m}, so this description is a compression of knowledge about F similar to that which results in the case of fractals generated as iterated function systems. In this case, the description of the fractal is encoded in the starting compact set C and in the coefficients of affine maps whose iterates applied to C generate the fractal in question. Our case here is more general as the compression in terms of upper approximations may be applied to any fractal regardless of the method of its generation.

As with the Collage Theorem, we ask for the first integer n such that sets F_n and F have the same upper approximation with respect to a_m. It follows from our discussion of the metric D' that we have $a_j^+ F_n = a_j^+ F$ for every $j \geq n$. It is a matter of a simple calculation to check that in case $a_m^+ F_n = a_m^+ F$ we also have $D_E(F_n, F) \leq 2^{-m+\frac{1}{2}}$ hence $(F_n)_n$ does converge to F in the metric D_E as well.

It follows that $a_m^+ F_n = a_m^+ F$ for a sufficiently large m does assure that sets F_n and F are sufficiently close to each other with respect to metrics d as well as D_E. The "roughification" $a_m^+ F_n$ may be thus taken as a satisfactory approximation to F.

Let us now refer to the case of fractals generated via iterated function systems. In this case in its simplest form we are given compact fractals $C_0, C_1, ..., C_k$ and an (usually) affine function $f : E \to E$ with the contraction coefficient $c \in (0, 1)$. The resulting fractal F is obtained as the limit of the sequence $(F_n)_n$ of sets defined inductively as follows

$$\begin{cases} F_0 = \bigcup_{i=1}^k C_i \\ F_{k+1} = f(F_k) \end{cases} \qquad (4.28)$$

It is easy to check that the resulting mapping on sets has also the contraction coefficient c. As we know, in the context of the Banach fixed–point theorem, the distance between the limit F and the set F_n may be evaluated as

$$D_E(F_n, F) \leq c^n \cdot (1 - c)^{-1} \cdot D_E(F_0, F_1) \qquad (4.29)$$

This general result implies,

Proposition 4.7. *(Polkowski [54]) Assume that $K = D_E(F_0, F_1)$. Then in order to satisfy the requirement*

$$D_E(F_n, F) \leq \varepsilon$$

it is sufficient to satisfy the requirement

$$a_m^+ F_n = a_m^+ F$$

with

$$m = \lceil \frac{1}{2} - log_2 \varepsilon \rceil$$

and

$$n \geq \lceil \frac{log[2^{-m+\frac{1}{2}} \cdot K^{-1} \cdot (1-c)]}{log c} \rceil$$

In addition to their role in inductive reasoning, rough sets allow also for deductive schemes of reasoning about knowledge.

4.7   A Rough Set Scheme for Reasoning about Knowledge

The setting of rough sets can be used in reasoning along classical lines about knowledge and in building logical systems reflecting the specific features of rough set context.

Consider an *approximation space* (U, E), where u is a set of objects, and E an equivalence relation on U; clearly, this notion is abstracted from the content of the notion of an information system, where E is the indiscernibility relation. In this context, the Rasiowa and Skowron [57] Rough Concept Logic (RCL) is constructed.

The relation E can be extended to powers U^n, $n = 1, 2, ...$, by letting $((u_1, ..., u_n), (u_1', ..., u_n')) \in E$ if and only if $(u_i, u_i') \in E$ for $i = 1, 2, ..., n$. By means of relations E^n, each relation $R \subseteq U^n$, $n = 1, 2, ...$, can be approximated in the rough set way as

$$(u_1, ..., u_n) \in \overline{E}(R) \Leftrightarrow (u_i, u_i') \in E \tag{4.30}$$

for a $(u_1', ..., u_n') \in R$.

Similarly,

$$(u_1, ..., u_n) \in \underline{E}(R) \Leftrightarrow (u_i, u_i') \in E \tag{4.31}$$

for each $(u_1', ..., u_n') \in R$.

$\overline{E}(R)$ is the upper approximation to R, and $\underline{E}(R)$ is the lower approximation to R; either of these sets is exact in U^n.

RCL is built of a variable set V, a predicate set $P = \{e, r_1, ..., r_m\}$, where e is interpreted as an equivalence relation, and each r_i is a predicate of arity k_i, propositional connectives $\{\vee, \wedge, \neg, \Rightarrow\}$, a unary connective $\underline{A}$, quantifiers $\forall, \exists$ and auxiliary symbols like parentheses, commas, etc. etc.

Elementary formulas are of the form $e(x, y), r_i(x_1, ..., x_{k_i})$, and formulas are built as in predicate calculus, cf., Ch.3, sect. 8.

Duality of lower and upper approximations is expressed by introducing a connective $\overline{A} = \neg\underline{A}\neg$.

An *interpretation frame* for RCL is a set X along with an equivalence relation E and relations $R_1, ..., R_m$, where arity of R_i is k_i for $i = 1, 2, ..., m$.

A valuation v in the frame $M=(X, E, R_1, ..., R_m)$ satisfies a formula α, in symbols $v \models_M \alpha$ when the following recursive conditions are satisfied

1. $v \models_M e(x, y)$ *if and only if* $(v(x), v(y)) \in E$;

2. $v \models_M r_i(x_1, ..., x_{k_i})$ *if and only if* $(v(x_1), ..., v(x_{k_i})) \in R_i$;

3. $v \models_M \alpha \vee \beta$ *if and only if* $v \models_M \alpha$ *or* $v \models_M \beta$;

4. $v \models_M \alpha \wedge \beta$ *if and only if* $v \models_M \alpha$ *and* $v \models_M \beta$;

5. $v \models_M \alpha \Rightarrow \beta$ *if and only if* $v \models_M \alpha$ *implies* $v \models_M \beta$;

6. $v \models_M \neg\alpha$ *if and only if it is not true that* $v \models_M \alpha$;

7. $v \models_M \exists x \alpha(x)$ *f and only if* $v(x/c) \models_M \alpha$ *for some* $c \in X$, *where* $v(x/c)(y)$ *is* $v(y)$ *for* $x \neq y$ *and* c *for* $y = x$;

8. $v \models_M \forall x \alpha(x)$ *if and only if* $v \models_M \alpha(x/c)$ *for each* $c \in X$;

9. $v \models_M \underline{A}\alpha(x_1, ..., x_n)$ *if and only if* $v \models_M \alpha(x_1/c_1, ..., x_n/c_n))$ *for each tuple* $(c_1, ..., c_n) \in X^n$ *such that* $(v(x_i), c_i) \in E$ *for* $i = 1, 2, ..., n$.

A formula α is true in a frame M if and only if $v \models_M \alpha$ for each valuation v in M. The formula α is a tautology if and only if α is true in every frame M. For a formula $\alpha(x_1, ..., x_n)$, and a frame M, one defines

$$[[\alpha]]_M = \{(c_1, ..., c_n) \in X^n : \exists v = v(x_1/c_1, ..., x_n/c_n).v \models_M \alpha\} \quad (4.32)$$

Then, Rasiowa and Skowron, op. cit., prove that α is true in M if and only if $[[\alpha]] = X^n$, as then α is satisfied by each valuation.

RCL is axiomatized by means of the following set of axioms

RCL1-3 *Axioms T1–T3 of propositional calculus*

RCL4 $\underline{A}(\alpha \Rightarrow \beta) \Rightarrow (\underline{A}\alpha \Rightarrow \underline{A}\beta$

RCL5 $\underline{A}\alpha \Rightarrow \alpha$

RCL6 $e(x, x)$

RCL7 $e(x, y) \Rightarrow [e(y, z) \Rightarrow e(x, z)]$

RCL8 $e(x, y) \Rightarrow e(y, x)$

RCL9 $e(x, y) \Rightarrow [\underline{A}\alpha(x) \Rightarrow \underline{A}\alpha(y)]$

RCL10 $e(x_1, y_1) \wedge ... \wedge e(x_n, y_n) \Rightarrow [\alpha(x_1, .., x_n) \Rightarrow \underline{A}\alpha(y_1, ..., y_n)$

RCL11 *If α has no free variables, then $\alpha \Rightarrow \underline{A}\alpha$*

Axioms RCL6–8 express the fact that e is interpreted as equivalence, RCL5, RCL9–11 do express the nature of the lower approximation $\underline{A}$.

Derivation rules are detachment (MP), substitution for free variables, rules for quantifier introduction and elimination IE, IU, EE, EU, see Ch. 3, sect. 8, and the rule of necessitation: if α, then $\underline{A}\alpha$.

Among formulas derived from axioms are the following, see Rasiowa and Skowron [57], Lemma 1.

Proposition 4.8. *The following are among theorems of RCL*

1. $(\alpha \Rightarrow \beta) \Rightarrow (\underline{A}\alpha \Rightarrow \underline{A}\beta)$;

2. $e(x_1, y_1) \Rightarrow ... \Rightarrow [(e(x_n, y_n) \Rightarrow (\alpha(x_1, ..., x_n) \Leftrightarrow \alpha(y_1, ..., y_n)]$;

3. $\underline{A}\alpha(x_1, ..., x_n) \Rightarrow \forall y_1, ..., y_n[e(x_1, y_1) \wedge ... \wedge e(x_n, y_n) \Rightarrow \alpha(y_1, ..., y_n)]$;

4. $\underline{A}\alpha \Rightarrow \underline{A}\underline{A}\alpha$;

5. $\alpha \Rightarrow \underline{A}\overline{A}\alpha$.

Let us observe that formulas 1, 4, 5 express the fact that when $\underline{A}$ is interpreted as the modality of necessity and dually $\overline{A}$ is interpreted as the modality of possibility, then RCL realizes the S5 system of modal logic, see Ch. 3, sect. 9. Completeness theorem follows from the main theorem in Rasiowa and Skowron, op. cit.

Proposition 4.9. *If a formula* α *is not derivable from a set of formulas* Γ, *non($\Gamma \vdash \alpha$), then there exists a frame M such that $M\models \Gamma$ and α is not true in M.*

The proof of this proposition goes on lines of the Rasiowa and Sikorski technique, cf. Ch. 3, sect. 8.

Modal character of rough set approximations was also revealed in Orlowska [42] and Vakarelov [67]. Epistemic logic for reasoning about knowledge was constructed in Orlowska [43].

4.8 Fuzzy Set Approach: Main Lines

We have witnessed the development of logical T–calculi, see Ch. 3, sect. 6, with truth values ranging continuously from 0 to 1 in the framework of *residuated lattices*, where an essential usage has been made of the adjoint pair $(T, \rightarrow_T)$ of a continuous t–norm and its residuum. For the first ideas of 'fuzziness before fuzzy', cf., Seising [64], see Fleck [18], Black [8], Menger [34]; Zadeh [70] introduced the subject of fuzzy set theory. t –norms and dual to them, t–*conorms* can be applied in the development of algebra of fuzzy sets, viz., t–norms determine intersections of fuzzy sets according to the formula,

$$\chi_{A \cap B}(x) = T(\chi_A(x), \chi_B(x)) \tag{4.33}$$

where T is a t–norm and χ_A is the fuzzy characteristic (membership) function of the fuzzy set A; whereas t–conorms may be used in determining unions of fuzzy sets, by means of

$$\chi_{A \cup B}(x) = S(\chi_A(x), \chi_B(x)) \tag{4.34}$$

where S is a t–conorm.

We focus in this chapter on deeper properties of t–norms (a fortiori, of co–norms) which were bypassed when discussing T–logics in Ch. 3, and are of interest for rough mereology, which is discussed in coming Ch. 6. We recall that a t–norm T is a function $T : [0,1]^2 \rightarrow [0,1]$, which satisfies the following postulates

1. T is *associative*: $T(T(x,y),z) = T(x,T(y,z))$;

2. T is *commutative*: $T(x,y) = T(y,x)$;

3. T is *non–decreasing in each coordinate*: $T(z,y) \geq T(x,y)$ *whenever* $z \geq x$;

4. $T(1, x) = x$;

5. $T(x, 0) = 0$.

Moreover, T may satisfy additional postulate,

6. T is continuous.

Known to us t–norms: the Łukasiewicz t–norm $L(x, y) = max\{0, x + y - 1\}$, the Product t–norm $P(x, y) = x \cdot y$, the minimum t–norm $M(x, y) = min\{x, y\}$ satisfy 1–6.

Let us observe that L and P do satisfy

7. $T(x, x) < x$ for each $x \in (0, 1)$.

Indeed, this property is obviously satisfied when $L(x, x) = 0$ and if $L(x, x) = 2x - 1$, then $x \geq \frac{1}{2}$ and $2x - 1 \geq x$ implies $x = 1$. For P, $P(x, x) = x^2 < x$ unless $x = 0, 1$.

A t – norm $T(x, y)$ which satisfies postulates 6, 7 is said to be an *Archimedean t–norm*.

Much effort was put into the task of recognizing the structure of t – norms.

Given a function $f : [a, b] \rightarrow [0, \infty]$ of which we assume it is continuous and increasing, we define its *pseudo – inverse* $g : [0, \infty] \rightarrow [a, b]$ by letting

$$g(x) = \begin{cases} a \ in \ case \ x \in [0, f(a)] \\ f^{-1}(x) \ in \ case \ x \in [f(a), f(b)] \\ b \ in \ case \ x \in [f(b), \infty] \end{cases} \qquad (4.35)$$

In the dual case when f is decreasing, the definition is similar, with the obvious changes, viz., b takes the place of a and vice versa.

The correctness of this definition follows from the fact that f is injective on $[a, b]$ and that by continuity of f, the image $f([a, b])$ is the interval $[f(a), f(b)]$ (or, $[f(b), f(a)]$).

We assume now that a function $T : [0, 1]^2 \rightarrow [0, 1]$, not necessarily a t – norm, satisfies postulates 1, 3, 4, 6, and 7.

We will say in this case that T is an *Archimedean function*. We recall a structure theorem about T due to Ling [29]. To this end, we establish first some properties of T. We introduce the symbol $T^n(x)$ defined inductively as follows

$$\begin{cases} T^0(x) = 1 \\ T^{n+1}(x) = T(T^n(x), x) \end{cases} \qquad (4.36)$$

Proposition 4.10. *(The auxiliary structure theorem, Ling [29]) Every archimedean function T has the following properties*

A1 The sequence $(T^n(x))_n$ is non–increasing for every $x \in [0,1]$

A2 Given $0 < y < x < 1$, we have $T^n(x) < y$ for some n; a fortiori, $\lim_n T^n(x) = 0$

A3 0 is the two–sided null element, i.e., $T(0,x) = 0 = T(x,0)$ for every $x \in [0,1]$

A4 $T(x,1) = x$ for every $x \in [0,1]$

A5 $y < x$, $0 < T^n(x)$ imply $T^n(y) < T^n(x)$

A6 $0 < x < 1$, $0 < T^n(x)$ imply $T^{n+1}(x) < T^n(x)$

A7 The set $T_n(x) = \{y \in [0,1] : T^n(y) = x\}$ is non–empty

We let $r_n(x) = \inf T_n(x)$.

A8 $r_n(x) \in T^n(x)$, i.e., $T^n(r_n(x)) = x$

A9 $r_n(x)$ is a continuous, decreasing function of x

A10 $r_n(x) < r_{n+1}(x)$ for every $0 < x < 1$

A11 $\lim_n r_n(x) = 1$ for every $0 < x < 1$

A12 The superposition $T^m \circ r_n$ depends on $\frac{m}{n}$ only, i.e., $T^m \circ r_n = T^{km} \circ r_{kn}$ for $k = 1, 2, \ldots$

Proof. For A1, given $x \in [0,1]$, one has $T^{n+1}(x) = T(T^n(x),x) \le T(T^n(x),1) = T^n(x)$.

For A2, assume to the contrary that there are x,y with $y < x, T^n(x) \ge y$ for each n; consider $z = \inf T^n(x) \ge y$. By postulates 6, 1,

$$T(z,z) = \lim_n T(T^n(x),T^n(x)) = \lim_n T^{2n}(x) = z$$

contradicting postulate 7.

In case of A3, indeed,

$$T(0,x) = T(\lim_n T^n(x),x) = \lim_n T(T^n(x),x) = \lim_n T^{n+1}(x) = 0$$

by A2; in case of $T(x,0)$ the argument is analogous.

For A4, consider the function $x \rightarrow T(x,1)$; by continuity, it takes all values from $0 = T(0,1)$ to $1 = T(1,1)$ so for a given x there is y with $x = T(y,1)$ and $T(x,1) = T(T(y,1),1) = T(y,T(1,1)) = T(y,1) = x$.

Concerning A5, first, observe $T(x,y) \leq T(1,y) = y$; if it was $T(x,y) = y$, we would have by induction that $T(T^n(x),y) = y$. But for some n_0 we have $T^{n_0}(x) < y$ by A2 and thus $y = T(T^{n_0}(x),y) \leq T(y,y) < y$, a contradiction. We have proved

(i) $T(x,y) < y, T(x,y) < x \text{ for } 0 < x, y < 1$

We consider now the function $z \rightarrow T(x,z)$ for a given x and let $y < x$; as this function is continuous and maps $[0,1]$ onto the interval $[0,x]$ so for some z_1 we have $T(x,z_1) = y$. Similarly, we prove that $T(z_2,x) = y$ for some z_2. We have verified

(ii) $x < y \text{ implies the existence of } z_1, z_2 \text{ such that } T(x,z_1) = y, T(z_2,x) = y$

We return to the proof of A5; assume to the contrary that for some x, y, n we have $y < x, 0 < T^n(x), T^n(x) = T^n(y)$. Then by (i), (ii), postulate 1, we have

$$0 < T^n(x) = T^n(y) = T(T^{n-1}(y),y)$$

$$=$$

$$T(T^{n-1}(y),T(x,z)) = T(T(T^{n-1}(y),x),z)$$
$$\leq T(T(T^{n-1}(x),x),z)$$

$$=$$

$$T(T^n(x),y) < T^n(x)$$

by (i), a contradiction.

A6 follows from A5(i). A8 follows as $T_n(x)$ is a closed set, hence, it contains its the least and the greatest bounds.

Concerning A7, as T^n is continuous and $T^n(0) = 0, T^n(1) = 1$, we have $T^n(y) = x$ with some y.

For A9, as, by A8, r_n is the right inverse to T^n which is continuous and decreasing by A5, r_n is also continuous and decreasing.

In case of A10 assume to the contrary that $0 < r_{n+1}(x) \leq r_n(x) < 1$ for some x, n, hence,

$$0 < T^n(r_{n+1}(x)) \leq T^n(r_n(x)) = x < 1 \tag{4.37}$$

and, by A5, one obtains

$$0 < x = T^{n+1}(r_{n+1}(x)) = T(T^n(r_{n+1}(x)), r_{n+1}(x)) \leq$$

$$T(x, r_{n+1}(x)) < x$$

a contradiction.

For A11, by A10, $(r_n(x))_n$ is increasing hence it has a limit $s = lim_n r_n(x)$; thus, $s > r_n(x)$ for each n. On the other hand, $x = T^n(r_n(x)) < T^n(s)$ for each n. Was $s < 1$, we would have $T^n(s) < 1$ for each n and we would find by A2 some m with $T^{mn}(s) = T^m(T^n(s)) < x$, a contradiction.

In case of A12, assume to the contrary that $T^m \circ r_n(x) \neq T^{km} \circ r_{kn}(x)$ for some x, m, n, k; then for $y = r_n(x)$ and $z = r_{kn}(x)$ we have $T^k(z) \geq y$ by definition of $r_n(x)$ and finally $T^k(z) > y$ by our assumption. By continuity of T^k we have $y = T^k(w)$ with some $w < z$ so $T^{kn}(w) = T^n(y) = x$, i.e., $w \in T_{kn}(x)$ contrary to $w < z$ and $z = r_{kn}(x)$ $\qquad\qquad\square$

We now are able to state and verify the principal structure theorem for archimedean functions.

Proposition 4.11. *(the principal structure theorem, Ling [29]) If a function $T : [0,1]^2 \rightarrow [0,1]$ is Archimedean, then there exists a continuous decreasing function f on $[0,1]$ with the property that*

$$(*) \quad T(x,y) = g(f(x) + f(y))$$

for each pair x, y where g is the pseudo – inverse to f. Moreover, every function $T(x,y)$ for which a representation of the form (*) exists is archimedean.

Proof. Selecting and fixing $a \in (0,1)$, we define by A12 a function $h(\frac{m}{n}) = T^m(r_n(a))$ on the set Q_0 of rational positive numbers. As both T^m, r_n are continuous decreasing, h is continuous and decreasing. Moreover, h does satisfy the functional equation

$$(**) \quad T(h(p), h(q)) = h(p+q)$$

on the set Q_0.

To verify (**), we consider $p = \frac{m}{n}, q = \frac{k}{l}$; finding a common denominator d we write down $x = \frac{c}{d}, y = \frac{f}{d}$ and compute

$$T(h(p), h(q)) = T(T^c(r_d(a)), T^f(r_d(a))) = T^{c+f}(r_d(a)) = h(\frac{c+f}{d}) = h(p+q)$$

Having (**) verified, we now extend h in a unique way by continuity and density of the set Q_0 in the interval $[0,1]$ to the continuous and decreasing function g. Clearly

$$(***) \quad T(g(x), g(y)) = g(x+y)$$

is satisfied on $[0,1]$.

Letting A to be the first argument with $g(A) = 0$, we have that g is a continuous decreasing function from $[0, A]$ onto $[0, 1]$ so it has the inverse f. Returning to the identity (***) and letting $g(x) = u, g(y) = v$, we have (***) in the form of

$$T(u, v) = g(f(u) + f(v))$$

for every pair $u, v \in [0, 1]$.

The second part, viz., that (*) implies that T is archimedean, may be verified by a straightforward verification that Postulates 1, 3, 4, 6, and 8 hold, cf., Schweizer and Sklar [62]   □

We may note as a direct consequence of the last theorem that, as shown by Proposition 4.11,

Proposition 4.12. *Each Archimedean function T is commutative.*

It follows that any continuous t – norm admits a representation of the form (*). The function f occurring in the representation (*) is called a *generator* of T. Let us return to the well – known to us the Łukasiewicz t–norm L and find its generator.

Example 4.1. We search for f with the property that $L(x, y) = g(f(x) + f(y))$. Assuming that $f(1) = a$, we have $g(x) = 1$ for $x \in [0, a]$ and thus $1 = L(1, 1) = g(f(1) + f(1)) = g(2a)$, hence, $2a \in [0, a]$ so $a = 0$.

As $f(1) = 0$, the candidate is $f(x) = 1 - x$; in this case the pseudo–inverse is $g(y) = 1 - y$ in case $y \le 1$ and $g(y) = 0$ in case $y > 1$. Combining f, g into (*), we have

$$g(f(x) + f(y)) = \begin{cases} 1 - (1 - x) - (1 - y) = x + y - 1 \ in \ case \ x + y \ge 1 \\ 0 \ in \ case \ x + y < 1 \end{cases}$$

$$(4.38)$$

i.e., $L(x, y) = g(f(x) + f(y))$ holds with the indicated f, g. Thus the generator for L is $f(x) = 1 - x$, i.e., the negation in the logic of Łukasiewicz.

It turns out that postulates 1, 3, 4, 6, 7 are essential; the example for that is provided by the t–norm *min* which does not satisfy Property 7. Actually,

Proposition 4.13. *(Ling [29]) In any representation*

$$(*) \ min(x, y) = g(f(x) + f(y))$$

the function f can neither be continuous nor decreasing.

Proof. Assuming (*), as $g \circ f$ is an identity, f is injective. For $f(0) = a, f(1) = b$, we may, e.g., assume $a > b$. As $0 = min(0, x) = g(f(0) + f(x))$ for every x, in case f is continuous, the value of g on $[a + b, 2a]$ is constantly 0. For

x close enough to 0 we have $a + b < f(x) + f(x) \le 2a$ so $x = min(x, x) = g(f(x) + f(x)) = 0$, a contradiction when $x \ne 0$. So f may not be continuous.

For the second part, assuming f decreasing, we may apply a well-known theorem of Riemann stating that any monotone real function on an interval has at most countably many points of discontinuity (these points may happen only where f experiences jumps and as each jump determines an open interval, the family of jumps may be at most countable). Thus there is a continuity point $a \in (0, 1)$ for f. Let $(a_n)_n$ be an increasing sequence with the limit a. It follows that $(f(a_n))_n$ is a decreasing sequence with the limit $f(a)$. As $f(a) > 0$, we have a_n with $f(a) < f(a_n) < 2 \cdot f(a)$ and applying g which is decreasing as the pseudo $-$ inverse to f, we get $a > a_n > g(f(a) + f(a)) = min(a, a) = a$, a contradiction. So there is no decreasing generator for min □

It will be useful to mention here the dual notion of a $t - co{-}norm$. Given a $t - norm$ T, the associated $t - co{-}norm$ S_T is produced via

$$S_T(x, y) = 1 - T(1 - x, 1 - y) \tag{4.39}$$

Clearly, associativity, commutativity, and continuity are retained by this operation. A closer inspection tells us that null elements for T (i.e., elements a such that $T(a, x) = a$ for every x) are changed into null elements $1 - a$ for S_T and the unit elements b for T (i.e., elements such that $T(b, x) = x$ for every x) are turned into unit elements $1 - b$ for S_T. Also S_T is non–decreasing as T is, and the inequality $T(x, x) < x$ is changed into inequality $S_T(x, x) > x$. Modifying accordingly postulates 1–7, we obtain dual properties for t–co–norms.

As an example we may calculate the t–co–norm S_L. We have

$$S_L(x, y) = 1 - L((1 - x), (1 - y)) = 1 - max\{0, 1 - x - y\} = min(x + y, 1)$$

The corresponding duality takes place also for representations of $t -$ norms and $t -$ co–norms. Letting

$$T(x, y) = g_T(f_T(x) + f_T(y))$$

as a representation for a t–norm T, we may produce the respective representation for the t–conorm S_T as

$$S_T(x, y) = 1 - T(1 - x, 1 - y) = 1 - g_T(f_T(1 - x) + f_T(1 - y))$$

and letting

$$f_S(x) = f_T(1 - x), g_S(x) = 1 - g_T(x)$$

we obtain the representation

$$S_T(x, y) = g_S(f_S(x) + f_S(y))$$

For instance, in case of S_L we have $f_S(x) = 1 - (1 - x) = x$ and $g_S(x) = 1 - g_T(x) = min(x, 1)$. Clearly, in the general case, f_S is increasing, and g_S is the pseudo–inverse to f_S. Having established basic properties of t–norms, we turn to residual implications, so essential for T–logics.

4.9   Residual Implications

Given a t–norm T, the *residual implication* $\Rightarrow_T$, cf., Ch. 3, sect. 6, is defined via the condition

$$RI \ x \Rightarrow_T y \geq z \Leftrightarrow T(x, z) \leq y \tag{4.40}$$

Let us collect in the following proposition the elementary properties of residual implications following from definition.

Proposition 4.14. *For every* $t - norm$ *T, the residual implication* $\Rightarrow_T$ *satisfies the following conditions*

$RI1 \ T(x, y) \leq z \Rightarrow_T u$ *if and only if* $x \leq T(y, z) \Rightarrow_T u$

$RI2 \ y \Rightarrow_T (z \Rightarrow_T u) = T(y, z) \Rightarrow_T u$

$RI3 \ y \Rightarrow_T z = 1$ *if and only if* $y \leq z$

$RI4 \ x \leq y \Rightarrow_T u$ *if and only if* $y \leq x \Rightarrow_T u$

Proof. Associativity of T implies

$$x \leq y \Rightarrow_T (z \Rightarrow_T u) \Leftrightarrow T(x, y) \leq z \Rightarrow_T u \Leftrightarrow$$

$$T(T(x, y), z) \leq u \Leftrightarrow T(x, T(y, z)) \leq u \Leftrightarrow x \leq T(y, z) \Rightarrow_T u$$

whence RI1, RI2 follow.

RI3 follows from definition of $\Rightarrow_T$ and the fact that $T(x, y) \leq min(x, y)$ following from $t(x, 1) = x$ and commutativity and monotonicity of T.

Finally, commutativity of T implies

$$x \leq y \Rightarrow_T u \Leftrightarrow T(x, y) \leq u \Leftrightarrow T(y, x) \leq u \Leftrightarrow y \leq x \Rightarrow_T u$$

which ends the proof   □

Passing to deeper properties of T and $\rightarrow_T$, let us observe that

$RI5$ *For a fixed* a, *the function* $T(x, a)$ *does preserve suprema*

To verify RI5, consider a set C with $c_0 = supC$; then

$$T(c_0, a) = T(supC, a) \geq T(c, a) \tag{4.41}$$

for every $c \in C$, hence, $T(supC, a) \geq sup_C T(c, a)$. On the other hand

$$T(supC, a) \leq u \Leftrightarrow supC \leq a \Rightarrow_T u \Leftrightarrow \forall c \in C.c \leq a \Rightarrow_T u \qquad (4.42)$$

which is equivalent to $\forall c \in C.t(c, a) \leq u$, hence, the equality $T(supC, a) = sup_C T(c, a)$ follows.

Clearly, the function $T(a, x)$ preserves suprema by commutativity of T.

Returning to $\Rightarrow_T$, let us observe that

Proposition 4.15. *The residual implication $\Rightarrow_T$ satisfies*

RI6 $\Rightarrow_T$ is non–increasing in the first coordinate and non–decreasing in the second coordinate

RI7 $a \Rightarrow_T x$ for a fixed a preserves infima; consequently, $x \Rightarrow_T a$ changes suprema into infima.

Proof. RI6 follows by monotonicity of T and definition of $\Rightarrow_T$, e.g., for $u \leq v$ from $x \leq y \Rightarrow_T u$, i.e., $T(x, y) \leq u$, it follows that $T(x, y) \leq v$, i.e., $x \leq y \Rightarrow_T v$ for every x, hence, $y \Rightarrow_T u \leq y \Rightarrow_T v$. The proof for the first coordinate follows on the same lines.

The argument for RI7 parallels that for RI5; given a set C, we have for every x

$$x \leq inf_C(a \Rightarrow_T c) \Leftrightarrow \forall c \in C.x \leq a \Rightarrow_T c$$

which is equivalent to

$$\forall c \in C.T(x, a) \leq c \Leftrightarrow T(x, a) \leq infC \Leftrightarrow x \leq a \Rightarrow_T infC$$

whence $a \Rightarrow_T infC = inf_C(a \Rightarrow_T c)$ follows $\square$

4.10 Topological Properties of Residual Implications

It turns out that properties mentioned in the preceding section may be equivalently stated in topological terms. We recall that a function f between metric spaces is continuous (cf., Ch. 2) if it satisfies the condition $f(lim_n x_n) = lim_n f(x_n)$ for every sequence $(x_n)_n$. This condition may be too strong for some nevertheless important functions and therefore it has been ramified into two *semi – continuity* conditions.

We introduce here two operations on sequences. Given a sequence $(x_n)_n$, we define its *lower limit* as

$$lim\ inf x_n = inf\{lim x_{n_k}\} \qquad (4.43)$$

where the infimum is taken over all convergent sub–sequences $(x_{n_k})_k$ of the sequence $(x_n)_n$, and, the *upper limit* as

$$lim\ sup x_n = sup\{lim x_{n_k}\},\qquad (4.44)$$

where the supremum is taken over all convergent sub–sequences $(x_{n_k})_k$ of the sequence $(x_n)_n$.

Two types of semi–continuity follow.

A function f is *lower–semicontinuous* at x_0 if and only if for every sequence $(x_n)_n$ with $x_0 = lim x_n$ the inequality $f(x_0) \leq lim\ inf f(x_n)$ holds. The function f is lower–semicontinuous if and only if it is lower–semicontinuous at every point.

A function f is *upper–semicontinuous* at x_0 if and only if for every sequence $(x_n)_n$ with $x_0 = lim x_n$ the inequality $f(x_0) \geq lim\ sup f(x_n)$ holds. The function f is upper–semicontinuous if and only if it is upper–semicontinuous at every point.

An important and not difficult to get characterization is the following

Proposition 4.16. *A function f is upper–semicontinuous (respectively, lower–semi–continuous) if and only if for every r, the set $\{x : f(x) \geq r\}$ is closed (respectively, for every r, the set $\{x : f(x) \leq r\}$ is closed).*

We may now characterize in topological terms t – norms and residual implications.

Proposition 4.17. *1. Any t – norm T is lower–semicontinuous; moreover, any associative and commutative function T with $T(0,x) = 0$, $T(1,x) = x$ which is lower–semicontinuos is a t–norm. In this case the residual implication $\Rightarrow_T$ is given by the condition $y \Rightarrow_T z = max\{x : T(x,y) \leq z\}$;*

2. Any residual implication $\Rightarrow_T$ is upper–semicontinuos and any function non–increasing in the first coordinate, non–decreasing in the second coordinate which is upper–continuous is of the form $\Rightarrow_T$ for some t–norm T.

Proof. We consider as a matter for an exemplary argument, a t–norm T about which we would like to prove upper–semicontinuity.

Let us fix a real number r and consider the set

$$M = \{(x,y) : T(x,y) \leq r\}\qquad (4.45)$$

We want to show that M is closed. Assume, to the contrary, that $(x_0, y_0) \in [0,1]^2$ and a sequence $(x_n)_n$ in M exist such that

(i) $T(x_0, y_0) > r$

(ii) $T(x_n, y_n) \leq r$ *for each n*

(iii) $(x_0, y_0) = lim_n (x_n, y_n)$

Then, by monotonicity of T, we have for each n that either $x_n < x$ or $y_n < y$; we may assume that the set $\{n : x_n < x_0\}$ is infinite and in consequence we may assume that $x_n < x_0$ for each n.

Letting $\bar{x}_n = max\{x_1, .., x_n\}$ and $\underline{y}_n = min\{y_1, ..., y_n\}$ we have $T(\bar{x}_n, \underline{y}_n) \leq r$, so we may assume that the sequence $(x_n)_n$ is increasing to x_0 and the sequence $(y_n)_n$ is decreasing to 0.

As $sup_n T(x_n, y_n) = T(x_0, y_1) \leq r$ it follows by monotonicity of T that $T(x_0, y_0) \leq r$, a contradiction. Thus M is closed and a fortiori T is lower–semicontinuous. In all remaining cases we may argue similarly.

The converse may be proved along the same lines. As a corollary to the lower–semicontinuity of a t – norm T, we have that the set $M = \{(x, y) : T(x, y) \leq z\}$ is closed hence the set $\{x : x \leq y \Rightarrow_T z\}$ is closed hence it contains its greatest element and thus $y \Rightarrow_T z = max\{x : T(x, y) \leq z\}$.

The proof of the second part follows along similar lines   □

The question which t–norms may have continuous residuals was settled in Menu and Pavelka [35], whose solution was based on results in Mostert and Shields [39] and Faucett [17], i.e., on the result that any continuous t–norm with $0, 1$ as the only solutions to the equation $T(x, x) = x$ and at least one nilpotent element x (i.e., $\exists n.T^n(x) = 0, x \neq 0$) is equivalent to L (Mostert and Shields [39]) and on the related result in (Faucett [17]) that any continuous t–norm with $0, 1$ as the only solutions to the equation $T(x, x) = x$ and no nilpotent element is equivalent to P.

We outline a proof of these results as consequences of the principal structure theorem in Ling [29], we produce an argument for the Mostert and Shields theorem in its dual form for the respective t–co–norm $S_L(x, y) = min\{x + y, 1\}$. The proof for L may be directly recovered from the given one via dualization, or it may be carried directly on parallel lines.

Let us comment briefly on nilpotent elements. We recall once more that an element $x \neq 0$ is nilpotent with respect to T if $T^n(x) = 0$ for some n (respectively $x \neq 1$ is nilpotent with respect to S_T in case $S_T^n(x) = 1$ for some n). For instance, with $S = S_L$, and $f_S(x) = x, g_S(x) = min\{x, 1\}$, we have by an easy calculation involving the identity $f(g(y)) = y$ (as g is the pseudo – inverse to f) that $S^n(x) = min\{n(x), 1\}$ hence $S^n(x) = 1$ for every $x \geq \frac{1}{n}$.

The difference between the existence and the non – existence of nilpotent elements may be captured in terms of the generator f and its pseudo–inverse g.

We may observe that regardless of the precise and specific formula for the t–co–norm S (or t – norm T), we have the inductively proved identity

$$N \; S^n(x, x) = g(f(2x) + (n - 2)f(x))$$

due to the identity $f(g(x)) = x$ which bounds the generator f and its pseudo – inverse g.

Hence, two possibilities arise

(i) *f is bounded, a fortiori g(x) returns to 1 sufficiently large values of x, i.e., from* N *it follows that S has nilpotent elements*

(ii) *f is unbounded so g is the inverse to f and there are no nilpotents*

We return to the problem of classification of t – norms and t – co–norms. We sketch a solution. First we need the notion of *equivalent t – norms*. It is manifest that given a t – norm T and an increasing function $\phi : [0, 1] \to [0, 1]$ the function

$$(E) \quad T'(x, y) = \phi^{-1}(T(\phi(x), \phi(y)))$$

is a t – norm. We will say that T, T' are *equivalent*. The relation (E) bears also on residual implications:

$$(ER) \quad y \Rightarrow_{T'} z = \phi^{-1}(\phi(y) \Rightarrow_T \phi(z))$$

We state the result due to Ling [29].

Proposition 4.18. *(Ling [29]) The theorem of Mostert and Shields:*

' *If $S(x, y)$ is associative and continuous on $[a, b]^2$ with values in $[a, b]$, with $S(a, x) = x$, and $S(b, x) = b$ for every x, and with at least one nilpotent element in the open interval (a, b), then S is equivalent to the t–co–norm S_L* '

is a consequence of the principal structure theorem for t–co–norms. The dual result follows for t–norms.

Proof. (after Ling [29]) We have to verify first that S satisfies the conditions of the dual principal structure theorem, i.e., that

1. *S is non – decreasing;*

2. *$S(x, x) > x$ for every $x \in (a, b)$.*

To this end, we may observe that if $S(x, z_1) = S(x, z_2)$ then by associativity we have

$$z_1 = S(a, z_1) = S(S(a, x), z_1) = S(a, S(x, z_1)) = S(a, S(x, z_2)) =$$

$$S(S(a, x), z_2) = S(a, z_2) = z_2$$

hence, $S(x, .)$ is an injective function for every x. As $S(x, .)$ is continuous with $S(x, a) = x, S(x, b) = b$ it is increasing. Similar proof shows that $S(., x)$ is increasing also. Property 1 holds.

By Property 1, we have $S(x, x) > S(a, x) = x$ proving Property 2. Hence, S satisfies assumptions of the principal structure theorem and there exist functions f, g with $S(x, y) = g(f(x) + f(y))$. As S has nilpotent elements, f is bounded, say, $f : [a, b] \to [0, M]$; then the pseudo $-$ inverse $g : [0, \infty] \to [a, b]$.

We define a new function $h : [a, b] \to [0, 1]$ via $h(x) = \frac{f(x)}{M}$ and we observe that $x = h(f^{-1}(Mx))$.

We denote by the symbol $g_L(x) = min\{x, 1\}$ the pseudo–inverse to the generator of $S_L(x, y) = min\{x + y, 1\}$. Then we have to verify

Claim. $g_L(x) = h(g(Mx))$ for every $x \in [0, \infty]$

But,

$$g_L(x) = h(f^{-1}(Mg_L(x))) = h(f^{-1}(Mmin(x, 1))) \qquad (4.46)$$

hence,

$$g_L(x) = \begin{cases} h(g(Mx)) \text{ in case } x \leq 1 \\ h(f^{-1}(M)) = h(b) = 1 \text{ in case } x \geq 1 \end{cases} \qquad (4.47)$$

whence it follows that $g_L(x) = h(g(Mx))$ for every $x \in [0, \infty]$.

By Claim,

$$h(S(x, y)) = h(g(f(x) + f(y))) = h(g(M \cdot (\frac{f(x)}{M} + \frac{f(y)}{M})))$$

which in turn equals

$$g_L(\frac{f(x)}{M} + \frac{f(y)}{M}) = g_L(h(x) + h(y))$$

which equals

$$min\{h(x) + h(y), 1\} = S_L(h(x), h(y))$$

and this means that S is equivalent to S_L □

Now, we pass to the problem of characterizing t–norms, which have continuous residual implications. We will need some additional properties of residuation.

Proposition 4.19. *The following properties hold for every* $t - norm$ T

RI8 $y \leq (y \Rightarrow_T z) \Rightarrow_T z$

RI9 $T(x, y) \Rightarrow_T 0 = x \Rightarrow_T (y \Rightarrow_T 0)$

Proof. As $y \Rightarrow_T z \leq y \Rightarrow_T z$ we have $T(y \Rightarrow_T z, y) \leq z$, hence, by commutativity of T it follows that $T(y, y \Rightarrow_T z) \leq z$ which implies $y \leq (y \Rightarrow_T z) \Rightarrow_T z$, i.e., RI8 follows.

Associativity of T implies that $z \leq T(x, y) \Rightarrow_T 0$ if and only if $T(z, T(x, y)) \leq 0$, i.e., $T(T(z, x), y) \leq 0$ so equivalently $T(z, x) \leq y \Rightarrow_T 0$ which is equivalent to $z \leq x \Rightarrow_T (y \Rightarrow_T 0)$. From the equivalence of the first and the last statements RI9 follows □

It turns out that continuity of $\Rightarrow_T$ implies continuity of T. This fact has remarkable consequences: if $\Rightarrow_T$ is continuous, then t–norm T is equivalent to L.

Proposition 4.20. *(Menu and Pavelka [35]) For every t – norm T, continuity of $\Rightarrow_T$ implies continuity of T. In this case, T is equivalent to L.*

Proof. We assume that $\Rightarrow_T$ is continuous; we first verify the following

Claim. *Consider the function $x \Rightarrow_T a$ with $0 \le a < 1$. Then*

$$y = (y \Rightarrow_T a) \Rightarrow_T a$$

for each $y \in [a, 1]$

To verify Claim, we recall that

$$(i)\ y \le (y \Rightarrow_T a) \Rightarrow_T a$$

by RI8; it remains to show that in case $y \ge a$, we have

$$(ii)\ y \ge (y \Rightarrow_T a) \Rightarrow_T a$$

As the function $q(x) = x \Rightarrow_T a$ is continuous decreasing with $q(a) = 1$, $q(1) = a$, there is $z \ge a$ with $y = z \Rightarrow_T a$. Then

$$\begin{cases} y = z \Rightarrow_T a \ge (((z \Rightarrow_T a) \rightarrow_T a) \Rightarrow_T a) = \\ (y \Rightarrow_T a) \Rightarrow_T a \end{cases} \tag{4.48}$$

by the fact that the superposition $(x \Rightarrow_T a) \Rightarrow_T a$ is increasing so (ii) follows, and (i) and (ii) together yield Claim.

From Claim it follows with $a = 0$ that

$$(iii)\ T(x, y) = (T(x, y) \Rightarrow_T 0) \Rightarrow_T 0 = ((x \Rightarrow_T (y \Rightarrow_T 0) \Rightarrow_T 0$$

where the last equality holds by RI9. Property (iii) shows that T is definable in terms of superpositions of $\Rightarrow_T$ and thus T is continuous.
We now solve the equation

$$(iv)\ T(x, x) = x$$

Assume that (iv) holds with $x \ne 0, 1$. Take $c \le x \le d$, so there is u with $c = T(x, u)$. Then

$$T(x, c) = T(x, T(x, u)) = T(T(x, x), u) = T(x, u) = c$$

and we have
$$c = T(x, c) \leq T(d, c) \leq T(1, c) = c$$
i.e., $T(d, c) = c$, hence, $d \rightarrow_T c = c$ for each pair d, c with $c \leq x \leq d$.

This means that for $c < x$, the function $\Rightarrow_T c$ is not any injection, contrary to Claim. We infer that the equation $T(x, x) = x$ has only $0, 1$ as solutions.

To conclude the proof one has to refer to the mentioned above results of Mostert and Shields [39] and Faucett [17]. By these results, T is equivalent either to L or to P; however, P has to be excluded as its residual implication (the Goguen implication), cf., Chapter 3, sect. 7, is discontinuous □

We now recall the notions of similarity and equivalence in the fuzzy universe.

4.11 Equivalence and Similarity in Fuzzy Universe

A fuzzy counterpart to set–theoretical notion of an equivalence relation is the T – *fuzzy similarity relation*, Zadeh [71] which in the most general formulation may be defined as a fuzzy set E_T on the square $X \times X$ satisfying the following requirements

FS1 $\chi_{E_T}(x, x) = 1$

FS2 $\chi_{E_T}(x, y) = \chi_{E_T}(y, x)$

FS3 $\chi_{E_T}(x, z) \geq T(\chi_{E_T}(x, y), \chi_{E_T}(y, z))$

FS1 models reflexivity of equivalence relations, FS2 models their symmetry whereas FS3 renders transitivity with respect to a t–norm T. Let us observe that for each idempotent a of the t–norm T (i.e., for a with $T(a, a) = a$), the relation $R_{a,E}$ defined via

$$(x, y) \in R_{a,E} \Leftrightarrow \chi_{E_T}(x, y) \geq a \qquad (4.49)$$

is an equivalence in the classical sense.

For instance, for *similarity relations*, Zadeh [71], defined with $T(x, y) = min(x, y)$ the relation $R_{a,E}$ is defined for every $a \in [0, 1]$ and relations $R_{a,E}$ form the descending chain of equivalence relations:

$$R_{1,E} \subseteq ... \subseteq R_{a,E} \subseteq .. \subseteq R_{0,E} \qquad (4.50)$$

On the other hand, for *probabilistic similarity relations* defined with $T(x, y) = P(x, y)$, Menger [33] , the only relations $R_{a,E}$ are $R_{1,E} \subseteq R_{0,E} = X \times X$.

The same happens to *likeness relations*, in Ruspini [60] defined with $T(x, y) = L(x, y)$.

In the classical case, an equivalence relation R does induce a *pseudo-metric* d_R on the underlying set X by $d_R(x, y) = 0$ in case $(x, y) \in R$ and $d_R(x, y) = 1$, otherwise. One could expect from a fuzzy similarity relation E_T a similar service; letting for $E = E_T$

$$\chi_{d_E}(x, y) = 1 - \chi_{E_T}(x, y) \tag{4.51}$$

defines a fuzzy set d_E which we may call the *fuzzy metric associated with E_T*. That d_E retains properties of a metric follows from the following

Proposition 4.21. *(de Mantaras and Valverde [32])*
The fuzzy set d_E induced by a fuzzy similarity relation $E = E_T$ has the following properties

FM1 $\chi_{d_E}(x, y) = 0$ *if and only if* $(x, y) \in R_{1,E}$

FM2 $\chi_{d_E}(x, y) = \chi_{d_E}(y, x)$

FM3 $\chi_{d_E}(x, z) \leq \chi_{d_E}(x, y) + \chi_{d_E}(y, z)$ *in case* $T = M, P, L$

Proof. FM1, FM2 follow from the definition of d_E, and concerning the triangle inequality FM3, we may paraphrase it to the equivalent inequality

FM4 $1 + \chi_{E_T}(x, z) \geq \chi_{E_T}(x, y) + \chi_{E_T}(y, z)$

which by FS3 follows from the inequality

$$1 + T(\alpha, \beta) \geq \alpha + \beta$$

satisfied by $T = M, P, L$ and verified directly $\square$

An interesting relation between fuzzy metrics and fuzzy similarities was discovered in de Mantaras and Valverde , op. cit., by dualization of the notion of fuzzy distance. Given a t–conorm $S(x, y)$, we may introduce the notion of the *residual dual implication* $x \leftarrow_S y$ via,

$$DRI \ x \Leftarrow_S y \leq z \Leftrightarrow S(z, x) \geq y \tag{4.52}$$

Then we define a fuzzy set d_S via

$$\chi_{d_S}(x, y) = f(max\{x, y\}) \Leftarrow_S f(min\{x, y\}) \tag{4.53}$$

where f is an arbitrarily chosen decreasing continuous function from the interval $[0, 1]$ onto itself. It turns out that

Proposition 4.22. *(de Mantaras and Valverde [32]) For each t–co–norm S, the fuzzy set d_S satisfies the following conditions*

SM1 $\chi_{d_S}(x,x) = 0$

SM2 $\chi_{d_S}(x,y) = \chi_{d_S}(y,x)$

SM3 $\chi_{d_S}(x,z) \le S(\chi_{d_S}(x,z), \chi_{d_S}(y,z))$

Let us remark for the proof that SM1, SM2 follow by definition of d_S and properties of S and SM3 follows by direct calculations involving the definition of $\Leftarrow_S$.

Hence, d_S is a pseudo–metric induced by the t–co–norm S. Involving once more a continuous decreasing $f : [0,1] \rightarrow [0,1]$, we may convert d_S into a fuzzy similarity relation $E = E_T$ with $T(x,y) = f^{-1}(S(f(x),f(y))$ by letting

$$\chi_{E_T}(x,y) = f(\chi_{d_S}(x,y)) \tag{4.54}$$

In this way we obtain the correspondence between fuzzy similarity relations and pseudo–metrics on X.

Now, we may address similarity (equivalence) classes for a fuzzy similarity relation $E = E_T$. In Zadeh [71], it is proposed to define the similarity class $[x]_E$ for $x \in X$ by letting

$$\chi_{[x]_E}(y) = \chi_E(x,y) \tag{4.55}$$

Then, the following properties are observed, see Höhle [25]

Proposition 4.23. *The similarity class $[x]_E$ satisfies the following conditions*

SC1 $\chi_{[x]_E}(x) = 1$

SC2 $T(\chi_{[x]_E}(y), \chi_E(y,z)) \le \chi_{[x]_E}(z)$

SC3 $T(\chi_{[x]_E}(y), \chi_{[x]_E}(z)) \le \chi_E(y,z)$

Proof. SC1 follows by FS1; SC2 is a direct paraphrase of FS3 and SC3 paraphrases FS3 via FS2 □

Properties SC1–SC3 reflect properties of classical equivalence relations, viz., SC1 corresponds to the property that $x \in [x]_R$, SC2 paraphrases the property that xRy implies $y \in [x]_R$, and SC3 paraphrases the property that $y, z \in [x]_R$ imply $(y,z) \in R$.

We may call a *fuzzy equivalence class* of a fuzzy similarity relation E any fuzzy set A satisfying SC1–SC3 in place of $[x]_E$. However, it turns out that we have already found all fuzzy equivalence classes. We denote with the symbol $A \times_T B$ the T – Cartesian product of fuzzy sets A, B defined via

$$\chi_{A \times_T B}(x,y) = T(\chi_A(x), \chi_B(y)) \tag{4.56}$$

Proposition 4.24. *(Höhle [25]) If A is a fuzzy equivalence class of a fuzzy similarity relation E_T then A is of the form $[x]_{E_T}$ with some $x \in X$. Moreover,*

1. $\bigcup_{x \in X}[x]_{E_T} \times_T [x]_{E_T} = E_T$;

2. In case X is finite, $\sup_y\{min\{\chi_{[x]_{E_T}}(y), \chi_{[z]_{E,T}}(y)\} < 1$ whenever $[x]_{E_T} \neq [z]_{E,T}$.

Proof. Assuming A a fuzzy equivalence class, we have

(i) $\chi_A(x) = 1$

with some x by SC1, and thus by RI5

$$\chi_A(y) = T(\chi_A(y), \sup_z T(\chi_{[x]_{E_T}}(z), \chi_{[x]_{E_T}}(z)))$$

which in turn is equal to

$$\sup_z T(\chi_A(y), T(\chi_{[x]_{E_T}}(z), \chi_{[x]_{E_T}}(z)))$$

As, by (i) and associativity and monotonicity of T, and our assumption about A

$$\sup_z T(\chi_A(y), T(\chi_{[x]_{E_T}}(z), \chi_{[x]_{E_T}}(z)))$$

is less or equal to

$$\sup_z T(T(\chi_A(y), \chi_A(z)), \chi_{[x]_{E_T}}(z)))$$

and moreover, by SC2

$$\sup_z T(T(\chi_A(y), \chi_A(z)), \chi_{[x]_{E_T}}(z)))$$

is less or equal to

$$\sup_z T(\chi_{E_T}(y, z), \chi_{[x]_{E_T}}(z))$$

it follows finally that, by SC3

$$\sup_z T(\chi_{E_T}(y, z), \chi_{[x]_{E_T}}(z)) \leq \chi_{[x]_{E_T}}(y)$$

Hence, $A \leq [x]_{E_T}$ and, replacing roles of A and $[x]_{E_T}$, one obtains by symmetry that $[x]_{E_T} \leq A$ so finally $A = [x]_{E_T}$.

To settle Property 1, we observe that, by FS3

$$\chi_{\bigcup_{[x]}[x]_{E_T} \times_T [x]_{E_T}}(y, z) = \sup_{[x]} T(\chi_{[x]_{E_T}}(y), \chi_{[x]_{E_T}}(z))$$

is equal to

$$sup_x T(\chi_{E_T}(x,y), \chi_{E_T}(x,z)) = \chi_{E_T}(y,z)$$

which verifies Property 1.

To verify Property 2, we may observe that by the already proved part, if $[x]_{E_T} \neq [z]_{E,T}$ then $min_y\{\chi_{[x]_{E_T}}(y), \chi_{[z]_{E_T}}(y)\} < 1$ $\qquad \square$

Properties 1 and 2 are counterparts to properties of classical equivalence relations: covering the universe by equivalence classes (1) and disjointness of equivalence classes (2).

It is justified to call collections $\mathcal{F}$ of fuzzy sets satisfying Properties 1 and 2, and additional property

3. $\exists x.\chi_A(x) = 1$ for every $A \in \mathcal{F}$.

by the name of *fuzzy partitions*.

Given a fuzzy partition $\mathcal{F} = \{A_i : i \in I\}$, we can ask about the existence of a fuzzy similarity relation E_T which could approximate fuzzy sets A_i with its equivalence classes and at the best give sets A_i as its equivalence classes.

In order to address this problem, we consider a t – norm T and we define a relation R_T on the universe X by the formula

$$\chi_{R_T}(x,y) = inf_i\{max\{\chi_{A_i}(x), \chi_{A_i}(y)\} \Rightarrow_T min\{\chi_{A_i}(x), \chi_{A_i}(y)\}\} \quad (4.57)$$

We have

Proposition 4.25. *(Valverde [68]) The relation R_T satisfies FS1–FS3 and thus R_T is a fuzzy similarity relation.*

Proof. Indeed, FS1 is satisfied as $x \to_T x = 1$ for every x hence $\chi_{R_T}(x,x) = 1$ for every x. Symmetry FS2 follows immediately from definition of R_T. Finally, transitivity property FS3 comes down after a suitable paraphrase to the property

RI10 $y \Rightarrow_T x \leq (z \Rightarrow_T y) \Rightarrow_T (z \Rightarrow_T x)$

of residual implications. To prove RI10, we begin with a string of equivalent inequalities following from the definition of residual implications starting with RI10.

(i) $T(y \Rightarrow_T x, z \Rightarrow_T y) \leq z \Rightarrow_T x$

(ii) $T(T(y \Rightarrow_T x, z \Rightarrow_T y), z) \leq x$

(iii) $T(y \Rightarrow_T x, T(z \Rightarrow_T y, z)) \leq x$

(iv) $T(y \Rightarrow_T x, y) \leq x$

(v) $T(z \Rightarrow_T y, z) \leq y$

and we obtain from (iv) and (v) the inequality

(vi) $T(y \Rightarrow_T x, T(z \Rightarrow_T y, z)) \leq T(y \Rightarrow_T x, y)) \leq x$

i.e., Property 3 is proved, verifying RI10 and a fortiori FS3.

As $\mathcal{F}$ was supposed to be a fuzzy partition, we have

$$\chi_{A_i}(x) = 1$$

with some x for every A_i. So SC1 is satisfied with $\mathcal{F}$.

We now verify that each A_i satisfies SC2. We need to verify that

(vii) $T(\chi_A(x), \chi_{R_T}(x, y)) \leq \chi_A(y)$ with $A = A_i$

Replacing $\chi_{R_T}(x, y)$ with its definition and using the generic symbol B for an element of $\mathcal{F}$, we have

$$T(\chi_A(x), inf_B\{max\{\chi_B(x), \chi_B(y)\} \Rightarrow_T min\{\chi_B(x), \chi_B(y)\}\})$$

is less or equal to

$$T(\chi_A(x), \{max\{\chi_A(x), \chi_A(y)\} \Rightarrow_T min\{\chi_A(x), \chi_A(y)\}\})$$

which in turn is less or equal to

$$T(max\{\chi_A(x), \chi_A(y)\}, \{max\{\chi_A(x), \chi_A(y)\} \Rightarrow_T min\{\chi_A(x), \chi_A(y)\}\})$$

which is less or equal to

$$\leq min\{\chi_A(x), \chi_A(y)\} \leq \chi_A(y)$$

proving SC2.

Now, for SC3, we should verify, again using generic symbols A, B that

(viii) $T(\chi_A(x), \chi_A(y)) \geq \chi_{R_T}(x, y)$

Let us look at

(ix) $\chi_{R_T}(x, y) = inf_B\{max\{\chi_B(x), \chi_B(y)\} \rightarrow_T min\{\chi_B(x), \chi_B(y)\}$

As far as we have no other information about $\mathcal{F}$, then that it is merely a fuzzy partition, we are not able to prove or disprove SC3   □

However, in case our fuzzy partition $\mathcal{F}$ is of the form $\{[x]_{E_T}\}$, i.e., it is generated by a fuzzy similarity relation E_T, we may say more, viz.,

Proposition 4.26. *(Valverde [68]) For the fuzzy partition $\mathcal{F} = \{[x]_{E_T} : x \in X\}$ induced by a fuzzy similarity relation $R = E_T$ with a t- norm T, the fuzzy similarity relation E_R induced by $\mathcal{F}$ according to the recipe of Proposition 4.25 coincides with E_T; a fortiori, $\mathcal{F}$ is a fuzzy partition with respect to E_R.*

Proof. Let us observe that by FS3

$$\chi_{E_T}(x,z) \geq T(\chi_{E_T}(x,y), \chi_{E_T}(y,z))$$

and we obtain

$$(i)\ \chi_{E_T}(x,y) \leq \chi_{E_T}(y,z) \Rightarrow_T \chi_{E_T}(x,z)$$

Going now to E_R, we consider the term

$$(ii)\ max\{\chi_B(x), \chi_B(y)\} \Rightarrow_T min\{\chi_B(x), \chi_B(y)\}$$

with $B = [b]_{E,T}$, for some b. By (i), we have

$$(iii)\ max\{\chi_B(x), \chi_B(y)\} \Rightarrow_T min\{\chi_B(x), \chi_B(y)\} \geq \chi_{E_T}(x,y)$$

and letting $b = x$, we have that

$$(iv)\ inf_B\{max\{\chi_B(x), \chi_B(y)\} \Rightarrow_T min\{\chi_B(x), \chi_B(y)\}\} = \chi_{E_T}(x,y)$$

which concludes our proof   □

It follows that the construction presented above is canonical: starting with a fuzzy similarity R it induces a fuzzy partition which in turn returns the initial fuzzy similarity R. We have here a counterpart to the classical correspondence between equivalence relations and partitions.

4.12 Inductive Reasoning: Fuzzy Decision Rules

Inductive reasoning by fuzzy sets is applied in inference about systems on basis of partial knowledge extracted from observing performance of a human operator or identifying essential parameters of a system and their mutual dependencies, see Zadeh [72]. In order to present basic features of this reasoning, we begin with predicated variant of fuzzy logic.

We introduce a set TYPE of *types* by conditions

(i) *there are finitely many elementary types* $X_1, ..., X_n$

(ii) *any finite sequence* (X_{i_j}) *of elementary types is a type*

(iii) *nothing else is a type*

For each elementary type X, there is given the *type domain* D_X. Variables of fuzzy predicate calculus are divided also: for each elementary type X, variables of type X form a subset Var_X of the variable set Var, and $Var_X \cap Var_Y = \emptyset$ for elementary types $X \neq Y$. Predicates are typed also and to each predicate $P(x_1, ..., x_n)$ there is type assigned $(X_1, ..., X_n)$ where each variable x_i is of elementary type X_i.

Open formulas are of the form $P(x_1, ..., x_n)$ where P is a predicate, and of the form $\alpha \vee \beta$, $\alpha \wedge \beta$, $\alpha \Rightarrow \beta$ where α, β are formulas.

The meaning of formulas is defined as follows; the meaning of the formula $P(x_1, ..., x_n)$ is the fuzzy membership function $\chi_P : \prod_i X_i \to [0, 1]$ of a fuzzy relation R_P in the Cartesian product $\prod_i X_i$, where each variable x_i is of elementary type X_i, the meaning of the formula $\alpha \vee \beta$ is $max\{\chi_\alpha, \chi_\beta\}$, the meaning of the formula $\alpha \wedge \beta$ is $min\{\chi_\alpha, \chi_\beta\}$, and the meaning of the formula $\alpha \Rightarrow \beta$ is $\chi_\alpha \Rightarrow_T \chi_\beta$, where χ_α, χ_β are meanings of formulas α and β. To reconcile the usage of possibly distinct t–norms in the last formula, one can stay with the basic fuzzy logic, see Chapter 3, sect. 3.6.

Quantifiers $\forall$, $\exists$ are added in usual form: given a formula $\alpha(x)$, expressions $\forall x.\alpha(x)$, $\exists x.\alpha(x)$ are formulas, where quantification occurs over variables of same type as the variable in the formula.

Meaning of quantified formulas is defined in accordance with the rule for classical predicate calculus, see Chapter 3, sect. 8, i.e., the meaning of the formula $\forall x.\alpha(x)$ is given as $inf\chi_{\alpha(a)} : a \in D_X$ where X is the elementary type of the variable x. Dually, the meaning of the formula $\exists x.\alpha(x)$ is given as $sup\chi_{\alpha(a)} : a \in D_X$.

In applications, elementary types come as *linguistic variables, attributes*, e.g., in the system of steam engine, attributes could be *temperature in the boiler, steam pressure, gauge angle*, in the system of the inverted pendulum attributes are *angle between pendulum and the vertical line, angular velocity, force applied to the car*. Each attribute X defines its domain D_X and predicates P_i^X. In applications, the notation 'P is X' is in use to denote the fact that input value from the domain D_X is characterized by the predicate P, e.g, in the phrase ' force is small '.

Rules induced from data are in the implicative form $\bigwedge_j P_j^{X_j} \Rightarrow Q^Y$ which in applications is written down as

If P_1 is X_1 AND ... AND P_k is X_k then Q is Y

with the meaning $min\{\chi_{P_j}(a_j)\} \Rightarrow_T \chi_Q(b)$, where a_j, b are input or output values from respective domains.

It should be noted that in practice many authors have applied solutions which have not obeyed any logical mechanism. For instance, often applied device is to regard the premise as a relation; we consider a simple case of two terms in premise: P is $X \wedge S$ is Z. The operation of *cylindrical extension* extends P to the predicate P^e of type (X, Y) by letting $\chi_{P^e}(a, b) = \chi_P(a)$ for $a \in D_X, b \in D_Y$. Similarly, S is extended to S^e. The resulting relation $P \times Q$ is defined then as the intersection $P^e \cap S^e$ with $\chi_{P \times Q}(a, b) = min\{\chi_P(a), \chi_Q(b)\}$.

The operation of *projection* produces from a predicate T of type (X, Y) a predicate of type X and a predicate of type Y; for instance, the predicate T_Y of type Y is produced with $\chi_{T_Y}(b) = sup\{\chi_T(a, b) : a \in D_X\}$. This formula can be reconciled with logic by observing that projection onto a coordinate space can be defined as $b \in proj_Y T \Leftrightarrow \exists a \in D_X.(a, b) \in D_T$ and applying the operator *sup* in the meaning of existential quantification.

4.13 On the Nature of Reductive Reasoning

We already have stated that reductive reasoning, including inductive reasoning, is performed along the Łukasiewicz [30] scheme,

$$\frac{q, p \Rightarrow q}{p} \tag{4.58}$$

Searching for an appropriate statement p in (4.58) can be termed a search for an *explanation*, or in case of inductive reasoning, for a *generalizing hypothesis*, cf., Mitchell [37]. It is instructive to measure this scheme against the contemporary framework of *non–monotonic reasoning* in order to give, taking rough sets as example, a brief recall of notions related to this area. We use the vehicle of *Scott consequence relations*, see Scott [63].

A Scott consequence relation with respect to a logic L, is defined as a binary relation $Cs(X, Y)$ between finite sets of statements in L, subject to requirements,

CS1 $Cs(\{\alpha\}, \{\alpha\})$

CS2 $Cs(X, Y), X \subseteq X', Y \subseteq Y'$ imply $Cs(X', Y')$

CS3 $Cs(X \cup \{\alpha\}, Y), Cs(X, Y \cup \{\alpha\})$ imply $Cs(X, Y)$

CS1 is the *reflexivity property*, CS2 is the *monotonicity property*, and CS3 is the *cut property*.

The usual interpretation of Cs is, see, e.g., Bochman [10], *if all statements in X are accepted, then at least one statement in Y is accepted.*

We depart from this standard semantic interpretation, in order to merge the Scott consequence relation with the Łukasiewicz scheme (4.58) for reductive (inductive, in particular) reasoning. Our proposed interpretation is

I *If all statements $q \in X$ are accepted, then at least one $p \in Y$ satisfies the scheme of* Łukasiewicz (4.58), *i.e., $p \Rightarrow q$ is true*

We need a proof that this interpretation does satisfy requirements CS1–CS3.

Proposition 4.27. *If the logic L does support theorems $\alpha \Rightarrow \alpha$ and $(\alpha \Rightarrow \beta) \Rightarrow [(\beta \Rightarrow \gamma) \Rightarrow (\alpha \Rightarrow \gamma)]$ as well as MP modus of deduction, then the interpretation I does satisfy CS1–CS3.*

Proof. CS1 holds in virtue of the assumption that $\alpha \Rightarrow \alpha$ is a theorem of L. For CS2, assume that $Cs(X, Y), X \subseteq X', Y \subseteq Y'$; if all $q \in X'$ are accepted, then all $q \in X$ are accepted, hence, by the assumption that $Cs(X, Y)$, at least one $p \in Y$ yields a true statement $p \Rightarrow q$, and of course, $p \in Y'$.

In case of CS3, the assumption that $Cs(X \cup \{\alpha\}, Y)$ implies that all $q \in X$ and α are accepted and at least one $p \in Y$ satisfies either (i) $p \Rightarrow q$ or (ii) $p \Rightarrow \alpha$. The assumption $Cs(X, Y \cup \{\alpha\})$ implies moreover that there exists $p \in Y$ such that (i) or (iii) $\alpha \Rightarrow q$ for some $q \in X$. To prove $Cs(X, Y)$, let us assume that there is no $p \in Y$ with $p \Rightarrow q$ true. Thus, by (iii) $\alpha \Rightarrow q$ and by (ii) $p \Rightarrow \alpha$ for some p. By theorem $(\alpha \Rightarrow \beta) \Rightarrow [(\beta \Rightarrow \gamma) \Rightarrow (\alpha \Rightarrow \gamma)]$, and double application of MP, $p \Rightarrow q$ follows, a contradiction. Hence, $Cs(X, Y)$ holds $\qquad\square$

We will call statements $q \in X$ *reductive queries*, $p \in Y$ *reductive hypotheses*, and, statements of the form $p \Rightarrow q$ *reductive explanations*. The adjective *reductive* will be often omitted for conciseness' sake.

Example 4.2. In the context of rough sets, as a query we can consider a descriptor formula $(d = v)$, where d is a decision and v its value. A hypothesis is then a formula, see sects. 3, 4, $\phi_B : \bigwedge_{a \in B}(a = v_a)$ such that the implication (decision rule) $\phi_B \Rightarrow (d = v)$ is true (exact) and it is an explanation for the query.

We know that important decision rules are often partially true; this is a specific property of logical schemes in all kinds of approximate reasoning. To account for them, we modify the consequence relations to *partial consequence relations*. First, we introduce a modified Łukasiewicz scheme

$$\frac{q, p \Rightarrow q}{p, r} \tag{4.59}$$

where $r \in [0, 1]$ is the truth state (degree) of the explanation $p \Rightarrow q$.

A partial Scott consequence relation, accordingly, is a relation $Cs_\pi(X, Y)$ subject to CS1–CS3 but with a new interpretation

I_π *If all statements $q \in X$ are accepted, then at least one $p \in Y$ satisfies the scheme of* Łukasiewicz (4.59), *i.e.,* $p \Rightarrow q$ *is true to a degree r for some $r \in [0, 1]$*

Proposition 4.27 extends to the partial case with a caveat, viz., as we know, composition of dependencies of the form $\alpha : p \Rightarrow q, \beta : q \Rightarrow r$ is true to degree $\geq L(r, s)$, where L is the Łukasiewicz t–norm, r, s are truth degrees of α, β, respectively. Thus, in order for the tautology $(\alpha \Rightarrow \beta) \Rightarrow [(\beta \Rightarrow \gamma) \Rightarrow (\alpha \Rightarrow \gamma)]$ to have a positive degree of truth, the particular implications should have high degrees of truth: denoting with r, s, t respective degrees of truth in order these implications appear in the tautology, we come at the condition $r + s + t > 2$; in case of rough sets, it is, hence, safe to restrict oneself to association rules. A *theory* for a reasoning scheme L is a set of statements X such that,

$$Cs(X, F \setminus X) \text{ is not true} \tag{4.60}$$

where F is the set of all well–formed statements in the language L. A theory then is a set of formulas in which each query is fully explained within the theory; one may say that X is *self–explainable*.

Example 4.3. For the case of rough sets, a theory will be a set of descriptor formulas X which with each descriptor query $(d = v)$ contains all exact decision rules $\phi_B \Rightarrow (d = v)$ in case of Cs, and all partially true decision rules $\phi_B \Rightarrow_r (d = v)$ in case of Cs_π.

It follows that in case of rough sets, a union of theories is again a theory, hence each theory extends to a maximal theory. A characterization of maximal theories for rough set decision rules poses itself.

Proposition 4.28. *Each maximal theory for the scheme L of descriptor logic of decision rules in the context of rough sets contains all non–void descriptors of the form $(d = v)$ and for each of these descriptors $(d = v)$, it contains all exact decision rules of the form $\phi_B \Rightarrow (d = v)$ in case of Cs and all partially true decision rules of the form $\phi_B \Rightarrow_r (d = v)$ in case of Cs_π.*

The notions of *belief* and of *knowledge* are formally introduced as follows. Belief is a set B of statements which is contained in each maximal theory. Knowledge is a set K of statements with the property that

$$Cs(\emptyset, K) \tag{4.61}$$

Example 4.4. In case of rough sets, by Proposition 4.28, the belief set consists of all non–void descriptors along with all possible explanations and explaining hypotheses, true or partially true, depending on the case. Knowledge is the set K of all true or partially true decision rules, depending on the case considered.

Contrary to the case of classical deductive reasoning, reductive reasoning has a *non–monotonic* character, meaning that when the evidence, i.e., available information increases, belief and knowledge can diminish. Non–monotonicity of reductive reasoning is usually exhibited by means of the notion of *inference*, see Van Benthem [6], Makinson [31]. We introduce the inference relation $\gg$ by letting,

$$\beta \gg \alpha \Leftrightarrow Cs(\alpha, \beta) \qquad (4.62)$$

Thus, α infers β if and only if $\beta$is an explaining hypothesis for α. From our assumptions about the logic L and from properties of the consequence relation Cn, the following properties of inference follow,

1. *$\beta \gg \alpha$ and $\gamma \Rightarrow \beta$ imply $\gamma \gg \alpha$;*

2. *In case of rough set decision systems, $\beta \gg \alpha$, $\gamma \gg \alpha$ imply $\beta \vee \gamma \gg \alpha$ and $\beta \wedge \alpha \gg \alpha$.*

Clearly, inference relation is relative to *evidence* available. We restrict ourselves to the case of decision systems; for a decision system D, we denote with the symbol $\gg_D$ the inference relation induced on basis of evidence D. A given decision system D is a subsystem of the *universal* decision system D^0 encompassing all possible objects and attributes. In case of Open World Assumption, we are provided input of new objects, and in some cases also of attributes, hence we can introduce a containment relation $\sqsubseteq$ on decision systems by means of

$$D = (U, A, d) \sqsubseteq D' = (U', A', d) \ \Leftrightarrow U \subseteq U', A \subseteq A' \qquad (4.63)$$

Non–monotonicity of inference can be expressed as follows: NM *For decision systems $D \sqsubseteq D'$, if $\Delta \gg_D \Theta$, then $\Delta \gg_{D'} \Theta$ need not hold* where $\Delta \gg_D \Theta$ is a synthetic notation for the fact that for each $\alpha \in \Theta$ there is $\beta \in \Delta$ with $\beta \gg_D \alpha$.

Inductive reasoning in the framework of fuzzy sets is discussed in Dubois and Prade [13] and Dubois, Lang and Prade [14]. The reader will find more information on various paradigms and methods of inductive reasoning in additional bibliography items listed below. Decision trees learning is covered in Quinlan [56]; evolutionary methods are introduced in Holland [24]. In Bishop [7] cognitive methods are described and Vapnik [69] discusses statistical learning. Dubois and Prade [15] give a survey on belief update, and Rissanen [58] introduces the minimal description length (MDL) principle. Muggleton [40] is a reference on inductive logic programming. Russell and Norvig [61] is a textbook covering many aspects of inductive learning.

References

1. Agrawal, R., Imieliński, T., Swami, A.: Mining association rules between sets of items in large databases. In: Proceedings ACM Sigmod Conf., Washington, DC, pp. 207–216 (1993)
2. Baader, F., Calvanese, D., McGuiness, D.L., Nardi, D., Patel–Schneider, P.F. (eds.): The Description Logic Handbook: Theory, Implementation and Applications. Cambridge University Press, Cambridge (2004)
3. Barnsley, M.F.: Fractals Everywhere. Academic Press, New York (1988)
4. Bazan, J.G.: A comparison of dynamic and non–dynamic rough set methods for extracting laws from decision tables. In: Polkowski, L., Skowron, A. (eds.) Rough Sets in Knowledge Discovery, vol. 1, pp. 321–365. Physica Verlag, Heidelberg (1998)
5. Bazan, J.G., Nguyen, H.S., Nguyen, S.H., Synak, P., Wróblewski, J.: Rough set algorithms in classification problems. In: Polkowski, L., Tsumoto, S., Lin, T.Y. (eds.) Rough Set Methods and Applications. New Developments in Knowledge Discovery in Information Systems, pp. 49–88. Physica Verlag, Heidelberg (2000)
6. Van Benthem, J.: Essays in Logical Semantics. D. Reidel, Dordrecht (1988)
7. Bishop, C.M.: Neural Networks for Pattern Recognition. Oxford University Press, Oxford (1995)
8. Black, M.: Vagueness: An exercise in logical analysis. Philosophy of Science 4, 427–455 (1937)
9. Bocheński, I.M.: Die Zeitgönossischen Denkmethoden. A. Francke AG, Bern (1954)
10. Bochman, A.: A Logical Theory of Nonmonotonic Inference and Belief Change. Springer, Berlin (2001)
11. Brown, F.M.: Boolean Reasoning: The Logic of Boolean Equations. Dover, New York (2003)
12. Carnap, R.: Logical Foundations of Probability. University of Chicago Press, Chicago (1950)
13. Dubois, D., Prade, H.: Fuzzy sets in approximate reasoning. Part 1: Inference with possibility distributions. Fuzzy Sets and Systems 40, 143–202 (1990)
14. Dubois, D., Lang, J., Prade, H.: Fuzzy sets in approximate reasoning. Part 2: Logical approaches. Fuzzy Sets and Systems 40, 203–244 (1991)
15. Dubois, D., Prade, H.: A survey of belief revision and updating rules in various uncertainty models. International Journal of Intelligent Systems 9(1), 61–100 (1994)
16. Duda, R.O., Hart, P.E., Stork, D.G.: Pattern Classification. John Wiley and Sons, New York (2001)

17. Faucett, W.M.: Compact semigroups irreducibly connected between two idempotents. Proc. Amer. Math.Soc. 6, 741–747 (1955)
18. Fleck, L.: O niektórych swoistych cechach myślenia lekarskiego (On some specific patterns of medical reasoning, in Polish). Archiwum Historji i Filozofji Medycyny oraz Historji Nauk Przyrodniczych 6, 55–64 (1927); Cohen R. S., Schnelle T.: Cognition and Fakt. Materials on Ludwik Fleck. D. Reidel, Dordrecht (1986)
19. Frege, G.: Grundgsetzte der Arithmetik II. Verlag Hermann Pohle, Jena (1903)
20. Grzymala–Busse, J.W.: LERS – a system for learning from examples based on rough sets. In: Słowiński, R. (ed.) Intelligent Decision Support: Handbook of Advances and Applications of the Rough Sets Theory, pp. 3–18. Kluwer, Dordrecht (1992)
21. Grzymała-Busse, J.W., Hu, M.: A comparison of several approaches to missing attribute values in data mining. In: Ziarko, W.P., Yao, Y. (eds.) RSCTC 2000. LNCS (LNAI), vol. 2005, pp. 378–385. Springer, Heidelberg (2001)
22. Hempel, C.G.: Aspects of Scientific Explanation. The Free Press, New York (1965)
23. Yu, H., Wang, G., Lan, F.: Solving the Attribute Reduction Problem with Ant Colony Optimization. In: Peters, J.F., Skowron, A., Chan, C.-C., Grzymala-Busse, J.W., Ziarko, W.P. (eds.) Transactions on Rough Sets XIII. LNCS, vol. 6499, pp. 240–259. Springer, Heidelberg (2011)
24. Holland, J.H.: Adaptation in Natural and Artificial Systems. University of Michigan Press, Ann Arbor (1975)
25. Höhle, U.: Quotients with respect to similarity relations. Fuzzy Sets and Systems 27, 31–44 (1988)
26. Kloesgen, W., Zytkow, J. (eds.): Handbook of Data Mining and Knowledge Discovery. Oxford University Press, Oxford (2002)
27. Kryszkiewicz, M., Rybiński, H.: Data mining in incomplete information systems from rough set perspective. In: Polkowski, L., Tsumoto, S., Lin, T.Y. (eds.) Rough Set methods and Applications. New Developments in Knowledge Discovery in Information Systems, pp. 567–580. Physica Verlag, Heidelberg (2000)
28. Leibniz, G.W.: Discourse on Metaphysics. In: Loemker, L. (ed.) Philosophical Papers and Letters, 2nd edn., D. Reidel, Dordrecht (1969); The Identity of Indiscernibles. In: Stanford Encyclopedia of Philosophy, http://plato.stanford.edu/entries/identity-indiscernible/ (last entered 01.04.2011)
29. Ling, C.-H.: Representation of associative functions. Publ. Math. Debrecen 12, 189–212 (1965)
30. Lukasiewicz, J.: W sprawie odwracalności stosunku racji i następstwa (Concerning the invertibility of the relation between the premise and the conclusion (in Polish)). Przegląd Filozoficzny 16 (1913)
31. Makinson, D.: General patterns in non–monotonic reasoning. In: Gabbay, D.M., Hogger, C.J., Robinson, J.A. (eds.) Handbook of Logic in Artificial Intelligence and Logic Programming 3. Non–monotonic and Uncertain Reasoning, vol. 2, pp. 35–110. Oxford University Press, Oxford (1994)
32. de Mántaras, L., Valverde, L.: New results in fuzzy clustering based on the concept of indistinguishability relation. IEEE Trans. on Pattern Analysis and Machine Intelligence 10, 754–757 (1988)
33. Menger, K.: Statistical metrics. Proc. Nat. Acad. Sci. USA 28, 535–537 (1942)

34. Menger, K.: Ensembles flous et fonctions alétoires. C. R. Académie des Sciences 37, 2001–2003 (1951)
35. Menu, J., Pavelka, J.: A note on tensor products on the unit interval. Comm. Univ. Carolinae 17, 71–83 (1976)
36. Michalski, R.S., Mozetic, I., Hong, J., Lavrac, N.: The multi–purpose incremental learning system AQ15 and its testing to three medical domains. In: Proceedings of AAAI 1986, pp. 1041–1045. Morgan Kaufmann, San Mateo (1986)
37. Mitchell, T.: Machine Learning. McGraw-Hill, Englewood Cliffs (1997)
38. Moshkov, M., Skowron, A., Suraj, Z.: Irreducible descriptive sets of attributes for information systems. In: Peters, J.F., Skowron, A. (eds.) Transactions on Rough Sets XI. LNCS, vol. 5946, pp. 92–105. Springer, Heidelberg (2010)
39. Mostert, P.S., Shields, A.L.: On the structure of semigroups on a compact manifold with boundary. Ann. Math. 65, 117–143 (1957)
40. Muggleton, S.H.: Inductive Logic Programming. Academic Press, New York (1992)
41. Novotny, M., Pawlak, Z.: Partial dependency of attributes. Bull. Pol. Ac.: Math. 36, 453–458 (1988)
42. Orłowska, E.: Modal logics in the theory of information systems. Z. Math. Logik u. Grund. d. Math. 30, 213–222 (1984)
43. Orłowska, E.: Logic for reasoning about knowledge. Z. Math. Logik u. Grund. d. Math. 35, 559–572 (1989)
44. Pawlak, Z.: Rough sets. International Journal of Computer and Information Sciences 11, 341–356 (1982)
45. Pawlak, Z.: On rough dependency of attributes in information systems. Bull. Pol. Ac.: Tech. 33, 551–559 (1985)
46. Pawlak, Z.: Rough Sets: Theoretical Aspects of Reasoning about Data. Kluwer, Dordrecht (1991)
47. Pawlak, Z., Skowron, A.: A rough set approach for decision rules generation. In: Proceedings of IJCAI 1993 Workshop W12, Warsaw University of Technology, Institute of Computer Science (1993)
48. Polkowski, L.: On convergence of rough sets. In: Słowinski, R. (ed.) Intelligent Decision Support. Handbook of Applications and Advances of Rough Sets Theory, pp. 305–311. Kluwer, Dordrecht (1992)
49. Polkowski, L.: Mathematical morphology of rough sets. Bull. Pol. Ac.: Math. 41, 241–273 (1993)
50. Polkowski, L.: Metric spaces of topological rough sets from countable knowledge bases. Foundations of Computing and Decision Sciences 18, 293–306 (1993)
51. Polkowski, L.: Concerning mathematical morphology of almost rough sets. Bull. Pol. Ac.: Tech. 42, 141–152 (1994)
52. Polkowski, L.: Hit–or–Miss topology. In: Hazewinkel, M. (ed.) Encyclopedia of Mathematics. Supplement 1, p. 293. Kluwer, Dordrecht (1998)
53. Polkowski, L.: Approximation mathematical morphology. In: Pal, S.K., Skowron, A. (eds.) Rough Fuzzy Hybridization. A New Trend in Decision Making, pp. 151–162. Springer, Singapore (1999)
54. Polkowski, L.: Rough Sets. Mathematical Foundations. Physica Verlag, Heidelberg (2002)
55. Popper, K.: The Logic of Scientific Discovery. Hutchinson, London (1959)
56. Quinlan, J.R.: C4.5: Programs for Machine Learning. Morgan Kauffmann, San Mateo (1993)

57. Rasiowa, H., Skowron, A.: Rough concept logic. In: Skowron, A. (ed.) SCT 1984. LNCS, vol. 208, pp. 288–297. Springer, Heidelberg (1985)
58. Rissanen, J.: Universal coding, information, prediction and estimation. IEEE Transactions on Information Theory IT-30(4), 629–636 (1984)
59. RSES. A system for data analysis, http://logic.mimuw.edu.pl/~rses/ (last entered 01. 04. 2011)
60. Ruspini, E.H.: On the semantics of fuzzy logic. Int. J. Approx. Reasoning 5, 45–88 (1991)
61. Russell, S., Norvig, P.: Artificial Intelligence. A Modern Approach, 3rd edn. Prentice Hall, Upper Saddle River (2009)
62. Schweizer, B., Sklar, A.: Probabilistic Metric Spaces. North – Holland, Amsterdam (1983)
63. Scott, D.: Completeness and axiomatizability in many–valued logics. Proc. Symp. in Pure Math. 25, 431–435 (1974)
64. Seising, R.: Fuzziness before fuzzy sets: Two 20th century philosophical approaches to vagueness – Ludwik Fleck and Karl Menger. In: Proceedings IFSA 2005, Beijing, pp. 1499–1504 (2005)
65. Skowron, A.: Boolean reasoning for decision rules generation. In: Komorowski, J., Raś, Z.W. (eds.) ISMIS 1993. LNCS, vol. 689, pp. 295–305. Springer, Heidelberg (1993)
66. Skowron, A., Rauszer, C.: The discernibility matrices and functions in decision systems. In: Słowiński, R. (ed.) Intelligent Decision Support. Handbook of Applications and Advances of the Rough Sets Theory, pp. 311–362. Kluwer, Dordrecht (1992)
67. Vakarelov, D.: Modal logics for knowledge representation systems. In: Meyer, A.R., Taitslin, M.A. (eds.) Logic at Botik 1989. LNCS, vol. 363, pp. 257–277. Springer, Heidelberg (1989)
68. Valverde, L.: On the structure of F – indistinguishability operators. Fuzzy Sets and Systems 17, 313–328 (1985)
69. Vapnik, V.N.: Statistical Learning Theory. John Wiley and Sons, New York (1998)
70. Zadeh, L.A.: Fuzzy sets. Information and Control 8, 338–353 (1965)
71. Zadeh, L.A.: Similarity relations and fuzzy orderings. Information Sciences 3, 177–200 (1971)
72. Zadeh, L.A.: Outline of a new approach to the analysis of complex systems and decision processes. IEEE Trans. Syst. Man and Cybern. 1, 28–44 (1973)

Chapter 5
Mereology

Mereology emerged in the beginning of XXth century due to independent efforts of S. Leśniewski and A. N. Whitehead. In the scheme of Leśniewski, the predicate of being a part was taken as the primitive notion whereas in the development of Whitehead's ideas the primitive notion was adopted as the predicate of being connected. Mereology presents an alternative, holistic, approach to concepts which is especially suited to reasoning about extensional objects, e.g., spacial ones as witnessed, e.g., by the Tarski axiomatization of geometry of solids, or recent applications to geometric information systems or analysis of statements about spatial objects and relations in natural language.

5.1 Mereology: The Theory of Leśniewski

Mereology due to Leśniewski arose from attempts at reconciling antinomies of naïve set theory, see Leśniewski [42], Sobociński [58]. Leśniewski [41] was the first presentation of foundations of this theory, see also Leśniewski [44], [45]; cf., Lejewski [38] and Sobociński [59].

The primitive notion of mereology due to Leśniewski is a notion of a *part*. Given some objects, a relation of a part is a binary relation π which is required to be

M1 *Irreflexive*: *For each* $x \in U$ *it is not true that* $\pi(x,x)$

M2 *Transitive*: *For each triple* x,y,z *of objects in* U, *if* $\pi(x,y)$ *and* $\pi(y,z)$, *then* $\pi(x,z)$

Remark. In the original scheme of Leśniewski, the relation of parts is applied to *individual objects* as defined in Ontology of Leśniewski, see Leśniewski [43], Iwanuś [30], Słupecki [54]. Ontology is founded on the predicate ϵ to be read "is" (in Greek, *ei* "you are", cf., Plutarch [48]) which is required to satisfy the Ontology Axiom AO, formulated by Leśniewski as early as 1920, see Słupecki [54], Lejewski [39].

L. Polkowski: Approximate Reasoning by Parts, ISRL 20, pp. 191–228.
springerlink.com © Springer-Verlag Berlin Heidelberg 2011

AO $x\epsilon\ y \Leftrightarrow \exists z.(z\epsilon\ x) \wedge \forall\ z.(z\epsilon\ x \Rightarrow\ z\epsilon\ y) \wedge \forall\ z,w.(z\epsilon\ x \wedge w\epsilon\ x \Rightarrow z\epsilon\ w)$

This axiom determines the meaning of the copula ϵ in the way adopted by Leśniewski: in spite of the copula occurring on either side of the equivalence, its meaning can be revealed by requiring the equivalence to be true. The three terms occurring on the right–hand side of the equivalence mean, respectively,

A $\exists z.(z\epsilon\ x)$ means that the name x is not empty and that some object responds to that name

B $\forall\ z,w.(z\epsilon\ x \wedge w\epsilon\ x \Rightarrow z\epsilon\ w)$: letting $z = w$ if and only if $z\epsilon\ w \wedge w\epsilon\ z$, we infer that the term B means that $z = w$ for each pair z,w of objects responding to the name of x. Thus, x is a singular name

C The term $\forall\ z.(z\epsilon\ x \Rightarrow\ z\epsilon\ y)$ means that each object responding to the name of x responds as well to the name of y: in a sense x is contained in y

From axiom AO, the following properties of the copula ϵ follow, cf., Słupecki [54]

1. $x\epsilon\ y \wedge z\epsilon x \Rightarrow z\epsilon\ y$;

2. $x\epsilon y \wedge z\epsilon x \Rightarrow x\epsilon\ z$;

3. $x\epsilon\ y \Rightarrow\ x\epsilon\ x$.

The phrase 'x is an object' is rendered as $x\epsilon\ V$ equivalent to $\exists y.x\epsilon\ y$. Then, the equivalence

4. $x\epsilon\ V \Leftrightarrow x\epsilon\ x$

does express the fact that x is an object.

In Mereology, the predicate *part* is applied to objects, called by singular names.

The relation of *part* induces the relation of an *ingredient*, *ingr*, defined as

$$ingr(x,y) \Leftrightarrow \pi(x,y)\ \vee x = y \qquad (5.1)$$

Clearly,

Proposition 5.1. *The relation of ingredient is a partial order on objects, i.e.,*

1. $ingr(x, x)$;

2. $ingr(x, y) \wedge ingr(y, x) \Rightarrow (x = y)$;

3. $ingr(x, y) \wedge ingr(y, z) \Rightarrow ingr(x, z)$.

We formulate the third axiom with a help from the notion of an ingredient.

M3 (*Inference*) *For objects* x, y, *the property*

$I(x, y)$: *For each object* t, *if* $ingr(t, x)$, *then there exist objects* w, z *such that* $ingr(w, t), ingr(w, z), ingr(z, y)$

implies $ingr(x, y)$.

The predicate of *overlap*, Ov in symbols, is defined by means of

$$Ov(x, y) \Leftrightarrow \exists z.ingr(z, x) \wedge ingr(z, y) \qquad (5.2)$$

Using the overlap predicate, one can write $I(x, y)$ down in the form

$I_{Ov}(x, y)$: *For each* t *with* $ingr(t, x)$, *there exists* z *such that* $ingr(z, y)$ *and* $Ov(t, z)$

The notion of a mereological class follows; for a non–vacuous property Φ of objects, the *class of* Φ, denoted $Cls\Phi$ is defined by the conditions

C1 *If* $\Phi(x)$, *then* $ingr(x, Cls\Phi)$

C2 *If* $ingr(x, Cls\Phi)$, *then there exists* z *such that* $\Phi(z)$ *and* $I_{Ov}(x, z)$

In plain language, the class of Φ collects in an individual object all objects satisfying the property Φ.

The existence of classes is guaranteed by an axiom.

M4 *For each non–vacuous property* Φ *there exists a class* $Cls\Phi$

The uniqueness of the class follows.

Proposition 5.2. *For each non–vacuous property* Φ, *the class* $Cls\Phi$ *is unique.*

Proof. Assuming that for some Φ there exist two distinct classes Y_1, Y_2, consider $ingr(t, Y_1)$. Then, by C2, and (5.2), there exists z such that $Ov(t, z)$ and $ingr(z, Y_2)$. It follows by M3 that $ingr(Y_1, Y_2)$. By symmetry, $ingr(Y_2, Y_1)$ holds and Proposition 5.1(2) implies that $Y_1 = Y_2$ □

Proposition 5.3. *For the non–vacuous property Φ, if for each object z such that $\Phi(z)$ it holds that $ingr(z, x)$, then $ingr(Cls\Phi, x)$.*

Proof. It follows directly from M3   □

The notion of an overlap allows for a succinct characterization of a class.

Proposition 5.4. *For each non–vacuous property Φ and each object x, it happens that $ingr(x, Cls\Phi)$ if and only if for each ingredient w of x, there exists an object z such that $Ov(w, z)$ and $\Phi(z)$.*

Remark. Proposition 5.2 along with existence of a class is an axiom in the Leśniewski [42] scheme, from which M3 is derived. Similarly, it is an axiom in the Tarski [60], [62] scheme.

Example 5.1. 1. The strict inclusion $\subset$ on sets is a part relation. The corresponding ingredient relation is the inclusion $\subseteq$. The overlap relation is the non–empty intersection. For a non–vacuous family F of sets, the class $ClsF$ is the union $\bigcup F$;

2. For reals in the interval $[0, 1]$, the strict order $<$ is a part relation and the corresponding ingredient relation is the weak order $\leq$. Any two reals overlap; for a set $F \subseteq [0, 1]$, the class of F is $supF$.

The notion of an element, Leśniewski [42], par. 6, Def. IV, $el(x, y)$ in symbols, is defined as follows

$$el(x, y) \Leftrightarrow \exists \Phi.y = Cls\Phi \wedge \Phi(x) \tag{5.3}$$

In plain words, $el(x, y)$ means that y is a class of some property and x responds to that property. To establish some properties of the notion of an element, we begin with

Proposition 5.5. *For each object x, and the property $INGR(x) = \{y : ingr(y, x)\}$, the identity $x = ClsINGR(x)$ holds.*

Proof. By Proposition 5.1(1), $INGR(x)$ is non–vacuous and

$$ingr(x, ClsINGR(x))$$

That $ingr(ClsINGR(x), x)$ follows by M3   □

Proposition 5.6. *For each pair x, y of objects, $el(x, y)$ holds if and only if $ingr(x, y)$ holds. Hence, $el = ingr$.*

Proof. By (5.3) and Proposition 5.5, if $ingr(x, y)$ then $el(x, y)$. The converse follows from the definition of an element (5.3) and class requirement C1   □

Corollary 1 *If $\pi(x,y)$, then $el(x,y)$.*

Corollary 2 *$el(x,x)$ holds for each object x.*

By Corollary 2, every object considered in mereology is non–empty in the sense of the element relation.

Corollary 3 *The property of objects of not being its own element is vacuous.*

Corollary 3 is one of means of expressing the impossibility of the Russell paradox within the mereology, cf., Leśniewski [42], Thms. XXVI, XXVII, see also Sobociński [58].

Extensionality of the overlap relation can be inferred from Proposition 5.5.

Proposition 5.7. *For each pair x,y of objects, $x = y$ if and only if for each object z, the equivalence $Ov(z,x) \Leftrightarrow Ov(z,y)$ holds.*

Proof. Assume the equivalence $Ov(z,x) \Leftrightarrow Ov(z,y)$ to hold for each z. By Proposition 5.5, if $ingr(t,x)$ then $Ov(t,x)$ and $Ov(t,y)$ hence by axiom M3 $ingr(t,y)$ and with $t = x$ we get $ingr(x,y)$. By symmetry, $ingr(y,x)$ and by Proposition 5.1(2), $x = y$ □

Concerning the class properties, we mention

Proposition 5.8. *For each pair of non–vacuous properties Φ, Ψ, from $\Phi \Rightarrow \Psi$ it follows that $ingr(Cls\Phi, Cls\Psi)$.*

Proof. The proposition is a direct consequence of the class definition C1, C2 and of M3   □

A corollary follows.

Proposition 5.9. *For each pair of non–vacuous properties Φ, Ψ, from $\Phi \Leftrightarrow \Psi$ it follows that $Cls\Phi = Cls\Psi$.*

The notion of a subset, $sub(x,y)$ is introduced in mereology, cf., Leśniewski [42], par. 10, Def. V, via the requirement

$$sub(x,y) \Leftrightarrow \forall z.[ingr(z,x) \Rightarrow ingr(z,y)] \tag{5.4}$$

It follows immediately that

Proposition 5.10. *For each pair x, y of objects, $sub(x,y)$ holds if and only if $el(x,y)$ holds if and only if $ingr(x,y)$ holds.*

For the property $Ind(x) \Leftrightarrow x\epsilon\, x$, one calls the class $ClsInd$, the *universe*, in symbols V, Leśniewski [42], par. 12, Def. VII.

It follows that

Proposition 5.11. *The following are properties of the universe V*

1. *The universe is unique;*

2. *$ingr(x, V)$ holds for each object x;*

3. *For each non–vacuous property Φ, it holds true that $ingr(Cls\Phi, V)$.*

The notion of an exterior object x to the object y, $extr(x, y)$ in symbols, Leśniewski [42], par. 13, Def. VIII, is the following

$$extr(x, y) \Leftrightarrow \text{ it is not true that } Ov(x, y) \qquad (5.5)$$

In plain words, x is exterior to y when no object is an ingredient both to x and y.

Clearly,

Proposition 5.12. *The operator of being exterior has properties*

1. *No object is exterior to itself;*

2. *$extr(x, y)$ implies $extr(y, x)$;*

3. *If for a non–vacuous property Φ, an object x is exterior to every object z such that $\Phi(z)$ holds, then $extr(x, Cls\Phi)$.*

The notion of a complement to an object, with respect to another object, is rendered as a ternary predicate $comp(x, y, z)$, Leśniewski [42], par. 14, Def. IX, to be read:'x is the complement to y with respect to z', and it is defined by means of the following requirements

1. $x = ClsEXTR(y, z)$;

2. $ingr(y, z)$,

where $EXTR(y, z)$ is the property which holds for an object t if and only if $ingr(t, z)$ and $extr(t, y)$ hold.

This definition implies that the notion of a complement is valid only when there exists an ingredient of z exterior to y.

Proposition 5.13. *For each triple x, y, z such that $comp(x, y, z)$,*

1. *$extr(x, y)$;*

2. *$\pi(x, z)$;*

3. *$comp(x, y, z)$ implies $comp(y, x, z)$.*

Proof. Property 1 follows from Proposition 5.12, 3. Property 2 follows from Proposition 5.3 which implies $ingr(x, z)$ and by Property 1, x is exterior to y hence distinct from z. To prove Property 3, we need to verify that $y = ClsEXTR(x, z)$. Consider t such that $ingr(t, z)$ and $extr(t, x)$; was $extr(t, y)$, it would be that $ingr(t, x)$, a contradiction. Hence, $Ov(t, y)$ and it follows by arbitrariness of t that $y = ClsEXTR(x, z)$ □

We let for an object x, $-x = ClsEXTR(x, V)$. It follows from Proposition 5.13 that

Proposition 5.14. *The operation* $-x$ *has properties*

1. $-(-x) = x$ for each object x;

2. $-V$ does not exist.

The operator $-x$ can be a candidate for the boolean complement in a structure of a Boolean algebra within mereology, constructed in Tarski [61], and anticipated in Tarski [60]; cf., in this respect Clay [13].

This algebra will be obviously rid of the null element, as the empty object is not allowed in mereology of Leśniewski, and the meet of two objects will be possible only when these objects overlap. Under this caveat, the construction of Boolean operators of join and meet proceeds on the following lines.

We let $Booladd(x, y)$ to denote a property of objects defined by

$$Booladd(x, y)(t) \Leftrightarrow ingr(t, x) \vee ingr(t, y) \qquad (5.6)$$

We denote the class $ClsBooladd(x, y)$ with the symbol $x + y$. As properties $ingr(t, x) \vee ingr(t, y)$ and $ingr(t, y) \vee ingr(t, x)$ are equivalent, we have by Proposition 5.9 that

Proposition 5.15. $x + y = y + x$.

Similarly, properties $(ingr(t, x) \vee ingr(t, y)) \vee ingr(t, z)$ and $ingr(t, x) \vee (ingr(t, y)) \vee ingr(t, z))$ are equivalent, hence

Proposition 5.16. $(x + y) + z = x + (y + z)$.

Consider now $x + (-x)$: the defining property is $\Phi(t) : ingr(t, x) \vee ingr(t, -x)$. For an arbitrary object z, either exists w such that $Ov(z, w) \wedge ingr(w, x)$ or exists v such that $Ov(z, v) \wedge ingr(v, -x)$, otherwise an easily perceived contradiction takes place, hence, $ingr(z, x + (-x))$ and a fortiori $x + (-x) = V$. We arrive at

Proposition 5.17. $x + (-x) = V$.

The object $x + y$ has the property

Proposition 5.18. *For each object* z, *the equivalence holds:* $Ov(z, x + y) \Leftrightarrow Ov(z, x) \vee Ov(z, y)$.

Proof. $Ov(z, x + y)$ means the existence of w such that $ingr(w, z)$ and $ingr(w, x + y)$ hence for some t such that $Ov(w, t)$ one has $ingr(t, x) \vee ingr(t, y)$ and thus for some u with $ingr(u, t), ingr(u, w)$ it follows that $ingr(u, z)$ and $ingr(u, x) \vee ingr(u, y)$ so finally $Ov(z, x) \vee Ov(z, y)$. The converse implication follows easily   □

For each pair x, y of overlapping objects, we introduce a property $Boolprod(x, y)$ defined as

$$Boolprod(x, y)(t) \Leftrightarrow ingr(t, x) \wedge ingr(t, y) \tag{5.7}$$

We denote the class $ClsBoolprod(x, y)$ with $x \cdot y$.

As with $+$ operation, one can prove properties of $\cdot$.

Proposition 5.19. *Under the assumption that* $Ov(x, y), Ov(y, z), Ov(x \cdot y, z)$ *the following properties hold*

1. $x \cdot y = y \cdot x$;

2. $(x \cdot y) \cdot z = x \cdot (y \cdot z)$;

3. $x \cdot (-x)$ *is not defined.*

The property counterparting 5.18 reads as follows

Proposition 5.20. *Assume the existence of non–empty complement. Then, for each object* z *it holds that* $ingr(z, x \cdot y) \Leftrightarrow ingr(z, x) \wedge ingr(z, y)$.

Proof. Assume $ingr(z, x \cdot y)$ so for each v with $ingr(v, z)$, there is w such that $Ov(v, w)$ and $ingr(w, x), ingr(w, y)$; hence there is u such that $ingr(u, v)$, $ingr(u, w)$, $ingr(u, x)$, $ingr(u, y)$ and $ingr(u, x \cdot y)$. However, was, e.g., $\neg ingr(z, x)$ we would have t such that $ingr(t, x)$ and $extr(t, x)$, contradicting the case $v = t$ considered in the preceding sentence. Thus, $ingr(z, x)$, $ingr(z, y)$ must hold and the proposition is proved as the converse implication follows directly by the class definition   □

Boolean relations hold between $x + y$ and $x \cdot y$, viz.,

Proposition 5.21. *The following relations between* $+$ *and* $\cdot$ *are valid*

1. $-(x + y) = (-x) \cdot (-y)$;

2. $-(x \cdot y) = (-x) + (-y)$.

Proof. Defining properties, respectively, for $-(x + y)$ and $(-x) \cdot (-y)$ are $\neg[(ingr(t, x) \vee ingr(t, y)]$ and $(\neg ingr(t, x)) \wedge (\neg ingr(t, y))$, respectively, and are equivalent, hence, classes of them are identical. Similarly, the second identity follows   □

For each object x, we construct a *set* $\sigma(x)$ by letting

$$\sigma(x) = \{y : ingr(y, x)\} \tag{5.8}$$

The meaning of $\sigma(x)$ is that this set consists of objects which are ingredients of x and it does correspond to class of ingredients of x.

Proposition 5.22. *Assume the existence of non–vacuous complement for all pairs a, b with $\pi(a, b)$. Objects satisfying $Booladd(x, y)$ make the set $\sigma(x) \cup \sigma(y)$. Then, $x + y$ corresponds to $\sigma(x) \cup \sigma(y)$.*

Proof. Assume $ingr(t, x + y)$; then, for each w, if $ingr(w, t)$, then $Ov(w, z)$, $ingr(z, x) \lor ingr(z, y)$ for some z. Assume that neither $ingr(t, x)$ nor $ingr(t, y)$. By complement assumption, there exist u, v such that $ingr(u, t)$, $ingr(v, t)$, $extr(u, x)$, $extr(v, y)$. The exteriority of u to x and v to y contradicts the preceding sentence $\qquad\square$

Objects satisfying $Boolprod(x, y)$ constitute the set $\sigma(x) \cap \sigma(y)$ by Proposition 5.20. The class $x \cdot y$ corresponds to the set $\sigma(x) \cap \sigma(y)$.

Finally, we consider the set $V \setminus \sigma(x)$ corresponding to the class $Cls\neg Ov(x)$ of the property $\neg Ov(x)$ which we denote $-x$. Then

Proposition 5.23. *The correspondence $\iota : x \rightarrow \sigma(x)$ is an isomorphism satisfying $ingr(u, v) \Leftrightarrow \sigma(u) \subseteq \sigma(v)$. At this correspondence, $x + y$ maps onto $\sigma(x) \cup \sigma(y)$, $x \cdot y$ maps onto $\sigma(x) \cap \sigma(y)$, and $-x$ maps onto $U \setminus \sigma(x)$. The Cls operator acts as the least upper bound operator, mapping a class of a property Φ onto the union $\bigcup \{\sigma(x) : \Phi(x)\}$.*

5.2 A Modern Structural Analysis of Mereology

An ex post analysis of the structure of Leśniewski's mereological theory has resulted in some stratification of the theory into more or less stronger sub–theories, e.g., in the way proposed in Casati–Varzi [10].

The basic notion is that of ingredient, subject to postulates

P1 $ingr(x, x)$

P2 $ingr(x, y) \land ingr(y, x) \Rightarrow (x = y)$

P3 $ingr(x, y) \land ingr(y, z) \Rightarrow ingr(x, z)$

The theory supporting P1–P3 is called *Ground Mereology* and denoted M. It does encompass a theory of ingredients and parts as partial and strict orders on a universe of objects. Clearly, parts are defined by means of

$$\pi(x, y) \Leftrightarrow ingr(x, y) \land \neg ingr(y, x) \tag{5.9}$$

The relation Overlap is defined as above and the dual predicate of *underlap* is defined by means of

$$U(x,y) \Leftrightarrow \exists z.ingr(x,z) \wedge ingr(y,z) \tag{5.10}$$

Combinations of these predicates for expressing various possible positions of each object relative to the other are possible, like *proper over–crossing* PO

$$PO(x,y) \Leftrightarrow Ov(x,y) \wedge \neg ingr(x,y) \wedge \neg ingr(y,x) \tag{5.11}$$

Further extensions are proposed by means of postulates which would secure the existence of complementary parts within objects , cf., the notion of a complement above; complementation is expressed by means of two postulates, see Casati–Varzi, op. cit., and Simons [52], a throughout discussion in Simons [53], viz., the *Postulate of Weak Supplementation*

P4 $\pi(x,y) \Rightarrow \exists z.ingr(z,y) \wedge \neg Ov(x,z)$

and the *Postulate of Strong Supplementation*

P5 $\neg ingr(y,x) \Rightarrow \exists z.ingr(z,y) \wedge \neg Ov(x,z)$

As from $\pi(x,y)$ it follows that $\neg ingr(y,x)$, P4 follows from P5. Since $\neg ingr(y,x)$ does not imply $\pi(x,y)$, the converse does not hold. The theory M+P5 denoted EM and called the *extensional mereology* is stronger than the theory M+P4 denoted MM and called the *minimal mereology*. In the latter, extensionality with respect to parts holds: If objects x,y have parts, then $\forall z.\pi(z,x) \Leftrightarrow \pi(z,y) \Rightarrow (x = y)$, see Casati–Varzi [10], p. 40.

Further postulates bring closer to fusion properties – the existence of sums (classes in the sense of Leśniewski) and intersections (called also products); a finitary version requires that objects in underlap relation should be encompassed in the smallest object which contains them

P6 $U(x,y) \Rightarrow \exists z.\forall w.[Ov(w,z) \Leftrightarrow Ov(w,x) \vee Ov(w,y)]$

and, dually, overlapping objects contain the greatest object contained in either of them

P7 $Ov(x,y) \Rightarrow \exists z.\forall w.[ingr(w,z) \Leftrightarrow ingr(w,x) \wedge ingr(w,y)]$

The theory M + P6, P7 is called the *closure mereology* denoted CM, MM+ P6,P7 is called the *closure minimal mereology* denoted CMM, and EM +P6, P7 is the *closure extensional mereology* denoted CEM. As M $\subseteq$ MM $\subseteq$ EM, it follows that M $\subseteq$ CM, MM $\subseteq$ CMM, EM $\subseteq$ CEM, CMM $\subseteq$ CEM. Actually, as pointed to in Casati–Varzi [10], p. 44, P4 $\Rightarrow$(P7 $\Rightarrow$ P5), hence CEM=CMM. Finally, GEM, the *general extensional mereology* is the mereology exposed above in sect. 5.1., i.e., the mereology as envisioned by Leśniewski, with full power of fusion, secured by the general notion of a class.

5.3 Mereotopology

Topological structures which arise in mereology due to Leśniewski, can be induced from overlap relations. As the first approximation to topology, let us define for each object x, its *closure* $c(x)$ by means of

$$c(x) = ClsOv(x) \tag{5.12}$$

where the property $Ov(x)$ is defined by $Ov(x)(y) \Leftrightarrow Ov(x, y)$, i.e., we build the closure $c(x)$ as the class of objects which overlap with x.

We have

Proposition 5.24. *The closure operator $c(.)$ has the following properties*

Cl1 $ingr(x, c(x))$

Cl2 If $ingr(x, y)$, *then* $ingr(c(x), c(y))$

Cl3 $ingr(c(x \cdot y), c(x) \cdot c(y))$

Cl4 $c(x + y) = c(x) + c(y)$

Proof. Cl1 and Cl2 follow from definition of the overlap relation and the class definition. For Cl3, if $ingr(t, c(x \cdot y))$, then there is z such that $Ov(t, z)$ and $Ov(z, x \cdot y)$ thus for some w one has $Ov(z, w)$ and $ingr(w, x)$, $ingr(w, y)$ which imply that $ingr(t, c(x))$, $ingr(t, c(y))$ and finally $ingr(t, c(x) \cdot c(y))$. By M3, $ingr(c(x \cdot y), c(x) \cdot c(y))$.

For Cl4, it suffices to observe that $Ov(z, x + y) \Leftrightarrow Ov(z, x) \vee Ov(z, y)$ □

Another possibility for a topology is in iteration of the operator c, viz, we let

$$Ov^{n+1}(x, y) \Leftrightarrow \exists z.Ov(x, z) \wedge Ov^n(z, y); Ov^1(x, y) \Leftrightarrow Ov(x, y) \tag{5.13}$$

and we define

$$OVLP(x)(y) \Leftrightarrow \exists n.Ov^n(x, y) \tag{5.14}$$

The closure $Cl(x)$ is defined as the class of the property $OVLP(x)$, i.e.,

$$Cl(x) = ClsOVLP(x) \tag{5.15}$$

The operator $Cl(x)$ has the following properties

Proposition 5.25. *Properties of $Cl(x)$ are*

$CL1\ Cl(Cl(x)) = Cl(x)$

$CL2\ ingr(x, Cl(x))$

$CL3\ ingr(x, y)\ implies\ ingr(Cl(x), Cl(y))$

$CL4\ Cl(x + y) = Cl(x) + Cl(y)$

Proof. CL2, CL3 follow straightforwardly from definitions. For CL1, observe that $ingr(t, Cl(x))$ if and only if $OVLP(x)(t)$. Thus, $ingr(t, Cl(Cl(x)))$ if and only if $OVLP(Cl(x))(t)$ if and only if $OVLP(x)(t)$ if and only if $ingr(t, Cl(x))$.

For CL4, assume first that $ingr(t, Cl(x + y))$ hence $OVLP(t, x + y)$ and thus $OVLP(t, x) \vee OVLP(t, y)$, i.e., $ingr(t, Cl(x)) \vee ingr(t, Cl(y))$ and thus $ingr(t, Cl(x) + Cl(y))$. Assume now that $ingr(t, Cl(x) + Cl(y))$, i.e., $OVLP(t, x) \vee OVLP(t, y)$, so there exists m such that $Ov^m(t, x) \vee Ov^m(t, y)$, i.e., $Ov^m(t, x+y)$, hence, $ingr(t, Cl(x+y))$ □

It follows that the operator Cl is a genuine closure operator; its properties are weak, as it in fact delineates components of objects with respect to the overlap property: it is not even a T_0–closure operator (a Kolmogoroff operator) which would mean that distinct objects should have distinct closures, as distinct objects in the same component have the same closure.

A definition of a *boundary* can be attempted on the lines of topological boundary concept. For an object x, let a property $\Upsilon(x)$ be defined as follows

$$\Upsilon(x)(t) \Leftrightarrow ingr(t, x) \wedge \forall z.[Ov(z, x) \wedge Ov(z, -x) \Rightarrow Ov(z, t)] \qquad (5.16)$$

We may define the *boundary of x*, $Fr(x)$, by letting

$$Fr(x) = Cls\Upsilon(x) \qquad (5.17)$$

Proposition 5.26. *Properties of $Fr(x)$ following directly from definitions above are*

1. $ingr(Fr(x), x);$

2. $\forall z.Ov(z, x) \wedge Ov(z, -x) \Rightarrow Ov(z, Fr(x)).$

The above notion of a boundary has a topological flavor; however, the notion of a boundary has a much wider scope. It can also support the idea of a *separator* between two objects within a third, which does encompass either, like a river flowing through a town separates parts on opposite banks. To implement this idea, for objects x, y, such that $extr(x, y)$, we define the property

$$\Omega(x,y)(t) \Leftrightarrow extr(t,x) \wedge extr(t,y) \tag{5.18}$$

and we let

$$Bd(x,y) = Cls\Omega(x,y) \tag{5.19}$$

Then

Proposition 5.27. *The boundary operation Bd has properties*

1. $Bd(x,y) = Bd(y,x)$.
2. $Bd(x+y,z) = Bd(x,z) \cdot Bd(y,z)$.

Proof. Property 1 is obvious. Property 2 follows from the equivalence $extr(x+y,z) \Leftrightarrow extr(x,z) \wedge extr(y,z)$ □

A relative variant can be defined; assuming that $ingr(x,z)$, $ingr(y,z)$ and $extr(x,y)$, a *boundary relative to* z *between* x *and* y, $Bd_z(x,y)$, is the class of objects t such that $ingr(t,z), extr(t,x), extr(t,y)$ provided this property is non–vacuous.

5.4 Timed Mereology

Timed component of mereology was introduced in Tarski [62], and presented in a systematic way in Woodger [68], [69]. The time component is introduced into the framework of mereology with a set of notions and postulates (axioms) concerning aspects of time like *momentariness*, coincidence in time, time slices.

Objects are considered as spatial only and their relevance to time is expressed as momentary or as spatial and extended in time and then the predicate of part is understood as a global descriptor covering spatio–temporal extent of objects whereas the temporal extension is described by the predicate Temp (T) with the intended meaning that $T(u,v)$ means that the object u *precedes* in time the object v (in terminology of Leśniewski, Tarski and Woodger: u *wholly precedes* v) meaning that, e.g., when u and v have some temporal extent, then u ends before or at the precise moment when v begins.

We follow in our exposition the presentation in Woodger [69], to whom all results belong. Proofs are mostly supplied by this author to make the exposition more accessible.

The property (predicate) *Mom* meaning *momentary being* is introduced to denote objects having only spatial aspect. This predicate is introduced by means of the following postulate

$$MOM \ Mom(x) \Leftrightarrow T(x,x) \tag{5.20}$$

Thus, x begins and ends at the same time, so its time aspect is like a spike in time; it renders the phrase 'to exist in a moment of time'.

The predicate T is required to satisfy postulates

TM1 $T(x,y) \wedge T(y,z) \Rightarrow T(x,z)$

TM2 $Mom(x) \wedge Mom(y) \Rightarrow T(x,y) \vee T(y,x)$

TM3 $T(x,y) \Leftrightarrow \forall u,v.ingr(u,x) \wedge ingr(v,y) \Rightarrow T(u,v)$

Postulate TM1 states that T is transitive, Postulate TM2 does state that of two momentary things, one precedes the other and Postulate TM3 relates T to the class operator, i.e., x precedes y if and only if each ingredient of x precedes each ingredient of y. Postulate TM3 provides a link between the part based mereology and the timed mereology, bonding spatial and temporal properties of objects.

As a consequence to postulates TM1–TM3, one obtains

Proposition 5.28. *The following properties result from postulates TM1–TM3*

1. $Mom(x) \Leftrightarrow \exists y.T(x,y) \wedge T(y,x);$

2. $ingr(u,x) \wedge T(x,y) \Rightarrow T(u,y);$

3. $T(Cls\Phi, x) \Leftrightarrow \forall y.\Phi(y) \Rightarrow T(y,x);$

4. $T(x, Cls\Phi) \Leftrightarrow \forall y.\Phi(y) \Rightarrow T(x,y);$

5. $Mom(Cls\Phi) \Leftrightarrow \forall x,y.\Phi(x) \wedge \Phi(y) \Rightarrow T(x,y).$

Proof. Property 1 follows by Postulate (5.20) and Postulate TM2: $T(x,y) \wedge T(y,x)$ implies $T(x,x)$ by Postulate TM2, hence, $Mom(x)$ by (5.20). Property 2 follows by Postulate TM3 as $ingr(u,x), ingr(y,y), T(x,y)$ imply $T(u,y)$. Property 3 follows from Postulate TM3: from $ingr(x,x), ingr(u,x)$ it follows that for some z, it happens that $Ov(u,z) \wedge \Phi(z)$, hence $ingr(z, Cls\Phi)$, hence, by Postulate TM3, $T(z,x)$ and by the class definition, $T(u,x)$.

Property 4 is dual to Property 3. Property 5 follows by (5.20) and either Property 3, or Property 4 applied to the pair $Cls\Phi, Cls\Phi$ □

Remark. Concisely, one can write down Property 2 in the form: $ingr \circ T = T$, hence, $T = T \circ ingr^{-1}$, where $ingr$ denotes the ingredient relation.

The notion of a *coincidence in time*, CT in symbols, needs a defining postulate

$$CT(x,y) \Leftrightarrow T(x,y) \wedge T(y,x) \qquad (5.21)$$

Hence, the relation of coincidence in time is symmetric and transitive.

Properties of coincidence in time can be summed up as follows

Proposition 5.29. *Operation CT has properties*

1. $Mom(x) \Leftrightarrow \forall y.ingr(y,x) \Rightarrow CT(y,x);$

2. $Mom(x) \wedge Mom(y) \wedge Ov(x,y) \Rightarrow CT(x,y);$

3. $Mom(Cls\Phi) \Leftrightarrow \exists x.[\Phi(x) \wedge \forall y.\Phi(y) \Rightarrow CT(y,x)].$

Proof. Property 1 follows by (5.20) implying $T(x,x)$ from which by Postulate TM3 it follows that if $ingr(y,x)$ then $T(y,x)$ and by Proposition 5.28(1) by which $T(x,y)$, hence, $CT(x,y)$. This proves the implication from left to right. The converse implication follows by definition (5.21). Property 2 follows by Property 1 and transitivity of CT. Property 3 follows from Property 1 with $Cls\Phi$ in place of x and transitivity of CT □

The notion of being *wholly before in time* is rendered as a predicate Z defined as follows

$$Z(x,y) \Leftrightarrow Mom(x) \wedge Mom(y) \wedge \neg T(y,x) \qquad (5.22)$$

Hence, $Z(x,y)$ means that the momentary object x is wholly before the momentary object y in time.

By Postulate TM2, $Z(x,y)$ implies $T(x,y)$. Hence,

Proposition 5.30. *For each pair of momentary objects x,y the equivalence*

$$T(x,y) \Leftrightarrow CT(x,y) \vee Z(x,y)$$

holds. Meaning that in case $T(x,y)$ holds, either x,y coincide in time or x is wholly before y in time.

Proof. By definitions of CT and Z, if $T(x,y)$ then either $T(y,x)$ and hence CT or $\neg T(y,x)$, hence, $Z(x,y)$ □

Proposition 5.31. *If x,y are momentary objects and $Z(x,y)$, then neither $ingr(x,y)$ nor $ingr(y,x)$.*

Proof. By Proposition 5.29, if either $ingr(x,y)$ or $ingr(y,x)$, then $CT(x,y)$ which excludes $Z(x,y)$ by virtue of respective definitions   □

Proposition 5.31 sets another important link between spatial and temporal aspects of objects. Actually in this proposition the condition that neither of objects is an ingredient of the other can be replaced by demanding that the objects do not overlap.

Finally, the notion of a *time–slice* is introduced, as a predicate $Slc(x,y)$ by means of

$$Slc(x,y) \Leftrightarrow Mom(x) \wedge ingr(x,y) \wedge \forall z.[ingr(z,y) \wedge C(z,x) \Rightarrow ingr(z,x)]$$
$$(5.23)$$

and thus a time–slice of an object y is an ingredient of y which is spatially so arranged that any ingredient of y coinciding with it in time is also its ingredient.Time slices are unique up to coincidence in time.

Proposition 5.32. *If x, y are time–slices of z, then x, y coincide in time if and only if $x = y$.*

Proof. By definition (5.23), one has $ingr(x, y), ingr(y, x)$, hence, $x = y$ □

Each object is the class of its time slices. To prove this statement, first we note that

Proposition 5.33. *Given a momentary ingredient x of an object y, the class $CT(y, x)$ of all ingredients of y which coincide in time with x is a time–slice of y coincident in time with x.*

Proof. Clearly, $CT(y, x)$ is an ingredient of y. By Postulate TM3, the class $CT(y, x)$ coincides in time with x. By transitivity of T, any ingredient of y which coincides in time with $CT(y, x)$ also coincides in time with x, hence, is an ingredient of $CT(y, x)$. It follows that $C(y, x)$ is a time–slice of y □

Proposition 5.33 suggests the way of embedding any ingredient of an object into a time–slice of that object. As any object is the class of its ingredients, it follows that

Proposition 5.34. *Each object x is the class of its ingredients which are its time–slices.*

Among time–slices of an object, one can distinguish the first (beginning) and the last (ending) time–slices.

A time–slice x of an object y is a *first time–slice* of y if and only if each time–slice z of y such that $T(z, x)$ is identical with x; similarly, a time–slice w of y is an *ending time–slice* of y if each time–slice z of y such that $T(w, z)$ is identical with w.

Proposition 5.35. *The important properties of first and last time–slices are*

1. *Each first time–slice x of an object y satisfies $T(x, y)$;*

2. *Each last time–slice w of an object y satisfies $T(y, w)$.*

Proof. As Statement 2 is dual to Statement 1, it suffices to verify the latter. Consider thus the first time–slice x of y. For any other time–slice $z \neq x$ of y, we have by Proposition 5.32 and by Postulate TM2 that $T(x, z)$, hence, by Proposition 5.34 and Postulate TM3 it follows that $T(x, y)$ □

One can define a *temporal interior of an object* y as the separator $Fr(x, w, y)$ between a first time–slice x and the last time–slice w of y relative to y.

5.5 Spatio–temporal Reasoning: Cells

A theory of cells, motivated by phenomenology of cell biology, but in fact an abstract timed mereology of time–evolving spatial structures, was developed in detail in Woodger [69], Ch. 2, see also Tarski [62].

By a *cell* an object y is understood for which an ingredient x exists with $T(x, y)$ and an ingredient w exists such that $T(y, w)$. Hence, a cell is time–bounded by some time–preceding object and some time–following object.

We denote the fact that y is a cell with the symbol $cell(y)$. Formally

TM4 $cell(y) \Leftrightarrow \exists x, w.[ingr(x, y) \wedge ingr(w, y) \wedge T(x, y) \wedge T(y, w)]$

The next postulate asserts that no cell is a momentary object,

TM5 $cell(y) \Rightarrow \neg Mom(y)$

We recall that the part relation is denoted with the symbol π. A first time–slice x of a cell y, respectively, a last time–slice w of y, will be denoted by $Beg(y)$, respectively $End(y)$. The last postulate about cells follows

TM6 $cell(y_1) \wedge cell(y_2) \wedge y_1 \neq y_2 \wedge Ov(y_1, y_2) \Rightarrow \pi(Beg(y_1), End(y_2)) \vee \pi(End(y_1), Beg(y_2))$

Properties of cells are collected in

Proposition 5.36. *The following are properties of cells*

1. *If $cell(y)$, then the class C of ingredients x of y for which $T(x, y)$ holds is a momentary ingredient of y such that $T(C, y)$ holds;*

2. *The class C of Statement 1 is a first time–slice of y;*

3. *If $cell(y)$, then the class D of ingredients w of y for which $T(y, w)$ holds is a momentary ingredient of y such that $T(y, D)$ holds;*

4. *The class D of Statement 2 is a last time–slice of y.*

Proof. As Statements 3 and 4 are dual to, respectively, Statements 1 and 2, it suffices to verify the latter. Concerning Statement 1, by Remark 1, for ingredients z, y of x such that $T(z, x), T(y, x)$ it follows (in view of $T \circ ingr^{-1} = T$) that $T(z, y)$ which by Postulate TM3 implies that $T(y, z)$ hence by Proposition 5.28, $T(x, x)$, hence, $Mom(x)$ follows by Postulate TM1. Clearly, $T(C, x)$ by Proposition 5.28, again. Concerning Statement 2, clearly, C is a first time–slice by its definition $\qquad \square$

Proposition 5.37. *Properties of first and last time–slices are*

1. For any cell x, the first time–slice $Beg(x)$ is not the last time–slice $End(x)$;

2. For any cell x, $T(Beg(x), End(x))$;

3. For any cell x, $Z(Beg(x), End(x))$.

Proof. By transitivity of T, in case 1, was it otherwise, we would have $T(x, x)$ contradicting Postulate TM5 in view of Postulate TM1. For Statement 2, it follows by transitivity of T from Proposition 5.36 1, 3. Statement 3 follows by definition (5.22) of the predicate Z and by Postulate TM4 about cells, and transitivity of T □

Postulate TM6 implies that

Proposition 5.38. *For two distinct cells x, y, if*

$$C(Beg(x), Beg(y)) \ or \ C(End(x), End(y))$$

then neither $ingr(Beg(x), End(y))$ nor $ingr(End(x), Beg(y))$.

Proof. Was, e.g., $C(Beg(x), Beg(y))$ and $ingr(Beg(x), End(y))$, one would have $Z(Beg(y), Beg(x))$ contradicting $C(Beg(x), Beg(y))$. Similarly for other combinations of terms in the premise and in the conclusion   □

The following is true in consequence

Proposition 5.39. *For cells x, y with the property that $C(Beg(x), Beg(y))$ or $C(End(x), End(y))$, if $Ov(x, y)$, then $x = y$.*

Proof. Premises imply by Proposition 5.38 that neither $ingr(Beg(x), End(y))$ nor $ingr(End(x), Beg(y))$ contradicting Postulate TM6 in case $x \neq y$ □

A corollary specializes Proposition 5.39
    Corollary 1. *For cells x, y, the following are equivalent*

1. $Beg(x) = Beg(y)$;

2. $End(x) = End(y)$;

3. $x = y$.

Proof. Assumptions of Proposition 5.39 are satisfied   □

Operations on cells include *division* and *fusion*.
    A cell x *arises by division from a cell y* if and only if $\pi(Beg(x), End(y))$. We write the fact down as $Div(x, y)$.
    Then, by already proven statements,

Proposition 5.40. *If $Div(x, y)$, then $Z(Beg(x), Beg(y))$, $T(y, x)$, and $x \neq y$.*

A dual operation of *fusion* is defined as follows.

A cell x *arises by fusion of a cell y with some other cell* if and only if $\pi(End(y), Beg(x))$. We write down the fact as $Fus(x, y)$.

A dual proposition to 5.40 is stated as follows

Proposition 5.41. *If $Fus(x, y)$, then $Z(Beg(y), Beg(x))$, $T(y, x)$, and $x \neq y$.*

It follows from definitions of division and fusion that

Proposition 5.42. *There exists no cell x which can divide and fuse with the same other cell y. There can be no cell which could arise by division and fusion from the same other cell.*

5.6 Mereology Based on Connection

A dual approach to parts, was initiated in Whitehead [64], [65], [66] in a form of propositions of axioms for the notion of 'x extends over y', dual to that of a part, and of vague proposals of desired properties of the notion. Th. de Laguna [37] published a variant of the Whitehead scheme, which led Whitehead [67] to another version of his approach, based on the notion of 'x is extensionally connected to y'. Connection Calculus based on the notion of a 'connection' was proposed in Clarke [12], and this version is presented here, see also Calculus of Individuals of Leonard and Goodman [40].

The relation/predicate of connection C is subject to basic requirements, see Clarke [12], A0.1, A0.2

CN1 $C(x, x)$ *for each object x*

CN2 *If $C(x, y)$, then $C(y, x)$ for each pair x, y of objects*

It follows that connection is reflexive and symmetric. This theory is sometimes called *Ground Topology* T, cf., Casati–Varzi [10]. Adding the extensionality requirement

CN3 *If $\forall z.[C(z, x) \Leftrightarrow C(z, y)]$, then $x = y$*

produces the *Extensional Ground Topology* ET., loc. cit.

Let us observe that within the mereology M, the predicate C can be realized by taking $C = Ov$; clearly, CN1–CN3 are all satisfied with Ov. We call this model of connection mereology, the *Overlap model*, denoted OVM.

In the universe endowed with C, satisfying CN1, CN2, one can define the notion of an ingredient $ingr_C$ by letting

$$IC \quad ingr_C(x, y) \Leftrightarrow \forall z.[C(z, x) \Rightarrow C(z, y)] \tag{5.24}$$

Then,

Proposition 5.43. *(Clarke [12]) The following properties of $ingr_C$ hold*

1. $ingr_C(x, x)$;

2. $ingr_C(x, y) \wedge ingr_C(y, z) \Rightarrow ingr_C(x, z)$;

3. *In presence of CN3*, $ingr_C(x, y) \wedge ingr_C(y, x) \Rightarrow x = y$;

4. $ingr_C(x, y) \Leftrightarrow \forall z.[ingr(z, x) \Rightarrow ingr(z, y)]$;

5. $ingr_C(x, y) \wedge C(z, x) \Rightarrow C(z, y)$;

6. $ingr_C(x, y) \Rightarrow C(x, y)$.

The notion of a part π_C can be introduced as

$$PC \quad \pi_C(x, y) \Leftrightarrow ingr_C(x, y) \wedge x \neq y \qquad (5.25)$$

By definition (5.25), π_C satisfies requirements of mereology for the notion of a part

Proposition 5.44. *(Clarke [12]) Properties of* part *are*

1. $\neg \pi_C(x, x)$;

2. $\pi_C(x, y) \wedge \pi_C(y, z) \Rightarrow \pi_C(x, z)$;

3. $\pi_C(x, y) \Rightarrow ingr_C(x, y)$;

4. $\pi_C(x, y) \Rightarrow \neg \pi_C(y, x)$.

The predicate of *overlapping*, $Ov_C(x, y)$ is defined by means of

$$OC \quad Ov_C(x, y) \Leftrightarrow \exists z.[ingr_C(z, x) \wedge ingr_C(z, y)] \qquad (5.26)$$

Basic properties of overlapping follow.

Proposition 5.45. *(Clarke [12]) Properties of overlapping are*

1. $Ov_C(x, x)$;

2. $Ov_C(x, y) \Leftrightarrow Ov_C(y, x)$;

3. $Ov_C(x, y) \Rightarrow C(x, y)$;

4. $ingr_C(x, y) \land Ov_C(z, x) \Rightarrow Ov_C(z, y)$;

5. $ingr_C(x, y) \Rightarrow Ov_C(x, y)$.

The counterpart of the notion of an *exterior* object, $extr_C$ is defined by means of

$$EC \ extr_C(x, y) \Leftrightarrow \neg Ov_C(x, y) \tag{5.27}$$

Proposition 5.46. *(Clarke [12]) The property*

$$[ingr_C(x, y) \land extr_C(z, y)] \Rightarrow extr_C(z, x).$$

holds by Property 4 in Proposition 5.45.

A new notion due to connectedness is the notion of *external connectedness*, EC in symbols, defined as follows

$$EC \ EC(x, y) \Leftrightarrow C(x, y) \land extr(x, y) \tag{5.28}$$

It is easy to see that in the model OVM, EC is a vacuous notion. Clearly, by definition (5.28),

Proposition 5.47. *(Clarke [12]) The following properties of external connectedness hold*

1. $\neg EC(x, x)$;

2. $EC(x, y) \Leftrightarrow EC(y, x)$;

3. $C(x, y) \Leftrightarrow EC(x, y) \lor Ov_C(x, y)$;

4. $Ov_C(x, y) \Leftrightarrow C(x, y) \land \neg EC(x, y)$;

5. $\neg EC(x, y) \Leftrightarrow [Ov_C(x, y) \Leftrightarrow C(x, y)]$: *This is a logical rendering of our remark that in OVM, no pair of objects is in EC, hence, $\neg EC(x, y) = TRUE$ for each pair of objects;*

6. $\neg \exists z. EC(z, x) \Rightarrow \{ingr_C(x, y) \Leftrightarrow [\forall w. Ov_C(w, x) \Rightarrow Ov_C(w, y)]\}$.

Proof. A comment in the way of proof. The implication

$$ingr_C(x, y) \Rightarrow [\forall w. Ov_C(w, x) \Rightarrow Ov_C(w, y)]$$

is always true. Thus, it remains to assume that

(i) $\neg \exists z. EC(z, x)$

and to prove that

$$(*) \ [\forall w. Ov_C(w, x) \Rightarrow Ov_C(w, y)] \Rightarrow ingr_C(x, y)$$

Assumption (i), can be written down as

(ii) $\forall z. \neg C(z, x) \vee Ov_C(z, x)$

To prove that $ingr_C(x, y)$ it should be verified that

(iii) $\forall z. (C(z, x) \Rightarrow C(z, y))$.

Consider an arbitrary object z'; either $\neg C(z', x)$ in which case implication in (iii) is satisfied with z', or, $Ov_C(z', x)$, hence, $Ov_C(z', y)$ by the assumed premise in (*), which implies that $C(z', y)$. The implication (iii) is proved and this concludes the proof   □

The richer structure of connection based calculus allows for some notions of a topological nature; the first is the notion of a *tangential ingredient*, in symbols: $Tingr_C(x, y)$ defined by means of

$$TI \ Tingr_C(x, y) \Leftrightarrow ingr_C(x, y) \wedge \exists z. EC(z, x) \wedge EC(z, y) \qquad (5.29)$$

Basic properties of tangential parts follow by definition TI

Proposition 5.48. *The following are basic properties of the predicate* $Tingr_C$

1. $\exists z. EC(z, x) \Rightarrow Tingr_C(x, x)$;

2. $\neg \exists z. EC(z, y) \Rightarrow \neg exists x. Tingr_C(x, y)$;

3. $Tingr_C(z, x) \wedge ingr_C(z, y) \wedge ingr_C(y, x) \Rightarrow Tingr_C(y, x)$.

Proof. For Property 3, some argument may be in order; consider w such that $EC(w, x), EC(w, z)$ existing by $Tingr_C(z, x)$. hence, $C(w, y)$. As $\neg Ov_C(w, x)$, it follows that $\neg Ov_C(w, y)$, hence, $EC(w, y)$, and $Tingr_C(y, x)$ □

These properties witness the fact that if there is something externally connected to x, then x is its tangential ingredient. This fact shows that the notion of a tangential ingredient falls short of the idea of a boundary. Dually, in absence of objects externally connected to y, no ingredient of y can be a tangential ingredient.

An object y is a *non–tangential ingredient* of an object x, $NTingr_C(y, x)$ in symbols, in case, it is an ingredient but not any tangential ingredient of x

$$NTI \ NTingr_C(y, x) \Leftrightarrow \neg Tingr_C(y, x) \wedge ingr_C(y, x) \qquad (5.30)$$

Proposition 5.49. *Basic properties of the operator NTI are*

1. $NTingr_C(y,x) \Rightarrow \forall z. \neg EC(z,y) \vee \neg EC(z,x)$;

2. $\neg \exists z. EC(z,x) \Rightarrow NTingr_C(x,x)$.

In absence of externally connected objects, each object is a non–tangential ingredient of itself.

Hence, in the model OVM, each object is its non–tangential ingredient and it has no tangential ingredients.

To produce models in which EC, $NTingr_C$, $Tingr_C$ will be exhibited, we may resort to topology; we recall, see Ch.2, sect. 6, that *a regular open set* in a topological space X, is a set $A \subseteq X$ such that $A = IntClB$ for some set $B \subseteq X$.

We define in the space $RO(X)$ of regular open sets in a regular space X (recall that a topological space is regular if for each element x, each neighborhood of x contains closure of another neighborhood of x, see Ch. 2, sect. 6) the connection C by demanding that $C(A,B) \Leftrightarrow ClA \cap ClB \neq \emptyset$. For simplicity sake, we assume that the regular space X is connected, so no set in it is clopen, equivalently, the boundary of each set is non–empty.

When investigating properties of a model ROM, we refer to Ch. 2, sect. 6.

First, we investigate what $ingr_C$ means in ROM. By definition IC in (5.24), for $A, B \in RO(X)$,

$$ingr_C(A,B) \Leftrightarrow \forall Z \in RO(X). ClZ \cap ClA \neq \emptyset \Rightarrow ClZ \cap ClB \neq \emptyset$$

This excludes the case when $A \setminus ClB \neq \emptyset$ as then we could find a $Z \in RO(X)$ with

$$Z \cap A \neq \emptyset = ClZ \cap ClB$$

(as our space X is regular). It remains that $A \subseteq ClB$, hence, $A \subseteq IntClB = B$.

It follows finally that

Proposition 5.50. *In model ROM, $ingr_C(A,B) \Leftrightarrow A \subseteq B$.*

It follows that in ROM ingredient means containment with connection C as intersection of closures.

Now, we can interpret overlapping in ROM. For $A, B \in RO(X)$, $Ov_C(A,B)$ means that there exists $Z \in RO(X)$ such that $Z \subseteq A$ and $Z \subseteq B$ hence $Z \subseteq A \cap B$, hence

$$A \cap B \neq \emptyset$$

This condition is also sufficient by regularity of X. We obtain

Proposition 5.51. *In ROM, $Ov_C(A,B) \Leftrightarrow A \cap B \neq \emptyset$.*

The status of EC in ROM is then

Proposition 5.52. *In ROM,*

$$EC(A, B) \Leftrightarrow ClA \cap ClB \neq \emptyset \wedge A \cap B = \emptyset$$

This means that closed sets ClA, ClB do intersect only at their boundary points.

We can address the notion of a tangential ingredient: $Tingr_C(A, B)$ means the existence of $Z \in RO(X)$ such that

$$ClZ \cap ClA \neq \emptyset \neq ClZ \cap ClB$$

and

$$Z \cap A = \emptyset = Z \cap B$$

along with $A \subseteq B$.
   In case

$$ClA \cap (ClB \setminus B) \neq \emptyset$$

letting $Z = X \setminus ClB$ we have

$$ClZ = Cl(X \setminus ClB)$$

and

$$BdZ = ClZ \setminus Z = Cl(X \setminus ClB) \setminus (X \setminus ClB)$$

which in turn is equal to

$$Cl(X \setminus ClB) \cap ClB = Cl(X \setminus B) \cap ClB = BdB$$

Hence, $ClB \setminus B \subseteq ClZ$, and $ClZ \cap ClA \neq \emptyset$; a fortiori, $ClB \cap ClZ \neq \emptyset$. As $Z \cap B = \emptyset$, a fortiori $Z \cap A = \emptyset$ follows.
   We know, then, that if

$$ClA \cap (ClB \setminus B) \neq \emptyset \Rightarrow Tingr_C(A, B)$$

Was to the contrary, $ClA \subseteq B$, from $Z \cap ClA \neq \emptyset$ it would follow that $Z \cap B \neq \emptyset$, negating $EC(A, B)$.
   It follows finally that

Proposition 5.53. *In the model ROM, $Tingr_C(A, B)$ if and only if $A \subseteq B$ and $ClA \cap (ClB \setminus B) \neq \emptyset$. From this analysis we obtain also that $NTingr_C(A, B)$ if and only if $ClA \subseteq B$.*

Further properties of the predicate $NTingr_C$ are collected in

Proposition 5.54. *(Clarke [12]) Properties of $NTingr_C$ are*

1. $NTingr_C(y, x) \wedge C(z, y) \Rightarrow C(z, x);$

2. $NTingr_C(y, x) \wedge Ov_C(z, y) \Rightarrow Ov_C(z, x);$

3. $NTingr_C(y, x) \wedge C(z, y) \Rightarrow Ov_C(z, x);$

4. $ingr_C(y, x) \wedge NTingr_C(x, z) \Rightarrow NTingr_C(y, z);$

5. $ingr_C(y, z) \wedge NTingr_C(x, y) \Rightarrow NTingr_C(x, z);$

6. $NTingr_C(y, z) \wedge NTingr_C(z, x) \Rightarrow NTingr_C(y, x).$

Proof. For Property 3, from already known $\forall z. \neg EC(z, y) \vee \neg EC(z, x)$, it follows

$$(i) \quad \forall w. \neg C(w, x) \vee Ov_C(w, x) \vee \neg C(w, y) \vee Ov_C(w, y)$$

As $C(z, y)$, one obtains $C(z, x)$. Thus, by (i), $Ov_C(z, y) \vee Ov_C(z, x)$ and $Ov_C(z, x)$.

For Property 4, assuming that $ingr_C(y, x)$, $ingr_C(x, z)$ and hence, $ingr_C(y, z)$ (else, there is nothing to prove), consider $\neg NTingr_C(y, z)$, i.e., for some w: $EC(w, z), EC(w, y)$. Thus, $C(w, z), \neg Ov_C(w, z), C(w, y), \neg Ov_C(w, y)$.

Then, $C(w, x)$ and $\neg Ov_C(w, x)$, hence, $EC(w, x)$ and $\neg NTingr_C(x, z)$, a contradiction. Similarly, one justifies Properties P5 and P6 $\qquad\qquad\square$

5.7 Classes in Connection Mereology

Definition of a class $Cls\Phi$ of a non–vacuous property Φ, is given in Clarke's version of connection mereology in terms of the primitive notion of the predicate C to satisfy the desired property

$$C(z, Cls_C\Phi) \Leftrightarrow \exists x. \Phi(x) \wedge C(z, x) \qquad (5.31)$$

The existence of a class is secured by an axiom

CN4 *For each non–vacuous property Φ of objects, there exists an object $Cls_C\Phi$ (a C–class of Φ) which satisfies (5.31)*

Proposition 5.55. *(Clarke [12])*

1. *By CN3 and CN4, there is a unique object $Cls_C\Phi$ for each non–vacuous property Φ;*

2. *If $\Phi(z)$, then $ingr_C(z, Cls_C\Phi)$;*

3. $ingr_C(z, Cls_C\Phi) \Rightarrow \exists w.\Phi(w) \wedge C(z, w)$;

4. $C(z, Cls\Phi) \wedge \neg\exists w.EC(z, w) \Rightarrow \exists w.\Phi(w) \wedge Ov_C(z, w)$;

Properties 3, 4 witness that

5. *In the model OVM, the class defined in connection mereology by Cls_C satisfies Le'sniewski's postulates for a class;*

6. $Ov_C(z, Cls_C\Phi) \Rightarrow \exists w.\Phi(w) \wedge C(z, w)$;

Indeed, $Ov_C(z, Cls_C\Phi)$ implies $\exists t.[ingr_C(t, z) \wedge ingr_C(t, Cls_C\Phi)]$, hence, $\exists w.[\Phi(w) \wedge C(t, w)]$. Thus, $C(z, w)$.

7. $\neg exists w.EC(z, w) \Rightarrow [Ov_C(z, Cls_C\Phi) \Rightarrow \exists t.\Phi(t) \wedge Ov_C(z, t)]$;

8. $\exists w.\Phi(w) \wedge C(z, w) \Rightarrow Ov_C(z, Cls_C\Phi)$.

As with the Leśniewski notion of a class, connection based classes have properties

9. *If $\Phi \Rightarrow \Psi$, then $ingr_C(Cls_C\Phi, Cls_C\Psi)$;*

10. *If $\Phi \Leftrightarrow \Psi$, then $Cls_C\Phi = Cls_C\Psi$;*

11. $x = Cls_C \ INGR(x)$.

5.8 C–Quasi–Boolean Algebra

By Tarski [62], [61], as with the Leśniewski mereology, in connection mereology a quasi–Boolean algebra arises. By Postulate CN4, there exists for each pair x, y of objects, an object denoted $x+y$ defined as the class of the property $+(x, y)$,

$$+ (x, y)(t) \Leftrightarrow ingr_C(t, x) \vee ingr_C(t, y) \qquad (5.32)$$

Then,

$$Cls + \ x + y = Cls + (x, y) \qquad (5.33)$$

Proposition 5.56. *(Clarke [12]) The operation $+$ has properties*

1. $ingr_C(x, x + y)$ *by 9, Proposition 5.55;*

2. $C(z, x + y) \Leftrightarrow \exists w.[(ingr_C(w, x) \vee ingr_C(w, y)] \wedge C(z, w)$ *by Cls_C;*

3. $C(z, x + y) \Leftrightarrow C(z, x) \vee C(z, y)$ by Property 2;

From Property 10 of Proposition 5.55, we obtain

4. $x + y = y + x$;

5. $x + (y + z) = (x + y) + z$;

6. $x + x = x$.

The dual property is defined as

$$\times (x, y)(t) \Leftrightarrow ingr_C(t, x) \wedge ingr_C(t, y) \tag{5.34}$$

Clearly, the definition (5.34) makes sense only if $Ov_C(x, y)$; in the sequel, we will tacitly assume this condition any time the following operation $x \cdot y$ is mentioned.

The Boolean product $x \cdot y$ is defined as

$$Cls \times x \cdot y = Cls \times (x, y) \tag{5.35}$$

It follows immediately by (5.35)

Proposition 5.57. *(Clarke [12]) The following are properties of the Boolean product*

1. $C(z, x \cdot y) \Leftrightarrow \exists w. [(ingr_C(w, x) \wedge ingr_C(w, y)] \wedge C(z, w)$;

2. $C(z, x \cdot y) \Rightarrow C(z, x) \wedge C(z, y)$ by Property 1;

3. $ingr_C(x \cdot y, x); ingr_C(x \cdot y, y)$ by Property 2;

4. $ingr_C(z, x \cdot y) \Leftrightarrow ingr_C(z, x) \wedge ingr_C(z, y)$ by Property 2 and Proposition 5.55;

By Property 10 of Proposition 5.55, the following are true

5. $x \cdot y = y \cdot x$;

6. $x \cdot (y \cdot z) = (x \cdot y) \cdot z$;

7. $x \cdot (y + z) = x \cdot y + x \cdot z$.

The universe V is defined as the class of the property Con

$$Con(x) \Leftrightarrow C(x, x) \tag{5.36}$$

$$V = ClsCon \qquad (5.37)$$

By (5.37)

Proposition 5.58. *For each object z, one has $ingr_C(z, V)$.*

The complement $-x$ is defined as the class of property

$$N(x)(t) \Leftrightarrow \neg C(t, x) \qquad (5.38)$$

by means of

$$-x = ClsN(x) \qquad (5.39)$$

Then, by Property 10 of Proposition 5.55

Proposition 5.59. *(Clarke [12]) The properties of complement are*

1. $-(-x) = x$;

2. $C(z, -x) \Leftrightarrow \exists w . \neg C(x, w) \wedge C(z, w)$;

3. $C(z, -x) \Leftrightarrow \neg ingr_C(z, x)$ *by Property 2;*

4. $ingr_C(x, y) \Leftrightarrow ingr_C(-y, -x)$;

> *Complement operator – flip–flops between boolean addition and multiplication*

5. $-(x \cdot y) = (-x) + (-y)$;

6. $(-x) \cdot (-y) = -(x + y)$

7. $\neg EC(x, -y) \Rightarrow [-x + y = V \Leftrightarrow ingr_C(x, y)]$.

Proof. As Property 5 obtains from Property 4 by means of Property 1, it suffices to prove Property 4. We have $C(t, -x \cdot y) \Leftrightarrow \neg ingr_C(t, x \cdot y)$ by 3, and by Property 3 of Proposition 5.57,

$$\neg ingr_C(t, x \cdot y) \Leftrightarrow \neg ingr_C(t, x) \vee \neg ingr_C(t, y)$$

which by the Property 3 is equivalent to $C(t, -x) \vee C(t, -y)$ and it is by Property 3 of Proposition 5.56, equivalent to $C(t, (-x) + (-y))$. By CN3, $-(x \cdot y) = (-x) + (-y)$.

For Property 6, premise $\neg EC(x, -y)$ is equivalent to $\neg C(x, -y) \vee Ov_C(x, -y)$ and in turn, by Property 3, to $ingr_C(x, y) \vee Ov_C(x, -y)$. The implication $ingr_C(x, y) \Rightarrow -x + y = V$ being obvious, we may observe that conditions $-x + y = V$ and $Ov_C(x, -y)$ are contradictory: hence, the premise $-x + y = V$ implies $ingr_C(x, y)$ □

In particular, in any model OVM, $-x + y = V \Leftrightarrow ingr_C(x, y)$.

5.9 C–Mereotopology

Topological operators are constructed in connection mereology under same caveat as quasi–Boolean operators: absence of the null object causes to make reservations concerning existence of some objects necessary for topological constructions. We will make this reservations not trying to add new axioms which would guarantee existence of some auxiliary objects.

The C–interior $Int_C(x)$ of an object x is defined as the class of non–tangential ingredients of x.

We define the property $NTP(x)$

$$NTP(x)(z) \Leftrightarrow NTP(z, x) \tag{5.40}$$

The interior $Int_C(x)$ is defined by means of

$$INT_C\ Int_C(x) = ClsNTP(x) \tag{5.41}$$

hence, properties follow

Proposition 5.60. *(Clarke [12]) properties of the operator Int_C are*

1. $C(z, Int_C(x)) \Leftrightarrow \exists w.NTingr_C(w, x) \wedge C(z, w)$ *by the class definition;*

2. $\neg\exists z.EC(z, x) \Rightarrow (Int_C(x) = x)$. *In particular, in the model OVM, $Int_C(x) = x$ for each object x;*

3. $ingr_C(Int_C(x), x)$ *as* $C(z, Int_C(x)) \Rightarrow C(z, x)$;

4. $C(z, Int_C(x)) \Rightarrow Ov_C(z, x)$;

5. $EC(z, x) \Rightarrow \neg C(z, Int_C(x))$;

6. $ingr_C(z, Int_C(x)) \Leftrightarrow NTingr_C(z, x)$;

7. $ingr_C(z, x) \Rightarrow ingr_C(Int_C(z), Int_C(x))$;

8. $Int_C(x) = x \Leftrightarrow C(z, x) \Rightarrow Ov_C(z, x)$;

9. $Int_C(x) = x \Leftrightarrow NTingr_C(x, x)$.

An *open* object is x such that $Int_C(x) = x$.

Under additional axiomatic postulate that the boolean product of any two open sets is open, see Clarke [12], A2.1, one can prove that $Int_C(x \cdot y) = Int_C(x) \cdot Int_C(y)$.

The notion of a topological *closure* $Cl_C(x)$ of x, can be introduced by means of the standard duality

$$Cl_C \; Cl_C(x) = -int(-x) \tag{5.42}$$

By properties of the interior and by duality (5.42), one obtains dual properties of closure

Proposition 5.61. *(Clarke [12]) Properties of the operator Cl_C are*

1. $ingr_C(x, Cl_C(x))$;

2. $Cl_C(Cl_C(x)) = Cl_C(x)$;

3. $ingr_C(x, y) \Rightarrow ingr_C(Cl_C(x), Cl_C(y))$;

4. $Int_C(x \cdot y) = Int_C(x) \cdot Int_C(y) \Leftrightarrow Cl_C(x + y) = Cl_C(x) + Cl_C(y)$;

5. $C(z, Cl_C(x)) \Leftrightarrow \exists w.NTingr_C(w, -x) \wedge C(z, w)$.

The notion of a *boundary* can be introduced along standard topological lines

$$Bd_C \; Bd_C(x) = -(int(x) + int(-x)) \tag{5.43}$$

We collect basic properties of the boundary in

Proposition 5.62. *The operator Bd_C has properties*

1. *Under Property 4 of Proposition 5.61, $Bd_C(x) = Cl_C(x) \cdot -Int_C(x)$, i.e., it can be expressed as the difference between the closure and the interior of the object;*

2. *$Bd(x) = Bd(-x)$ by Property 1 of Proposition 5.59;*

3. *$ingr_C(Bd_C(x), Cl_C(x))$.*

We now include a section on applications of reasoning schemes based on mereology in spatial reasoning.

5.10 Spatial Reasoning: Mereological Calculi

Spatial reasoning belongs to the oldest known to humankind forms of reasoning: the space around man and orientation in it were doubtless the earliest experiences they had had. The strict relations among various objects, e.g., segments of various length, especially those forming figures in the plane were among the first scientific observations of man: for many historians of civilization, see Bronowski [9], the Pythagorean Theorem is the most important mathematical fact known to man and one can argue plausibly the same about the Tales Theorem most probably known to Imhotep, and other pyramid builders, see Shaw [51].

It comes thus as no surprise that axiomatic system of geometry due to Euclid, see Joyce [31], is one of the two earliest formal systems besides the Syllogistic of Aristotle, see Ch. 1. Formal questions posed by the Euclidean system, like the famous Vth Postulate have led to many fundamental discoveries like non–Euclidean geometries. In XX century, axiomatization of geometry was pursued among others by Tarski [63].

With Riemann, geometry entered the realm of abstract spaces and work of Cantor alongside early anticipations by Riemann brought forth topology – a theory of invariants of continuous mappings, cf., Ch. 2. Topology created a more flexible and adaptable to various contexts reasoning framework which have grown to a many–faceted theory permeating the whole of mathematical insight into reality, mostly due to the universal and simple notion of a neighborhood.

Formal discussion of topology have been carried out within the framework of set theory, initiated by Cantor, whose elementary notion is the relation of being an element, in symbols, $\in$.

The philosophical assumption concerning the nature of a set was therefore that sets are built of elements and identical sets possess of the same elements. This assumption brought forth some paradoxes like the well-known Russell paradox whether there exists a set of all sets. The emergence of paradoxes demonstrated the difficulty with the notion of a set; mathematicians and philosophers of the turn of XX century found seemingly the way out of difficulty by imposing on the notion of a set some strict requirements which allowed the existence of some sets and prohibited the construction of some too abstract sets, see Kanamori [32]. Yet, some questions have turned out to be very difficult to answer and even undecidable, so currently there exist many models of set theory to choose from, cf., Kanamori, op. cit.

Qualitative Spatial Reasoning developed from the above mentioned ideas of Leśniewski and Whitehead and it has become a basic ingredient in a variety of problems, e.g., in mobile robotics, see, e.g., Kuipers [34], Kuipers and Byun [35], Kuipers and Levitt [36]. Spatial reasoning which deals with objects like solids, regions etc., by necessity refers to and relies on mereological theories of concepts based on the opposition part–whole, see Gotts et al. [27].

For expressing relations among entities, mathematics proposes two basic languages: the language of set theory, based on the opposition element–set, where distributive classes of entities are considered as sets consisting of (discrete) atomic entities, and languages of mereology, for discussing entities continuous in their nature, based on the opposition part–whole. Due to continuous nature of spatial real objects, Spatial Reasoning relies to great extent on mereological theories of part, cf., Asher and Vieu [5], Asher et al. [4], Aurnague and Vieu [6], Cohn and Gotts [15], Gotts and Cohn [26], Cohn et al. [16], Galton [24], Smith [57], [56], Masolo and Vieu [47].

Qualitative Reasoning aims at studying concepts and calculi on them that arise often at early stages of problem analysis when one is refraining from qualitative or metric details, cf., Cohn [14]; as such it has close relations with design, cf., Booch [8] as well as planning stages, cf., Glasgow [25] of the model synthesis process.

Classical formal approaches to spatial reasoning, i.e., to representing spatial entities (points, surfaces, solids) and their features (dimensionality, shape, connectedness degree) rely on Geometry or Topology, i.e., on formal theories whose models are spaces (universes) constructed as sets of points; contrary to this approach, qualitative reasoning about space often exploits pieces of space (regions, boundaries, walls, membranes) and argues in terms of relations abstracted from a commonsense perception (like *connected, discrete from, adjacent, intersecting*). In this approach, points appear as ideal objects, e.g., ultrafilters of regions/solids as in Tarski [60].

Qualitative Spatial Reasoning has a wide variety of applications, among them, to mention only a few, representation of knowledge, cognitive maps and navigation tasks in robotics, see Kuipers, Kuipers and Byun, Kuipers and Levitt op. cit., op. cit., op. cit., as well as AISB97 [1], Arkin [3], Dorigo and Colombetti [18], Kortenkamp [33], Freksa [23], Geographical Information Systems and spatial databases including *Naive Geography*, see Frank and Campari [21], Frank and Kuhn [22], Hirtle and Frank [29], Egenhofer and Golledge [20], Mark [46], and in studies in semantics of orientational lexemes and in semantics of movement in Asher et al. [4] and Aurnague and Vieu [6].

Spatial Reasoning establishes a link between Computer Science and Cognitive Sciences, cf., Freksa [23] and it has close and deep relationships with philosophical and logical theories of space and time, cf., Reichenbach [50] , vanBenthem [7], Allen [2].

Any formal approach to Spatial Reasoning requires Ontology, which differs from the above exposed Ontology due to Leśniewski, and is based on established hierarchies of concepts, cf., Guarino [28], Smith [55], Casati et al. [11]. In reasoning with spatial objects, of primary importance is to develop an ontology of spatial objects, taking into account complexity of these objects.

The scheme for Connection calculus, presented above, sect. 5, has inspired many authors toward creation of a calculus on specified geometric objects which would implement connection predicate, although in a modified according to a given context form. A good example of such approach is the calculus of regions RCC (Region Connection Calculus), see Randell et al. [49]. Primitive geometric objects considered here are regions; this standpoint distinguishes RCC from the original setting by Clarke in [12] where points were primitives and connection C was interpreted as having a point in common. In RCC authors tend to interpret connection as the property of having a common point in closures of regions. Thus, RCC is essentially a calculus of regularly open regions, i.e., regions R such that $R = IntClR$: each region is the interior of its closure, see sect. 5. Hence, the framework adopted by authors of RCC is that of the model ROM.

5.10.1 On Region Connection Calculus

RCC adopts requirements CN1–CN3 of Connection calculus in Clarke [12], and defines other predicates in a similar way; we recall them here preserving the notation of RCC.

DISCONNECTED FROM(x)(y): $DC(x, y)$ *if and only if not* $C(x, y)$

IMPROPER PART OF(x)(y): $P(x, y)$ *if and only if for each* z, $C(z, y) \rightarrow$ $C(z, x)$ *(y is an improper part of x)*

PROPER PART OF(x)(y): $PP(x, y)$ *if and only if* $P(x, y)$ *and not* $P(y, x)$ *(y is a proper part of x)*

EQUAL(x)(y): $EQ(x, y)$ *if and only if* $P(x, y)$ *and* $P(y, x)$ *(y is identical to x)*

OVERLAP(x)(y): $Ov(x, y)$ *if and only if there exists* z *such that* $P(x, z)$ *and* $P(y, z)$ *(y overlaps x)*

DISCRETE FROM(x)(y): $DR(x, y)$ *if and only if not* $Ov(x, y)$ *(y is discrete from x)*

PARTIAL OVERLAP(x)(y): $POv(x, y)$ *if and only if* $Ov(x, y)$ *and not* $P(x, y)$, *and not* $P(y, x)$ *(y partially overlaps x)*

EXTERNAL CONNECTED(x)(y): $EC(x, y)$ *if and only if* $C(x, y)$ *and not* $Ov(x, y)$ *(y is externally connected to x)*

TANGENTIAL PART OF(x)(y): $TPP(x, y)$ *if and only if* $PP(x, y)$ *and there exists* z *such that* $EC(x, z)$ *and* $EC(y, z)$ *(y is a tangential proper part of x)*

NON–TANGENTIAL PART OF(x)(y): $NTPP(x, y)$ *if and only if* $PP(x, y)$ *and not* $TPP(x, y)$ *(y is a non–tangential proper part of x)*

Each non–symmetric predicate X among the above is also accompanied by its inverse Xi (e.g., $TPP(x, y)$ by $TPPi(y, x)$).

Of these predicates, the eight: DC, EC, PO, EQ, TPP, NTPP, TPPi, NTPPi are shown to have the JEPD property (Jointly Exclusive and Pairwise Disjoint) and they form the fragment of RCC called the RCC8 calculus.

Due to topological assumptions, RCC has some stronger properties than Clarke's calculus of C. Witness, the two properties, not mentioned by us when discussing the model ROM.

The first is the extensionality of the Overlap predicate

If for each z $Ov(x, z) \leftrightarrow Ov(y, z)$, then $x = y$

Indeed, for regular open sets A, B, the condition that $Ov(x, z) \Leftrightarrow Ov(y, z)$ means that $ClA = ClB$, hence, $A = IntClA = IntClB = B$.

The second property concerns complementarity

If $PP(A, B)$, then there exists C such that $P(C, B)$ and $DR(C, A)$

Indeed, was $B \subseteq ClA$, we would have $ClB = ClA$, a fortiori, $B = A$, contrary to the assumptions. Hence, $B \setminus ClA \neq \emptyset$, and there exists, by regularity of X, a $C \in RO(X)$ such that $ClC \subseteq B \setminus ClA \neq \emptyset$.
One more consequence of topological assumptions is that the region and its complement are externally connected, as their closures do intersect,

$EC(x, -x)$ for each region x

For practical reasons, RCC8 is presented in the form of the *transition table*: a table in which for entries $R_1(x, y)$ and $R_2(y, z)$ a result $R_3(x, z)$ is given, see Egenhofer [19]. The transition table for RCC8 is shown in the table of Fig. 5.1.

-	DC	EC	PO	TPP	NTPP	TPPi	NTPPi
DC	-	DR,PO,PP	DR,PO,PP	DR,PO,PP	DC	DC	
EC	DR,PO,PPi	DR,PO,TPP,TPi	DR,PO,P	EC,PO,PP	PO,PP	DR	DC
PO	DR,PO,PPi	DR,PO,PPi	-	PO,PP	PO,P	DR,PO,PPi	DR,PO,PPi
TPP	DC	DR	DR,PO,PP	PP	NTPP	DR,PO,PP	-
NTPP	DC	DC	DR,O,PP	NTPP	NTPP	DR,PO,PP	-
TPPi	DR,PO,PPi	EC,PO,PPi	PO,PPi	PO,TPP,TPi	PO,PP	PPi	NTPPi
NTPPi	DR,PO,PPi	PO,PPi	PO,PPi	PO, PPi	0	NTPPi	NTPPi

Fig. 5.1 Transition table for RCC8 calculus

RCC8 allows for additional predicates characterizing shape, connectivity, see Gotts et al. [27] and regions with vague boundaries ("the egg–yolk" approach), see Cohn and Gotts [15].

References

1. AISB–1997: Spatial Reasoning in Mobile Robots and Animals. In: Proceedings AISB 1997 Workshop. Manchester Univ., Manchester, UK (1997)
2. Allen, J.: Towards a general theory of action and time. Artificial Intelligence 23(20), 123–154 (1984)
3. Arkin, R.C.: Behavior-Based Robotics. MIT Press, Cambridge (1998)
4. Asher, N., Aurnague, M., Bras, M., Sablayrolles, P., Vieu, L.: De l'espace-temps dans l'analyse du discours. Rapport interne IRIT/95-08-R. Institut de Recherche en Informatique, Univ. Paul Sabatier, Toulouse (1995)
5. Asher, N., Vieu, L.: Toward a geometry of commonsense: a semantics and a complete axiomatization of mereotopology. In: Proceedings IJCAI 1995, pp. 846–852. Morgan Kauffman, San Mateo (1995)
6. Aurnague, M., Vieu, L.: A theory of space-time for natural language semantics. In: Korta, K., Larrazábal, J.M. (eds.) Semantics and Pragmatics of Natural Language: Logical and Computational Aspects. ILCLI Series I, pp. 69–126. Univ. Pais Vasco, San Sebastian (1995)
7. Van Benthem, J.: The Logic of Time. D. Reidel, Dordrecht (1983)
8. Booch, G.: Object–Oriented Analysis and Design with Applications. Addison–Wesley Publ., Menlo Park (1994)
9. Bronowski, J.: The Ascent of Man. BBC Paperbacks, BBC (1976)
10. Casati, R., Varzi, A.C.: Parts and Places. The Structures of Spatial Representations. MIT Press, Cambridge (1999)
11. Casati, R., Smith, B., Varzi, A.C.: Ontological tools for geographic representation. In: Guarino, N. (ed.) Formal Ontology in Information Systems, pp. 77–85. IOS Press, Amsterdam (1998)
12. Clarke, B.L.: A calculus of individuals based on connection. Notre Dame Journal of Formal Logic 22(2), 204–218 (1981)
13. Clay, R.: Relation of Leśniewski's Mereology to Boolean Algebra. The Journal of Symbolic Logic 39, 638–648 (1974)
14. Cohn, A.G.: Calculi for qualitative spatial reasoning. In: Pfalzgraf, J., Calmet, J., Campbell, J. (eds.) AISMC 1996. LNCS, vol. 1138, pp. 124–143. Springer, Heidelberg (1996)
15. Cohn, A.G., Gotts, N.M.: Representing spatial vagueness: a mereological approach. In: Proceedings of the 5th International Conference KR 1996, pp. 230–241. Morgan Kaufmann, San Francisco (1996)
16. Cohn, A.G., Randell, D., Cui, Z., Bennett, B.: Qualitative spatial reasoning and representation. In: Carrete, N., Singh, M. (eds.) Qualitative Reasoning and Decision Technologies, Barcelona, pp. 513–522 (1993)

17. Cohn, A.G., Varzi, A.C.: Connections relations in mereotopology. In: Prade, H. (ed.) Proceedings of 13th European Conference on Artificial Intelligence, ECAI 1998, pp. 150–154. Wiley and Sons, Chichester (1998)
18. Dorigo, M., Colombetti, M.: Robot Shaping. An Experiment in Behavior Engineering. MIT Press, Cambridge (1998)
19. Egenhofer, M.J.: Reasoning about binary topological relations. In: Günther, O., Schek, H.-J. (eds.) SSD 1991. LNCS, vol. 525, pp. 143–160. Springer, Heidelberg (1991)
20. Egenhofer, M.J., Golledge, R.G. (eds.): Spatial and Temporal Reasoning in Geographic Information Systems. Oxford University Press, Oxford (1997)
21. Campari, I., Frank, A.U. (eds.):Spatial Information Theory: A Theoretical Basis for GIS. LNCS, vol. 716. Springer, Heidelberg (1993)
22. Kuhn, W., Frank, A.U. (eds.): Spatial Information Theory: A Theoretical Basis for GIS. LNCS, vol. 988. Springer, Heidelberg (1995)
23. Freksa, C., Habel, C.: Repraesentation und Verarbeitung raeumlichen Wissens. Informatik-Fachberichte. Springer, Berlin (1990)
24. Galton, A.: The mereotopology of discrete space. In: Freksa, C., Mark, D.M. (eds.) COSIT 1999. LNCS, vol. 1661, pp. 251–266. Springer, Heidelberg (1999)
25. Glasgow, J.: A formalism for model–based spatial planning. In: Kuhn, W., Frank, A.U. (eds.) COSIT 1995. LNCS, vol. 988, Springer, Heidelberg (1995)
26. Gotts, N.M., Cohn, A.G.: A mereological approach to representing spatial vagueness. In: Working papers. The Ninth International Workshop on Qualitative Reasoning, QR 1995 (1995)
27. Gotts, N.M., Gooday, J.M., Cohn, A.G.: A connection based approach to commonsense topological description and reasoning. The Monist 79(1), 51–75 (1996)
28. Guarino, N.: The ontological level. In: Casati, R., Smith, B., White, G. (eds.) Philosophy and the Cognitive Sciences. Hoelder–Pichler–Tempsky, Vienna (1994)
29. Frank, A.U. (ed.): Spatial Information Theory A Theoretical Basis for GIS. LNCS, vol. 1329. Springer, Heidelberg (1997)
30. Iwanuś, B.: On Leśniewski's elementary Ontology. Studia Logica XXXI, pp. 73–119 (1973)
31. Joyce D.: Euclid: Elements (1998),
    http://aleph0.clarku.edu/~djoyce/java/elements/elements.html
    (last entered 01. 04. 2011)
32. Kanamori, A.: The Mathematical Development of Set Theory from Cantor to Cohen. The Bulletin of Symbolic Logic 2(1), 1–71 (1996)
33. Kortenkamp, D., Bonasso, R.P., Murphy, R. (eds.): Artificial Intelligence and Mobile Robots. AAAI Press/MIT Press, Menlo Park, CA (1998)
34. Kuipers, B.J.: Qualitative Reasoning: Modeling and Simulation with Incomplete Knowledge. MIT Press, Cambridge (1994)
35. Kuipers, B.J., Byun, Y.T.: A qualitative approach to robot exploration and map learning. In: Proceedings of the IEEE Workshop on Spatial Reasoning and Multi-Sensor Fusion, pp. 390–404. Morgan Kaufmann, San Mateo (1987)
36. Kuipers, B.J., Levitt, T.: Navigation and mapping in large-scale space. AI Magazine 9(20), 25–43 (1988)
37. de Laguna, T.: Point, line and surface as sets of solids. The Journal of Philosophy 19, 449–461 (1922)

38. Lejewski, C.: A contribution to Leśniewski's mereology. Yearbook for 1954–1955 of the Polish Society of Arts and Sciences Abroad V, pp. 43–50 (1954–1955)
39. Lejewski, C.: On Leśniewski's Ontology. Ratio I (2), 150–176 (1958)
40. Leonard, H., Goodman, N.: The calculus of individuals and its uses. The Journal of Symbolic Logic 5, 45–55 (1940)
41. Leśniewski, S.: Podstawy Ogólnej Teoryi Mnogości, I (Foundations of General Set Theory, I, in Polish). Prace Polskiego Koła Naukowego w Moskwie, Sekcya Matematyczno–przyrodnicza, No. 2, Moscow (1916)
42. Leśniewski, S.: O podstawach matematyki (On foundations of mathematics, in Polish). (1927) Przegląd Filozoficzny XXX, pp 164–206; (1928) Przegląd Filozoficzny XXXI, pp 261–291; (1929) Przegląd Filozoficzny XXXII, pp 60–101; (1930) Przegląd Filozoficzny XXXIII, pp 77–105 (1930); (1931) Przegląd Filozoficzny XXXIV, pp 142–170 (1927–1931)
43. Leśniewski, S.: Über die Grundlagen der Ontologie. C.R. Soc. Sci. Lettr. Varsovie Cl. III, 23 Anneé, pp. 111–132 (1930)
44. Leśniewski, S.: On the foundations of mathematics. Topoi 2, 7–52 (1982)
45. Leśniewski, S., Srzednicki, J., Surma, S.J., Barnett, D., Rickey, V.F. (eds.): Collected Works of Stanisł aw Leśniewski. Kluwer, Dordrecht (1992)
46. Mark, D.M. (ed.): Spatial Information Theory. In: Freksa, C., Mark, D.M. (eds.) COSIT 1999. LNCS, vol. 1661, pp. 205–220. Springer, Heidelberg (1999)
47. Masolo, C., Vieu, L.: Atomicity vs. Infinite divisibility of space. In: Freksa, C., Mark, D.M. (eds.) COSIT 1999. LNCS, vol. 1661, pp. 235–250. Springer, Heidelberg (1999)
48. Plutarch: The E at Delphi. In: Moralia (ed.), vol. 5, Harvard University Press, Cambridge (1936)
49. Randell, D., Cui, Z., Cohn, A.G.: A spatial logic based on regions and connection. In: Proceedings of the 3rd International Conference on Principles of Knowledge Representation and Reasoning KR 1992, pp. 165–176. Morgan Kaufmann, San Mateo (1992)
50. Reichenbach, H.: The Philosophy of Space and Time repr. Dover, New York (1957)
51. Shaw, I.: The Oxford History of Ancient Egypt. Oxford U. Press, Oxford (2000)
52. Simons, P.: Free part–whole theory. In: Lambert, K. (ed.) Philosophical Applications of Free Logic, pp. 285–306. Oxford University Press, Oxford (1991)
53. Simons, P.: Parts. A Study in Ontology. Clarendon Press, Oxford (2003)
54. Słupecki J, S. Leśniewski's calculus of names. Studia Logica III, pp. 7–72 (1955)
55. Smith, B.: Logic and formal ontology. In: Mohanty, J.N., McKenna, W. (eds.) Husserl's Phenomenology: A Textbook, pp. 29–67. University Press of America, Lanham (1989)
56. Smith, B.: Boundaries: an essay in mereotopology. In: Hahn, L. (ed.) The Philosophy of Roderick Chisholm, pp. 534–561. Library of Living Philosophers. La Salle: Open Court (1997)
57. Smith, B.: Agglomerations. In: Freksa, C., Mark, D.M. (eds.) COSIT 1999. LNCS, vol. 1661, pp. 267–282. Springer, Heidelberg (1999)
58. Sobociński, B.: L'analyse de l'antinomie Russellienne par Leśniewski. Methodos I, II, 94–107, 220–228, 308–316,237–257 (1949–1950)
59. Sobociński, B.: Studies in Leśniewski's Mereology. Yearbook for 1954-1955 of the Polish Society of Art and Sciences Abroad V, pp. 34–43 (1954–1955)
60. Tarski, A.: Les fondements de la géométrie des corps. Supplement to Annales de la Société Polonaise de Mathématique 7, 29–33 (1929)

61. Tarski, A.: Zur Grundlegung der Booleschen Algebra. I. Fundamenta Mathe-
    maticae 24, 177–198 (1935)
62. Tarski, A.: Appendix E. In: Woodger, J.H. (ed.) The Axiomatic Method in
    Biology, p. 160. Cambridge University Press, Cambridge (1937)
63. Tarski, A.: What is elementary geometry? In: Henkin, L., Suppes, P., Tarski, A.
    (eds.) The Axiomatic Method with Special Reference to Geometry and Physics.
    Studies in Logic and Foundations of Mathematics, pp. 16–29. North-Holland,
    Amsterdam (1959)
64. Whitehead, A.N.: La théorie relationniste de l'espace. Revue de Métaphysique
    et de Morale 23, 423–454 (1916)
65. Whitehead, A.N.: An Enquiry Concerning the Principles of Human Knowledge.
    Cambridge University Press, Cambridge (1919)
66. Whitehead, A.N.: The Concept of Nature. Cambridge University Press,
    Cambridge (1920)
67. Whitehead, A.N.: Process and Reality: An Essay in Cosmology. Macmillan,
    New York (1929)
68. Woodger, J.H.: The Axiomatic Method in Biology. Cambridge University Press,
    Cambridge (1937)
69. Woodger, J.H.: The Technique of Theory Construction. In: International En-
    cyclopedia of Unified Science, vol. II, 5, pp III+81. Chicago University Press,
    Chicago (1939)

Chapter 6
Rough Mereology

A scheme of mereology, introduced into a collection of objects, see Ch. 5, sets an exact hierarchy of objects of which some are (exact) parts of others; to ascertain whether an object is an exact part of some other object is in practical cases often difficult if possible at all, e.g., a robot sensing the environment by means of a camera or a laser range sensor, cannot exactly perceive obstacles or navigation beacons. Such evaluation can be done approximately only and one can discuss such situations up to a degree of certainty only. Thus, one departs from the exact reasoning scheme given by decomposition into parts to a scheme which approximates the exact scheme but does not observe it exactly.

Such a scheme, albeit its conclusions are expressed in an approximate language, can be more reliable, as its users are aware of uncertainty of its statements and can take appropriate measures to fend off possible consequences.

Imagine two robots using the language of connection mereology for describing mutual relations; when endowed with touch sensors, they can ascertain the moment when they are connected; when a robot has as a goal to enter a certain area, it can ascertain that it connected to the area or overlapped with it, or it is a part of the area, and it has no means to describe its position more precisely.

Introducing some measures of overlapping, in other words, the extent to which one object is a part to the other, would allow for a more precise description of relative position, and would add an expressional power to the language of mereology. Rough mereology answers these demands by introducing the notion of a *part to a degree* with the degree expressed as a real number in the interval $[0,1]$. Any notion of a part by necessity relates to the general idea of *containment*, and thus the notion of a part to a degree is related to the idea of *partial containment* and it should preserve the essential intuitive postulates about the latter.

The predicate of a part to a degree stems ideologically from and has as one of motivations the predicate of an element to a degree introduced by L. A. Zadeh as a basis for fuzzy set theory [24]; in this sense, rough mereology is to

L. Polkowski: Approximate Reasoning by Parts, ISRL 20, pp. 229–257.
springerlink.com   © Springer-Verlag Berlin Heidelberg 2011

mereology as the fuzzy set theory is to the naive set theory. To the rough set theory, owes rough mereology the interest in concepts as objects of analysis.

The primitive notion of rough mereology is the notion of a *rough inclusion* which is a ternary predicate $\mu(x, y, r)$ where x, y are *objects* and $r \in [0, 1]$, read '*the object x is a part to degree at least of r to the object y*'. Any rough inclusion is associated with a mereological scheme based on the notion of a part by postulating that $\mu(x, y, 1)$ is equivalent to $ingr(x, y)$, where the ingredient relation is defined by the adopted mereological scheme, see Ch. 5, sect. 1. Other postulates about rough inclusions stem from intuitions about the nature of partial containment; these intuitions can be manifold, a fortiori, postulates about rough inclusions may vary. In our scheme for rough mereology, we begin with some basic postulates which would provide a most general framework. When needed, other postulates, narrowing the variety of possible models, can be introduced.

6.1 Rough Inclusions

We have already stated that a rough inclusion is a ternary predicate $\mu(x, y, r)$. We assume that a collection of objects is given, on which a part relation π is introduced with the associated ingredient relation $ingr$. We thus apply inference schemes of mereology due to Leśniewski, presented in Ch. 5, sect. 1.

Predicates $\mu(x, y, r)$ were introduced in Polkowski and Skowron [18], [19]; they satisfy the following postulates, relative to a given part relation π and the induced by π relation $ingr$ of an ingredient, on a set U of entities

RINC1 $\mu(x, y, 1) \Leftrightarrow ingr(x, y)$

This postulate asserts that parts to degree of 1 are ingredients.

RINC2 $\mu(x, y, 1) \Rightarrow \forall z[\mu(z, x, r) \Rightarrow \mu(z, y, r)]$

This postulate does express a feature of partial containment that a 'bigger' object contains a given object 'more' than a 'smaller' object. It can be called a *monotonicity condition* for rough inclusions.

RINC3 $\mu(x, y, r) \wedge s < r \Rightarrow \mu(x, y, s)$

This postulate specifies the meaning of the phrase 'a part to a degree at least of r'.

From postulates RINC1–RINC3, and known properties of the ingredient predicate, see Ch. 5, sect. 1, some consequences follow.

Proposition 6.1. *The immediate consequences of postulates RINC1–RINC3 are*

1. $\mu(x,x,1)$;

2. $\mu(x,y,1) \wedge \mu(y,z,1) \Rightarrow \mu(x,z,1)$;

3. $\mu(x,y,1) \wedge \mu(y,x,1) \Leftrightarrow x = y$;

4. $x \neq y \Rightarrow \neg\mu(x,y,1) \vee \neg\mu(y,x,1)$;

5. $\forall z \forall r [\mu(z,x,r) \Leftrightarrow \mu(z,y,r)] \Rightarrow x = y$.

Proof. Property 1 follows by by RINC1 and Property 1 of Proposition 1, Ch. 5, Property 2 is implied by transitivity of ingredient, Property 2 of Proposition 1, Ch. 5, and RINC1. Property 3 follows by RINC1 and Property 3 of Proposition 1, Ch. 5, Property 4 holds by Property 3 of Proposition 1, Ch. 5. Lastly, Property 5 is true by Property 3 of Proposition 1, Ch. 5   □

Property 5 above may be regarded as an *extensionality postulate* in rough mereology.

By a *model* for rough mereology, we mean a quadruple

$$M = (V_M, \pi_M, ingr_M, \mu_M)$$

where V_M is a set with a part relation $\pi_M \subseteq V_M \times V_M$, the associated ingredient relation $ingr_M \subseteq V_M \times V_M$, and a relation $\mu_M \subseteq V_M \times V_M \times [0,1]$ which satisfies RINC1–RINC3.

We now describe some models for rough mereology which at the same time give us methods by which we can define rough inclusions, see Polkowski [10]–[15].

6.2   Rough Inclusions: Residual Models

We begin with continuous t–norms on the unit interval [0,1], see Ch. 4, sect. 7.

We recall that it follows from results in Mostert and Shields [7] and Faucett [2], see Ch. 4, sect. 9, cf., Hájek [3], that the structure of a continuous t–norm T depends on the set $F(T)$ of idempotents of T, i.e, values x such that $T(x,x) = x$; we denote with O_T the countable family of open intervals $A_i \subseteq [0,1]$ with the property that $\bigcup_i A_i = [0,1] \setminus F(T)$.

Then, see Ch. 4, Proposition 18,

Proposition 6.2. $T(x,y)$ *is an isomorph to either* $L(x,y)$ *or* $P(x,y)$ *when* $x,y \in A_i$ *for some* i, *and* $T(x,y) = min\{x,y\}$, *otherwise.*

We recall, see Ch. 4, sect. 10, that, for a continuous t–norm $T(x, y)$, the
residual implication, residuum, $x \Rightarrow_T y$ is defined by the condition

$$x \Rightarrow_T y \geq z \Leftrightarrow T(x, z) \leq y \tag{6.1}$$

It follows that $x \Rightarrow_T y = 1$ if and only if $x \leq y$, as $T(x, x) \leq x$ for each
continuous t–norm T.

For a continuous t–norm T, we define a relation $\mu_T \subseteq [0, 1]^3$ by means of

$$RIT \ \mu_T(x, y, r) \Leftrightarrow x \Rightarrow_T y \geq r \tag{6.2}$$

Proposition 6.3. *The quadruple* $M(T) = ([0, 1], <, \leq, \mu_T)$ *is a model for
rough mereology induced by the residuum of the t–norm* T.

Proof. First, let us make positive that μ_T satisfies RINC1–RINC3. For
RINC1, $\mu_T(x, y, 1)$ means that $x \Rightarrow_T y = 1$, hence, $x \leq y$, i.e., $ingr_M(x, y)$.
For RINC2, assume that $\mu_T(x, y, 1)$ and $\mu_T(z, x, r)$, hence (i) $x \leq y$ (ii) $z \Rightarrow_T$
$x \geq r$, i.e., by (1), (iii) $T(z, r) \leq x$. By (i), (iii), $T(z, r) \leq y$, hence, by (1),
$z \Rightarrow_T y \geq r$. RINC3 follows by (6.2) $\square$

Clearly, the underlying part relation in the above proposition is the strict
ordering $<$ and the ingredient relation is $\leq$.

In particular important cases, of t–norms L, P, M, one obtains the specific
models M_L, M_P, M_M. In each model $M(T)$, $\mu(x, y, 1) \Leftrightarrow x \leq y$, hence, we
recall below only the case when $x > y$, see Ch. 4. sect. 10.

In case of the Lukasiewicz t–norm L, we have, $x \Rightarrow_L y = min\{1, 1 - x + y\}$;
accordingly,

$$\mu_L(x, y, r) \Leftrightarrow min\{1, 1 - x + y\} \geq r \tag{6.3}$$

equivalently for $x > y$

$$\mu_L(x, y, r) \Leftrightarrow x - y \leq 1 - r \tag{6.4}$$

From (6.4), we can extract a transitivity rule

Proposition 6.4. *From* $\mu_L(x, y, r), \mu_L(y, z, s)$ *it follows that* $\mu_L(x, z, L(r, s))$.

Proof. It suffices to consider the case when $x > y$ and $y > z$, by (6.4), we
have $x - y \leq 1 - r$ and $y - z \leq 1 - s$, hence, $x - z \leq 1 - (r + s - 1)$, i.e.,
$\mu_L(x, y, L(r, s))$ $\square$

The proposition does encompass as well cases when $x \leq y$ or $y \leq z$.

For the product t–norm P, if $x > y$ then we have $x \Rightarrow_P y = \frac{y}{x}$; hence

$$\mu_P(x, y, r) \Leftrightarrow \frac{y}{x} \geq r \tag{6.5}$$

The transitivity rule follows.

Proposition 6.5. $\mu_P(x,y,r), \mu_P(y,z,s)$ imply $\mu_P(x,zP(r,s))$.

Proof. The conclusion follows by (6.5) from $\frac{y}{x} \geq r$, $\frac{z}{y} \geq s$, which imply $\frac{z}{x} \geq r \cdot s$. Cases when either $x \leq y$ or $y \leq z$, or $x \leq z$ are discussed analogously □

Finally, we consider the minimum t–norm M with $x \Rightarrow_P y = y$, hence

$$\mu_M(x,y,r) \Leftrightarrow y \geq r \qquad (6.6)$$

Proposition 6.6. If $\mu_M(x,y,r), \mu_M(y,z,s)$, then $\mu_M(x,z,min\{r,s\})$.

Proof. From $\mu_M(x,y,r), \mu_M(y,z,s)$ it follows that $min\{x,r\} \leq y$, $min\{y,s\} \leq z$, hence, $min\{x,min\{r,s\}\} \leq z$, i.e., $\mu_M(x,z,M(r,s))$ □

The partial results in Propositions 6.3, 6.5, 6.6, can be generalized to,

Proposition 6.7. *For each continuous t–norm T, the transitivity rule is obeyed by the rough inclusion μ_T: if $\mu_T(x,y,r), \mu_T(y,z,s)$, then $\mu_T(x,z, T(r,s))$.*

Proof. $\mu_T(x,y,r)$ is equivalent to $T(x,r) \leq y$, and, $\mu_T(y,z,s)$ is equivalent to $T(y,s) \leq z$. By coordinate–wise monotonicity of T, it follows that $T(T(x,r),s) \leq z$, and, by associativity of T, one obtains $T(x,T(r,s)) \leq z$, hence, $\mu_T(x,z) \geq T(r,s)$ □

Let us also put for the record the observation

Proposition 6.8. *For each r, and every continuous t–norm T, the set $D_T(r) = \{(x,y) : \mu_T(x,y,r)\}$ is a closed subset of the unit square $[0,1]^2$; sets $D_L(r)$,*
$D_M(r)$ are moreover convex.

Proof. $\mu(x,y,r)$ is equivalent to $T(x,r) \leq y$ and the result follows by continuity of T □

To carry further a topological analysis of rough inclusions of the form μ_T, we consider for a given $r \in [0,1]$, the set $D_T(r)$. It is equal, by RINC3, to the set $D_T(r^+) = \{(x,y) : \exists s \geq r\mu(x,y,s)\}$. The structure of $D_T(r^+)$ can be revealed by RINC2.

Proposition 6.9. *If $(x,y) \in D_T(r^+)$, then the segment $\{x\} \times [0,y] \subseteq D_T(r^+)$.*

Proof. By RINC2, if $y' \leq y$, and $\mu(x,y',r)$ then $\mu(x,y,r)$ □

We know, by Proposition 7, that each μ_T is transitive. As such it satisfies

Proposition 6.10. *If* $\mu_T(x,y,r)$ *and* $ingr_M(z,x)$, *then* $\mu_T(z,y,r)$.

Proof. As $ingr_M(z,x)$, it follows that $\mu_T(z,x,1)$ by RINC1, hence, by transitivity, $\mu_T(z,y,T(1,r))$, i.e., $\mu_T(z,y,r)$ □

The structure theorem for μ_T can be strengthened,

Proposition 6.11. *If* $(x,y) \in D_T(r^+)$, *then the Cartesian product of segments* $[0,x] \times [0,y] \subseteq D_T(r^+)$.

As for each r, the set $D_T(r^+)$ is closed, the topological characterization of μ_T follows,

Proposition 6.12. μ_T *is upper–semicontinuous, in the sense that the set* $\{(x,y) : \mu(x,y,r)$ *is closed for each* $r \in [0,1]$.

We may observe that μ_T is in fact a ternary relation, hence, the continuity property of μ_T can be expressed – more accurately even – when μ_T is regarded as a many–valued mapping, i.e., given (x,y), the value of $\mu_T(x,y,r)$, by RINC3, is an interval $[0, r_{max}(x,y)]$. Given $s \in [0,1]$, the following holds

Proposition 6.13. *The set* $E_T(s) = \{(x,y) \in [0,1]^2 : r_{max}(x,y) < s\}$ *is open for each* s, *i.e.,* μ_T *is upper–semicontinuous as a many–valued mapping.*

Proof. $E_T(s)$ is open as the complement to the closed set $D_T(s^+)$ □

Proposition 17 in Ch. 4, implies the converse to the above Proposition 13, see Ch. 2, sect. 12, for the notion of semi–continuity for multi–valued mappings,

Proposition 6.14. *Each rough inclusion* $\mu(x,y,r)$, *with* $x,y,r \in [0,1]$, *which is non–increasing in the first coordinate, non–decreasing in the second coordinate and upper semi–continuous as a multi–valued mapping is of the form* $\mu_T(x,y,r)$ *for some t–norm* T.

Proof. It follows from assumptions that the function $f_\mu(x,y) = r_{max}$ satisfies assumptions of Proposition 17 in Ch. 4, hence, $f_\mu(x,y) = x \Rightarrow_T y$ for some t–norm T, and our thesis follows □

We now turn to Archimedean t–norms of Ch. 4 in search for new rough inclusions.

6.3   Rough Inclusions: Archimedean Models

We recall that a continuous t–norm T is *Archimedean*, see Ch. 4, sect. 8, when $T(x,x) < x$ for each $x \in (0,1)$. Thus, the only idempotents of T are $0, 1$.

We also recall, see Ch. 4, sect. 9, that Archimedean t–norms admit a functional characterization, a very special case of the general Kolmogorov

[5] theorem, viz., for any Archimedean t–norm T, the following functional equation holds

$$T(x,y) = g_T(f_T(x) + f_T(y)) \qquad (6.7)$$

where the function $f_T : [0,1] \to R$ is continuous decreasing with $f_T(1) = 0$, and $g_T : R \to [0,1]$ is the pseudo–inverse to f_T, i.e., $g \circ f = id$, see Ling [6].

We consider two Archimedean t–norms: L and P. Their representations are

$$f_L(x) = 1 - x; \; g_L(y) = 1 - y \qquad (6.8)$$

and

$$f_P(x) = exp(-x); \; g_P(y) = -ln \; y \qquad (6.9)$$

For an Archimedean t–norm T, we define the rough inclusion μ^T on the interval $[0,1]$ by means of

$$ARI \; \mu^T(x,y,r) \Leftrightarrow g_T(|x - y|) \geq r \qquad (6.10)$$

equivalently,

$$\mu^T(x,y,r) \Leftrightarrow |x - y| \leq f_T(r) \qquad (6.11)$$

It follows from (6.11), that

Proposition 6.15. *The relation μ^T satisfies conditions RINC1–RINC3 with ingr as identity* $=$.

Proof. For RINC1: $\mu^T(x,y,1)$ if and only if $|x-y| \leq f_T(1) = 0$, hence, if and only if $x = y$. This implies RINC2. In case $s < r$, and $|x - y| \leq f_T(r)$, one has $f_T(r) \leq f_T(s)$ and $|x - y| \leq f_T(s)$ □

Specific recipes are: for μ^L

$$\mu^L(x,y,r) \Leftrightarrow |x - y| \leq 1 - r \qquad (6.12)$$

and for μ^P,

$$\mu^P(x,y,r) \Leftrightarrow |x - y| \leq -ln \; r \qquad (6.13)$$

The counterpart of Proposition 6.7 obeys for Archimedean rough inclusions,

Proposition 6.16. *For each Archimedean t–norm T, if $\mu^T(x,y,r)$ and $\mu^T(y,z,s)$, then $\mu^T(x,z,T(r,s))$.*

Proof. Assume $\mu^T(x,y,r)$ and $\mu^T(y,z,s)$, i.e., $|x - y| \leq f_T(r)$ and $|y - z| \leq f_T(s)$. Hence, $|x - z| \leq |x - y| + |y - z| \leq f_T(r) + f_T(s)$, hence, $g_T(|x - z|) \geq g_T(f_T(r) + f_T(s)) = T(r,s)$, i.e., $\mu^T(x,z,T(r,s))$ □

It may be worth–while to relate the residual and archimedean approaches to rough inclusions. For an Archimedean t–norm $T(x,y) = g(f(x) + f(y))$, one can easily calculate that in case $x > y$, the residual implication $x \Rightarrow_T y$ can be expressed as $g(f(y) - f(x))$, which, e.g., for the t–norm L comes down to $g(y - x)$, i.e., a formula (6.10), albeit with different mereological context.

From Proposition 6.2, a general result on structure of μ_T can be inferred

Proposition 6.17. *For a continuous t–norm T, in notation of Proposition 6.2, a rough inclusion μ_T can be defined as follows,*

$$\mu_T(x,y,r) \text{ if and only if } \begin{cases} \mu_T(x,y,r) \text{ for } x,y \in A_i, T = L \text{ or } T = P \\ \mu^M(x,y,r) \text{ otherwise} \end{cases}$$

$$(6.14)$$

For future applications, we need rough inclusions on sets.

6.4 Rough Inclusions: Set Models

Consider now a finite set X along with the family 2^X of its subsets. We define a rough inclusion $\mu^S \subseteq 2^X \times 2^X \times [0,1]$, by letting

$$\mu^S(A,B,r) \Leftrightarrow \frac{|A \cap B|}{|A|} \geq r \qquad (6.15)$$

where $|X|$ denotes the cardinality of X.

Then, we observe

Proposition 6.18. *The relation μ^S is a rough inclusion with the associated ingredient relation of containment $\subseteq$ and the part relation being the strict containment $\subset$.*

Proof. Clearly, $\mu^S(A,B,1)$ if and only if $A \subseteq B$; in that case, for every $Z \subseteq X, Z \cap A \subseteq Z \cap B$, hence, $\mu^S(Z,A,r)$ implies $\mu^S(Z,B,R)$ for every r. RINC3 is obviously satisfied $\square$

For containment on sets, there is no transitivity rule.

6.5 Rough Inclusions: Geometric Models

The set model can be modified in a geometric context; consider, e.g., a Euclidean space of a finite dimension E with objects as compact convex regions; this environment is usually applied, e.g., in modeling problems of intelligent mobile robotics, where compact convex regions model robots as well as obstacles.

For regions A, B, we define a rough inclusion μ^G by means of

$$\mu^G(A,B,r) \Leftrightarrow \frac{||A \cap B||}{||A||} \geq r \qquad (6.16)$$

where $||A||$ denotes the area (the Lebesgue measure) of the region A.

We have

Proposition 6.19. μ^G *is a rough inclusion with the containment $\subseteq$ as the associated ingredient relation.*

Proof. Clearly, if $||A \cap B|| = ||A||$, then $A \subseteq B$, as $A \setminus B \neq \emptyset$ implies $||A \setminus B|| \neq 0$ for compact convex regions A, B. The rest is already standard   □

Again, as with μ^S, no transitivity rule can be given here.

6.6 Rough Inclusions: Information Models

An important domain where rough inclusions will play a dominant role in our analysis of reasoning by means of parts is the realm of information systems, see Ch. 4., sect. 2. We will define information rough inclusions denoted with a generic symbol μ^I. We assume that *indiscernibility = identity*, i.e., each indiscernibility class, see Ch. 4., sect. 2, is represented by a unique object.

In order to define μ^I, for each pair $u, v \in U$, we define the set

$$DIS(u,v) = \{a \in A : a(u) \neq a(v)\} \tag{6.17}$$

We begin with an Archimedean rough inclusion T, and we define a rough inclusion μ_T^I by means of

$$\mu_T^I(u,v,r) \Leftrightarrow g_T(\frac{|DIS(u,v)|}{|A|}) \geq r \tag{6.18}$$

Then, it is true that

Proposition 6.20. μ_T^I *is a rough inclusion with the associated ingredient relation of identity and the part relation empty.*

Proof. We have $g_L(y) = 1 - y$ and $g_P(y) = exp(-y)$, i.e, $g^{-1}(1) = 0$ in either case, so $\mu_T^I(u,v,1)$ implies $DIS(u,v) = \emptyset$, hence,$IND_A(u,v)$ which, by our assumption, is the identity $=$. This verifies the condition RINC1, the rest follows along standard lines   □

Rough inclusions defined by means of (6.18) will be called *archimedean information rough inclusions* (shortened to *airi's*).

In specific cases, for the Łukasiewicz t–norm, L, the *airi* μ_L^I is given by means of the formula

$$\mu_L^I(u,v,r) \Leftrightarrow 1 - \frac{|DIS(u,v)|}{|A|} \geq r \tag{6.19}$$

We introduce the set $IND(u,v) = A \setminus DIS(u,v)$. With its help, we obtain a new form of (6.19)

$$\mu_L^I(u,v,r) \Leftrightarrow \frac{|IND(u,v)|}{|A|} \geq r \qquad (6.20)$$

The formula (6.20) witnesses that the reasoning based on the rough inclusion μ_L^I is the probabilistic one. At the same time, we have given a logical proof for formulas like (6.20) that are very frequently applied in Data Mining and Knowledge Discovery, also in rough set methods in those areas, see, e.g., Kloesgen and Zytkow [4]. It also witness that μ_L^I is a generalization of indiscernibility relation to the relation of partial indiscernibility.

In case of the product t–norm P, the formula (6.18) specifies to

$$\mu_P^I(u,v,r) \Leftrightarrow exp(-\frac{|DIS(u,v)|}{|A|}) \geq r \qquad (6.21)$$

We can prove, for any *airi* μ_T^I, the transitivity property in the form, see Polkowski [12]

$$If\ \mu_T^I(u,v,r)\ and\ \mu_T^I(v,w,s),\ then\ \mu_T^I(u,w,T(r,s)) \qquad (6.22)$$

Proof. We begin with the observation that

$$DIS(u,w) \subseteq DIS(u,v) \cup DIS(v,w) \qquad (6.23)$$

hence

$$\frac{|DIS(u,w)|}{|A|} \leq \frac{|DIS(u,v)|}{|A|} + \frac{|DIS(v,w)|}{|A|} \qquad (6.24)$$

We let

$$\begin{cases} g_T(\frac{|DIS(u,v)|}{|A|}) = r \\ g_T(\frac{|DIS(v,w)|}{|A|}) = s \\ g_T(\frac{|DIS(u,w)|}{|A|}) = t \end{cases} \qquad (6.25)$$

Then

$$\begin{cases} \frac{|DIS(u,v)|}{|A|} = f_T(r) \\ \frac{|DIS(v,w)|}{|A|} = f_T(s) \\ \frac{|DIS(u,w)|}{|A|} = f_T(t) \end{cases} \qquad (6.26)$$

Finally, by (6.24)

$$f_T(t) \leq f_T(r) + f_T(s) \qquad (6.27)$$

hence

$$t = g_T(f_T(t)) \geq g_T(f_T(r) + f_T(s)) = T(r,s) \qquad (6.28)$$

witnessing $\mu_T(u,w,T(r,s))$. This concludes the proof $\square$

We would like as well to exploit, in the context of information systems, residual implications that served us well in case of the unit interval. The formalism of descriptor logic, see Ch. 4, sect. 2, however, gives us not many possibilities for characterization of objects, save sets DIS and IND. To come as close as possible to requirements for a rough inclusion, we select an object $s \in U$ referred to as a *standard object, or, a pattern*. From application point of view, s may be, e.g., the best classified case, or, the pattern set as ideal.

For any object $x \in U$, we let

$$IND(x, s) = \{a \in A : a(x) = a(s)\} \qquad (6.29)$$

For a continuous t–norm T, we define a rough inclusion $\mu_T^{IND,s}$, under generic name of *indri*, by letting

$$INDRI \ \mu_T^{IND,s}(u, v, r) \Leftrightarrow \frac{|IND(u, s)|}{|A|} \Rightarrow_T \frac{|IND(v, s)|}{|A|} \geq r \qquad (6.30)$$

We obtain a rough inclusion, indeed

Proposition 6.21. *The indri* $\mu_T^{IND,s}$ *satisfies conditions RINC1–RINC3 with ingredient relation* $ingr_{IND}(u, v) \Leftrightarrow |IND(u, s)| \leq |IND(v, s)|$ *and identity defined as* $u =_{IND} v \Leftrightarrow |IND(u, s)| = |IND(v, s)|$.

Proof. $\mu_T^{IND,s}(u, v, 1)$ is equivalent to $|IND(u, s)| \leq |IND(v, s)|$, i.e., to $ingr_{IND}(u, v)$. That is for RINC1. For RINC2, assume that $|IND(u, s)| \leq |IND(v, s)|$, and $\frac{|IND(w,s)|}{|A|} \Rightarrow_T \frac{|IND(u,s)|}{|A|} \geq r$. Then, by monotonicity of T, $\frac{|IND(w,s)|}{|A|} \Rightarrow \frac{|IND(v,s)|}{|A|} \geq r$ follows. RINC3 is obviously satisfied $\qquad \square$

The rough inclusion $\mu_T^{IND,s}$ specifies to distinct formulas for three basic t–norms, L, P, M. The specific formulas are, for $r < 1$,

$$\mu_L^{IND,s}(u, v, r) \Leftrightarrow 1 - |IND(u, s)| + |IND(v, s)| \geq r \cdot |A| \qquad (6.31)$$

$$\mu_P^{IND,s}(u, v, r) \Leftrightarrow |IND(v, s)| \geq r \cdot |IND(u, s)| \qquad (6.32)$$

and,

$$\mu_M^{IND,s}(u, v, r) \Leftrightarrow |IND(v, s)| \geq r \cdot |A| \qquad (6.33)$$

We may observe that the quotient $\frac{|IND(u,s)|}{|A|}$ is the value of the *(reduced modulo A) Hamming distance* between u and s, and $ingr_{IND}(u, v)$ means in this context that v is 'closer' to the standard s than u.

This, as well as property (6.4), suggests usage of metrics in definitions of rough inclusions; actually, metrics were used in Poincaré [9] to give an example of a tolerance relation, see Ch. 1, sect. 13. We will build on this idea.

6.7   Rough Inclusions: Metric Models

We consider a metric space (X, ρ) and we let

$$\mu_\rho(x, y, r) \Leftrightarrow \rho(x, y) \leq 1 - r \qquad (6.34)$$

We check that μ_ρ is a rough inclusion.

Proposition 6.22. *The relation* μ_ρ *satisfies conditions RINC1–RINC3 with the ingredient relation of identity $=$ and the part relation empty.*

Proof. $\mu_\rho(x, y, 1)$ means that $\rho(x, y) \leq 0$, i.e., $x = y$. This proves RINC1 and RINC2, RINC3 follow   □

The rough inclusion μ_ρ obeys a transitivity law

Proposition 6.23. *If* $\mu_\rho(x, y, r)$ *and* $\mu_\rho(y, z, s)$, *then* $\mu_\rho(x, z, L(r, s))$.

Proof. From $\rho(x, y) \leq 1 - r$ and $\rho(y, z) \leq 1 - s$, by the triangle inequality for ρ, it follows that $\rho(x, z) \leq (1 - r) + (1 - s)$, i.e., $\rho(x, z) \leq 1 - (1 - r + s) = 1 - L(r, s)$, hence, $\mu_\rho(x, z, L(r, s))$ □

In particular, we may consider the *discrete metric* $D(x, y)$ defined as

$$D(x, y) = \begin{cases} 1 \text{ in case } x \neq y \\ 0 \text{ in case } x = y \end{cases} \qquad (6.35)$$

The rough inclusion μ_D satisfies the following

Proposition 6.24. *1.* $\mu_D(x, y, 1) \Leftrightarrow x = y$.
2. For $r < 1$, $\mu_D(x, y, r) \Leftrightarrow \mu_D(x, y, 0) \Leftrightarrow x \neq y$.

We have produced a *two–valued* rough inclusion, which we may justly call the *discrete rough inclusion*.

   We conclude our review of known to us types of rough inclusions with a 3–valued rough inclusion.

6.8   Rough Inclusions: A 3–Valued Rough Inclusion on Finite Sets

We define a 3–valued rough inclusion μ_3 by formulas

$$\mu_3(A, B, 1) \Leftrightarrow A \subseteq B \qquad (6.36)$$

$$\mu_3(A, B, 1/2) \Leftrightarrow A \triangle B \neq \emptyset, \qquad (6.37)$$

and

$$\mu_3(A, B, 0) \Leftrightarrow A \cap B = \emptyset \tag{6.38}$$

Symmetric rough inclusions offer technical advantage over non–symmetric ones; the possibility of inverting roles of objects allows for better control of mutual relationships between pairs of objects and from technical point of view offers a considerable advantage. Symmetric by definitions are rough inclusions of sects. 4, 6, 8.

6.9 Symmetrization of Rough Inclusions

Assume μ a transitive non–symmetric rough inclusion; by μ^{sym}, we denote the symmetrized version of μ defined by letting

$$\mu^{sym}(x, y, r) \Leftrightarrow \mu(x, y, r) \wedge \mu(y, x, r) \tag{6.39}$$

Proposition 6.25. *Properties of μ^{sym} are summed in*

1. μ^{sym} is a rough inclusion with the ingredient relation of identity $=$;

2. μ^{sym} is a transitive symmetric rough inclusion.

Proof. $\mu^{sym}(x, y, 1)$ means $\mu(x, y, 1)$ and $\mu(y, x, 1)$, hence, $ingr(x, y)$ and $ingr(y, x)$ thus $x = y$. RINC2, RINC3 follow easily. Symmetry is obvious by definition and transitivity follows easily $\square$

Here our discussion of basics of rough inclusions ends and we proceed to the analysis of structures which can be defined in rough mereological universes.

6.10 Mereogeometry

Elementary geometry was defined by Alfred Tarski in His Warsaw University lectures in the years 1926–27 as a part of Euclidean geometry which can be described by means of 1st order logic. Alfred Tarski proposed an axiomatization of elementary geometry, and many others including David Hilbert, Moritz Pasch, Eugenio Beltrami proposed also some axiomatizations of geometry.

There are two main aspects in formalization of geometry: one is metric aspect dealing with the distance underlying the space of points which carries geometry and the other is affine aspect taking into account linear relations.

In Tarski axiomatization, Tarski [23], the metric aspect is expressed as a relation of equidistance (congruence) and the affine aspect is expressed by means of the betweenness relation. The only logical predicate required is the identity $=$.

We recall here the Tarski formalism, although it is not our goal to discuss elementary geometry; we include this for completeness' sake as well as to be used in the sequel.

Equidistance relation denoted $Eq(x, y, u, z)$ (or, as a congruence: $xy \equiv uz$) means that the distance from x to y is equal to the distance from u to z (pairs x, y and u, z are equidistant) is subject to requirements

1. *Eq–reflexivity:* $Eq(x, y, y, x)$;

2. *Eq–identity:* If $Eq(x, y, z, z)$, then $x = y$;

3. *Eq–transitivity:* If $Eq(x, y, u, z)$ and $Eq(x, y, v, w)$, then $Eq(u, z, v, w)$.

Betweenness relation, denoted $B(x, y, z)$, (y is between x and z) is required to satisfy the requirements

1. *B–identity:* If $B(x, y, x)$, then $x = y$;

2. *B–Pasch axiom:* If $B(x, u, z)$ and $B(y, v, z)$, then there is some a such that $B(u, a, y)$ and $B(v, a, x)$;

3. *B–continuity:* Let $\phi(x)$ and $\psi(y)$ be first–order formulas in which objects a, b do not occur as free, and, x is not free in $\psi(y)$ and y is not free in $\phi(x)$. If there is a such that for each pair x, y [from $\phi(x)$ and $\psi(y)$ it follows that $B(a, x, y)$], then there is b such that for each pair x, y [from $\phi(x)$ and $\psi(y)$ it follows that $B(b, x, y)$];

4. *B–lower dimension:* For some triple a, b, c [not $B(a, b, c)$ or not $B(b, c, a)$ or not $B(c, a, b)$].

Requirements concerning mutual relations between the two predicates are following

1. *B, Eq–upper dimension:* If $Eq(x, u, x, v)$ and $Eq(y, u, y, v)$ and $Eq(z, u, z, v)$ and $u \neq v$, then $B(x, y, z)$ or $B(y, z, x)$, or $B(z, x, y)$;

2. *B, Eq–parallel postulate:* If $B(x, y, w)$ and $Eq(x, y, y, w)$, $B(x, u, v)$, $Eq(x, u, u, v)$, $B(y, u, z)$ and $Eq(y, u, z, u)$, then $Eq(y, z, v, w)$;

3. *B, Eq–five segment postulate*: If $(x \neq y)$ and $B(x, y, z)$, $B(x', y', z')$, $Eq(x, y, x', y')$, $Eq(y, z, y', z')$, $Eq(x, u, x', u')$, $Eq(y, u, y', u')$, then $Eq(z, u, z', u')$;

4. *B, Eq–segment extension postulate*: There exists a, such that $B(w, x, a)$ and $Eq(x, a, y, z)$.

Van Benthem [1] took up the subject proposing a version of betweenness predicate based on the nearness predicate which was a departing point for rough mereological geometry, see Polkowski and Skowron [20], and has served us in our definition of robot formations, see Ch. 8.

We are interested in introducing into the mereological world defined by μ of a geometry in whose terms it will be possible to express spatial relations among objects; a usage for this geometry has been found in navigation and control tasks of mobile robotics, see Polkowski and Osmialowski [16], [17], Osmiałowski [8], Polkowski and Szmigielski [21], Szmigielski [22].

We first introduce a notion of a distance κ, induced by a rough inclusion μ, see Polkowski and Skowron [20], and applications in Polkowski and Osmialowski [16], [17], Osmiałowski [8], Polkowski and Szmigielski [21], Szmigielski [22]

$$\kappa(X, Y) = min\{max\ r, max\ s :\ \mu(X, Y, r), \mu(Y, X, s)\} \qquad (6.40)$$

Observe that the mereological distance differs essentially from the standard distance: the closer are objects, the greater is the value of κ: $\kappa(X, Y) = 1$ means $X = Y$ whereas $\kappa(X, Y) = 0$ means that X, Y are either externally connected or disjoint, no matter what is the Euclidean distance between them.

The notion of *betweenness in the Tarski sense* $T(Z, X, Y)$ (read: Z is between X and Y), due to Tarski [23], is

$$T(Z, X, Y) \Leftrightarrow \text{for each region W, } \kappa(Z, W) \in [\kappa(X, W), \kappa(Y, W)] \quad (6.41)$$

Here, $[a, b]$ means the non–oriented interval with endpoints a, b.

Proposition 6.26. *The relation T satisfies the basic properties resulting from axioms of elementary geometry of Tarski [23] for the notion of betweenness*

1. *TB1 $T(Z, X, X)$ if and only if $Z = X$ (identity);*

2. *TB2 $T(V, U, W)$ and $T(Z, V, W)$ imply $T(V, U, Z)$ (transitivity);*

3. *TB3 $T(V, U, Z)$, $T(V, U, W)$ and $U \neq V$ imply $T(Z, U, W)$ or $T(W, U, Z)$ (connectivity).*

Proof. Indeed, by means of κ, the properties of betweenness in our context are translated into properties of betweenness in the real line which hold by the Tarski theorem, Tarski [23], Thm. 1   □

We apply κ to define in our context the relation N of *nearness* proposed in Van Benthem [1]

$$N(X, U, V) \;\Leftrightarrow\; \kappa(X, U) > \kappa(V, U) \tag{6.42}$$

Here, $N(X, U, V)$ means that X *is closer to* U *than* V *is to* U.

Then, N does satisfy all axioms for nearness in Van Benthem [1]

Proposition 6.27. *The relation N does satisfy the following postulates*

NB1 $N(Z, U, V)$ and $N(V, U, W)$ imply $N(Z, U, W)$ (transitivity)

NB2 $N(Z, U, V)$ and $N(U, V, Z)$ imply $N(U, Z, V)$ (triangle inequality)

NB3 $N(Z, U, Z)$ is false (irreflexivity)

NB4 $Z = U$ or $N(Z, Z, U)$ (selfishness)

NB5 $N(Z, U, V)$ implies $N(Z, U, W)$ or $N(W, U, V)$ (connectedness)

Proof. For NB1, assumptions are $\kappa(Z, U) > \kappa(V, U)$ and $\kappa(V, U) > \kappa(W, U)$; it follows that $\kappa(Z, U) > \kappa(W, U)$ i.e. the conclusion $N(Z, U, W)$ follows.

For NB2, assumptions $\kappa(Z, U) > \kappa(V, U)$, $\kappa(V, U) > \kappa(Z, V)$ imply $\kappa(Z, U) > \kappa(Z, V)$, i.e., $N(U, Z, V)$.

For NB3, it cannot be true that $\kappa(Z, U) > \kappa(Z, U)$.

For NB4, $Z \neq U$ implies in our world that $\kappa(Z, Z) = 1 > \kappa(Z, U) \neq 1$.

For NB5, assuming that neither $N(Z, U, W)$ nor $N(W, U, V)$, we have $\kappa(Z, U) \leq \kappa(W, U)$ and $\kappa(W, U) \leq \kappa(V, U)$ hence $\kappa(Z, U) \leq \kappa(V, U)$, i.e., $N(Z, U, V)$ does not hold   □

In addition to betweenness T, we make use of a *betweenness* relation in the sense of Van Benthem T_B introduced in Van Benthem [1]

$$T_B(Z, U, V) \;\Leftrightarrow\; [\text{for each } W \; (Z = W) \text{ or } N(Z, U, W) \text{ or } N(Z, V, W)] \tag{6.43}$$

The principal example bearing, e.g., on our approach to robot control deals with rectangles in 2D space regularly positioned, i.e., having edges parallel to coordinate axes. We model robots (which are represented in the plane as discs of the same radii in 2D space) by means of their safety regions about robots; those regions are modeled as rectangles circumscribed on robots. One of advantages of this representation is that safety regions can be always implemented as regularly positioned rectangles.

Given two robots a, b as discs of the same radii, and their safety regions as circumscribed regularly positioned rectangles A, B, we search for a proper choice of a region X containing A, and B with the property that a robot C contained in X can be said to be between A and B. In this search we avail ourselves with the notion of betweenness relation T_B.

Taking the rough inclusion μ^G defined in (6.16), sect. 6, for two disjoint rectangles A, B, we define the *extent*, $ext(A, B)$ of A and B as the smallest rectangle containing the union $A \cup B$. Then we have the claim, obviously true by definition of T_B.

Proposition 6.28. *We consider a context in which objects are rectangles positioned regularly, i.e., having edges parallel to axes in R^2. The measure μ is μ^G. In this setting, given two disjoint rectangles C, D, the only object between C and D in the sense of the predicate T_B is the extent $ext(C, D)$ of C, D, , i.e., the minimal rectangle containing the union $C \cup D$.*

Proof. As linear stretching or contracting along an axis does not change the area relations, it is sufficient to consider two unit squares A, B of which A has (0,0) as one of vertices whereas B has (a,b) with $a, b > 1$ as the lower left vertex (both squares are regularly positioned). Then the distance κ between the extent $ext(A, B)$ and either of A, B is $\frac{1}{(a+)(b+1)}$.

For a rectangle $R : [0, x] \times [0, y]$ with $x \in (a, a + 1), y \in (b, b + 1)$, we have that

$$\kappa(R, A) = \frac{(x - a)(y - b)}{xy} = \kappa(R, B) \tag{6.44}$$

For $\phi(x, y) = \frac{(x-a)(y-b)}{xy}$, we find that

$$\frac{\partial \phi}{\partial x} = \frac{a}{x^2} \cdot (1 - \frac{b}{y}) > 0 \tag{6.45}$$

and, similarly, $\frac{\partial \phi}{\partial y} > 0$, i.e., ϕ is increasing in x, y reaching the maximum when R becomes the extent of A, B.

An analogous reasoning takes care of the case when R has some (c,d) with $c, d > 0$ as the lower left vertex   □

Further usage of the betweenness predicate is suggested by the Tarski axiom of B, Eq–upper dimension, which implies collinearity of x, y, z. Thus,a line segment may be defined via the auxiliary notion of a pattern; we introduce this notion as a relation Pt.

We let $Pt(u, v, z)$ if and only if $T_B(z, u, v)$ or $T_B(u, z, v)$ or $T_B(v, u, z)$.

We will say that a finite sequence $u_1, u_2, ..., u_n$ of objects *belong in a line segment* whenever $Pt(u_i, u_{i+1}, u_{i+2})$ for $i = 1, ..., n-2$; formally, we introduce the functor *Line* of finite arity defined by means of

$$Line(u_1, u_2, ..., u_n) \text{ if and only if } Pt(u_i, u_{i+1}, u_{i+2}) \text{ for } i < n - 1$$

For instance, any two disjoint rectangles A, B and their extent $ext(A, B)$ form a line segment.

Proposition 6.29. *The relation T_B does satisfy the Tarski properties TB1–TB2, in the case of regular rectangles. The condition TB3 is easily falsified with very simple examples.*

Proof. For the case of rectangles, TB1 follows by the fact that $ext(X, X) = X$ (we interpret $Z \subseteq X$ as $Z = X$). As for TB2, assume that $T_B(V, U, W)$, $T_B(Z, V, W)$ hold, i.e., $V = ext(U, W)$, and $Z = ext(V, W)$ hence $Z = V$ and $V = ext(U, Z)$, i.e. $T_B(V, U, Z)$ holds.

In the case of robots, assume robots are modeled as squares of side length 1. Then TB1 follows as for rectangles, and for TB2, assume that $T_B(a, b, c)$, $T_B(d, a, c)$ hold. We describe this situation by assuming that in the extent $ext(b, c)$, b is at the left upper vertex of $ext(b, c)$, and c is in the right lower corner of $ext(b, c)$.

We will say: a is down and right to b and a is up and left to c. Similarly for $T_B(d, a, c)$: d is down and right to a and d is up and left to c. Hence, a is down and right to b, and a is up and left to d, meaning that $T_B(a, b, d)$ holds   □

We now return to the theme of mereotopology discussed in Ch. 5 for mereology.

6.11   Rough Mereotopology

We analyze now topological structures in rough mereological framework. We consider separately some cases depending on types of rough inclusions.

6.11.1   *The Case of Transitive and Symmetric Rough Inclusions*

Here we have rough inclusions of the form μ^T induced by Archimedean t–norms in sect. 6.3, rough inclusions of the form of *airi's*, see sect. 6.6 (18), and rough inclusions of the form μ_ρ induced by a metric ρ, see sect. 6.7 . We will use the general notation of $\mu_{s,t}$ to denote either of these forms. We thus assume that a rough inclusion $\mu_{s,t}$ is given, on a collection of objects, which obeys the transitivity law

$$\mu_{s,t}(x, y, r), \mu_{s,t}(y, z, s) \Rightarrow \mu_{s,t}(x, z, T(r, s)) \tag{6.46}$$

and is symmetric, i.e.

$$\mu_{s,t}(x, y, r) \Leftrightarrow \mu_{s,t}(y, x, r) \tag{6.47}$$

For each object x, we define an object $O_r(x)$ as the class of property $M(x, r)$, where

$$M(r, x)(y) \Leftrightarrow \mu_{s,t}(y, x, r), \tag{6.48}$$

and,

$$O_r(x) = ClsM(x, r) \tag{6.49}$$

Hence,

Proposition 6.30. $ingr(z, O_r(x))$ *if and only if* $\mu_{s,t}(z, x, r)$.

Proof. By Ch. 5, Proposition 4, $ingr(z, O_r(x))$ if and only if there exists t such that $Ov(z, t)$ and $\mu_{s,t}(t, x, r)$, hence, there exists w such that $ingr(w, z)$, $ingr(w, t)$, hence $w = z = t$, and finally $\mu_{s,t}(z, x, r)$ ☐

We regard the object $O_r(x)$ as an analogue of the notion of the 'closed ball about x of the radius r'.

To define the analogue of an open ball, we consider the property

$$M_r^+(x)(y) \Leftrightarrow \exists q > r.\mu_{s,t}(y, x, q) \tag{6.50}$$

The class of the property $M_r^+(x)$ will serve as the open ball analogue

$$Int(O_r(x)) = ClsM_r^+(x) \tag{6.51}$$

The counterpart of Proposition 6.30 for the new property is

Proposition 6.31. $ingr(z, Int(O_r(x)))$ *if and only if* $\exists q > r.\mu_{s,t}(z, x, q)$.

Proof. We follow the lines of the preceding proof. It is true that

$$ingr(z, Int(O_r(x)))$$

if and only if there exists t such that $Ov(z, t)$ and there exists $q > r$ for which $\mu_T(t, x, q)$ holds, hence, there exists w such that $ingr(w, z)$, $ingr(w, t)$, which implies that $w = z = t$, and finally $\mu_T(z, x, q)$ ☐

From Propositions 6.30 and 6.31, we infer that

Proposition 6.32. $ingr(Int(O_r(x)), O_r(x))$.

Proposition 6.33. *If* $s < r$, *then* $ingr(O_r(x), O_s(x))$, $ingr(Int(O_r(x)), Int(O_s(x)))$.

Consider z with $ingr(z, Int(O_r(x)))$. By Proposition 6.31, $\mu_T(z, x, s)$ holds with some $s > r$. We can choose $\alpha \in [0, 1]$ with the property that $T(\alpha, s) > r$. For any object w with $ingr(w, O_\alpha(z))$, we can find an object u such that $\mu_T(u, z, \alpha)$ and $Ov(w, u)$.

For an object t such that $ingr(t, u)$ and $ingr(t, w)$, we have $\mu_T(t, w, 1)$, $\mu_T(t, u, 1)$, hence, $\mu_T(t, x, T(\alpha, s))$, i.e, $ingr(t, Int(O_r(x)))$. As $t = w$, we find that $ingr(w, Int(O_r(x)))$. We have verified

Proposition 6.34. *For each object z with $ingr(z, Int(O_r(x)))$, there exists $\alpha \in [0,1]$ such that $ingr(O_\alpha(z), Int(O_r(x)))$.*

For any object z, what happens when $ingr(z, Int(O_r(x))$ and $ingr(z, O_s(y))$? We would like to find some $O_q(z)$ which would be an ingredient in either of $O_r(x), O_s(y))$.

Proposition 6.34 answers this question positively: we can find $\alpha, \beta \in [0,1]$ such that $ingr(O_\alpha(z), Int(O_r(x)))$ and $ingr(O_{beta}(z), Int(O_s(y)))$. By Proposition 6.33, $ingr(O_q(z), Int(O_r(x)))$, $ingr(O_q(z), Int(O_s(y)))$, for $q = max\{\alpha, \beta\}$.

We can sum up the last few facts

Proposition 6.35. *The collection $\{Int(O_r(x)) : x \text{ an object}, r \in [0,1]\}$ is an open basis for a topology on the collection of objects.*

We will call an object x *open*, $Open(x)$ in symbols, in case it is a class of some property of objects of the form $Int(O_r(x))$.

$$Open(x) \Leftrightarrow \exists \Psi, \text{ a non} - -\text{vacuous property of basic open sets. } x = Cls\Psi\} \tag{6.52}$$

Hence,

Proposition 6.36. *In consequence of (6.52)*

1. *If Φ is any non–vacuous property of objects of the form $Open(x)$, then $Open(Cls\Phi)$;*

2. *If $Ov(Open(x), Open(y))$, then $Open(Open(x) \cdot Open(y))$.*

We define closures of objects, and to this end, we introduce a property $\Phi(x)$ for each object x

$$\Phi(x)(y) \Leftrightarrow \forall s < 1.Ov(O_s(y), x) \tag{6.53}$$

Closures of objects are defined by means of

$$Cl(x) = Cls\Phi(x) \tag{6.54}$$

We find what it does mean to be an ingredient of the object $Cl(x)$.

Proposition 6.37. *$ingr(z, Cl(x))$ if and only if $Ov(O_r(z), x)$ for every $r < 1$.*

Proof. By definition, there exists w such that $Ov(w, z)$ and $\Phi(w)$. For t which is an ingredient of z, w, hence, $z = t = w$, we have $\Phi(z)$, i.e., $Ov(O_r(z), x)$ for all $r < 1$ $\qquad\qquad\square$

In particular

Proposition 6.38. *$ingr(z, Cl(O_w(x)))$ if and only if $Ov(O_r(z), O_w(x))$ for every $r < 1$.*

We follow this line of analysis. $Ov(O_r(z), O_w(x))$ means that we find q such that $ingr(q, O_r(z))$, $ingr(q, O_w(x))$, hence, by Proposition 6.30, $\mu_{s,t}(q, z, r)$ and $\mu_{s,t}(q, x, w)$, hence, by symmetry of $\mu_{s,t}$, we have $\mu_{s,t}(z, x, T(r, w))$, and, by continuity of T, with $r \to 1$, we obtain $\mu_{s,t}(z, x, w)$, i.e., by Proposition 6.30, $ingr(z, O_w(x))$. We have proved

Proposition 6.39. $ingr(z, Cl(O_w(x)))$ *if and only if* $ingr(z, O_w(x))$.

A corollary follows

Proposition 6.40. $Cl(O_w(x)) = O_w(x)$.

We can also address the notion of the *interior of an object*. We define $Int(x)$, the *interior of* x as

$$ingr(z, Int(x)) \Leftrightarrow \exists w.[Ov(z, w) \land \exists r < 1.ingr(O_r(w), x)] \qquad (6.55)$$

A standard by now reasoning shows

Proposition 6.41. $ingr(z, Int(x))$ *if and only if there exists* $r < 1$ *such that* $ingr(O_r(z), x)$.

This implies that the notion of an interior is valid for objects of the form $O_r(x)$, where $r < 1$ and x is an object.

We can now address the problem of a *boundary* of any object of the form $O_r(x)$. We define the boundary $Bd(O_r(x))$ as

$$Bd(O_r(x)) = O_r(x) \cdot -Int(O_r(x)) \qquad (6.56)$$

We have a characterization of boundary ingredients

Proposition 6.42. $ingr(z, Bd(O_r(x)))$ *if and only if*

$$\mu_{s,t}(z, x, r) \land \neg \exists q > r.\mu_{s,t}(z, x, q)$$

Proof. $ingr(z, Bd(O_r(x)))$ if and only if

$$ingr(z, O_r(x)), \neg ingr(z, Int(O_r(x)))$$

hence, $\mu_{s,t}(z, x, r)$ and $\mu_{s,t}(z, x, q)$ for no $q > r$ □

We introduce a symbol $\overline{\mu}_{s,t}(z, x, r) = sup\{q : \mu_T s, t(z, x, q)\}$. With its help, we write the last result down as

$$ingr(z, Bd(O_r(x))) \Leftrightarrow \mu_T(z, x, r) \land \overline{\mu(z, x)} = r \qquad (6.57)$$

Summing up our discussion, we may state that in case of transitive symmetric rough inclusions, objects to whom notions of closure as well as interior can be assigned are 'collective' objects of the form $O_r(x)$ but not object of the form x; in one sense we obtain a 'pointless' topology, on the other hand, this topology is like the orthodox topology as open objects are 'neighborhoods' of the form $O_r(x)$.

6.11.2 The Case of Transitive Non–symmetric Rough Inclusions

This case is more difficult as lack of symmetry of μ prohibits some inferences; on the other hand, it is more interesting, as, e.g., no longer $ingr(x,y)$ implies $x = y$. We use a symbol μ to denote a rough inclusion induced by a continuous t–norm T; this time, μ is transitive but not symmetric, e.g., it is a residual rough inclusion of the form $x \Rightarrow_T y \geq r$, see sect. 6.2.

We preserve definitions of objects $O_r(x)$, $Int(O_r(x))$ and we analyze these notions for a transitive, non–symmetric rough inclusion μ. When transitivity issue comes into play, we assume that it is due to a continuous t–norm T.

We begin with $O_r(x)$, and we prove

Proposition 6.43. $ingr(z, O_r(x))$ if and only if there exists t such that $ingr(t, z)$ and $\mu(t, x, r)$.

Proof. $ingr(z, O_r(x))$ means that $Ov(z, w)$ and $\mu(w, x, r)$. For t with $ingr(t, w)$ and $ingr(t, z)$, we have $\mu(t, w, 1)$ and $\mu(t, z, 1)$, hence $\mu(t, x, T(r, 1))$, i.e., $\mu(t, x, r)$ □

We will call t in the conclusion of Proposition 6.43, an 'O–*witness*' for the property $ingr(z, O_r(x))$ of z.

Similarly, we characterize interiors of O–objects

Proposition 6.44. $ingr(z, Int(O_r(x)))$ if and only if there is t such that $ingr(t, z)$ and $\mu(t, x, s)$ for some $s > r$.

By analogy, we call t in the conclusion of Proposition 6.44, an 'O^+–*witness*' for the property $ingr(z, Int(O_r(x)))$ of z.

The witness has good neighborhood properties.

Proposition 6.45. For an O^+–*witness* t for $ingr(z, Int(O_r(x)))$, there exists $s < 1$ such that $ingr(O_s(t), Int(O_r(x)))$ holds.

Proof. We have $\mu(t, x, s)$ with $s > r$, and, $ingr(t, z)$. We choose $q < 1$ such that $T(q, s) > r$. Consider y with $ingr(y, O_q(t))$, so there is w with properties $ingr(w, y)$, $\mu(w, t, q)$.

Hence, $\mu(w, x, T(s, q))$ and $ingr(w, Int(O_r(x)))$. By mereology axiom M3, Ch. 5, sect. 1, $ingr(O_q(t), Int(O_r(x)))$ □

As an upshot to Proposition 6.45, we will call an object x *open*, $Open(x)$ in symbols, when the condition is satisfied

$$Open(x) \Leftrightarrow [ingr(z, x) \Leftrightarrow \exists t, 0 < r < 1. ingr(O_r(t), x) \wedge ingr(t, z)] \quad (6.58)$$

A corollary follows

$$Open(Int(O_r(x))) \quad (6.59)$$

for each object x and each $r < 1$.

Following this line of reasoning, we define the *interior* of an object x, $Int(x)$ in symbols, letting

$$ingr(z, Int(x)) \Leftrightarrow \exists t.[ingr(t, z) \land \exists r < 1.ingr(O_r(t), x)] \qquad (6.60)$$

Clearly

Proposition 6.46. *Properties of the operator Int are*

1. $ingr(Int(x), x)$;

2. $ingr(Int(Int(x)), Int(x))$;

3. $ingr(x, y) \Rightarrow ingr(Int(x), Int(y))$.

Dually, we define the *closure* of x, $Cl(x)$ in symbols, by means of

$$ingr(z, Cl(x)) \Leftrightarrow \exists t.[Ov(t, z) \land \forall r < 1.Ov(O_r(t), x)] \qquad (6.61)$$

This condition can be disentangled; there exists w with $ingr(w, t), ingr(w, z)$ and there exists u_r such that $ingr(q_r, x), \mu(q_r, z, r)$ for each $r < 1$.

This intricate requirement can be simplified in *discrete* case when the number of objects is *finite*. Then, for r sufficiently close to 1, the only q_r satisfying the condition $\mu(q_r, z, r)$ is z itself, hence, $ingr(z, x)$ and finally, $Ov(z, x)$. We obtain

Proposition 6.47. *In discrete case, $ingr(z, Cl(x))$ if and only if $Ov(z, x)$.*

Hence

Proposition 6.48. *In discrete case,*

1. $ingr(x, Cl(x))$;

2. $ingr(Cl(x), Cl(Cl(x)))$;

3. $ingr(x, y) \Rightarrow ingr(Cl(x), Cl(y))$.

We obtain a *quasi–Čech* closure operation. Finally, we take up the boundary question. Again, we define the boundary of x, $Bd(x)$ as $Cl(x) \cdot -Int(x)$. The specific condition can be revealed as

Proposition 6.49. *In discrete case, for each object x, the boundary $Bd(x)$ satisfies $ingr(z, Bd(x))$ if and only if $Ov(z, x)$ and $\exists t.[ingr(t, z) \land \forall r < 1.\neg ingr(O_r(t), x)]$.*

6.12   Connections from Rough Inclusions

We will explore now the possibility of inducing a connection in an environment endowed with a rough inclusion μ. As before, we consider cases when μ is symmetric or not. We still retain the symbol μ as generic for rough inclusions.

6.12.1   The Case of Transitive and Symmetric Rough Inclusions

For a transitive and symmetric rough inclusion μ, we define a connection C_μ by letting

$$C_\mu(x,y) \Leftrightarrow \forall r < 1.Ov(O_r(x), O_r(y)) \tag{6.62}$$

We need to make sure that C_μ is a connection, see Ch. 5, 5

Proposition 6.50. C_μ *does satisfy conditions CN1–CN3 for connections.*

Proof. CN1, CN2 are satisfied obviously. For CN3, observe that when $x \neq y$ then, e.g., $\neg ingr(x,y)$, i.e., $\mu(x,y,r)$ implies $r < 1$, hence, there is $s < 1$ that $\mu(x,y,s)$ does not hold,i.e., it is not true that $Ov(O_q(x), O_q(y))$ holds for q large enough to satisfy $T(q,q) > s$ □

We can explore the form of other predicates induced from C_μ. We know that $Ov(x,y)$ means $x = y$, hence, external connectedness is expressed by means of

$$EC_\mu(x,y) \Leftrightarrow (x \neq y) \wedge C_\mu(x,y). \tag{6.63}$$

This implies the form of $Tingr_{C_\mu}$ and $NTingr_{C_\mu}$.

In discrete case, for r sufficiently close to 1, $O_r(x) = x$ for each x, hence, $C_\mu(x,y)$ means simply $x = y$, $EC_\mu(x,y)$ is not defined, hence, $Tingr_{C_\mu}$ is not defined and $NTingr_{C_\mu}(x) = x$ for each x. In continuous case, the situation depends on additional properties of μ: if μ is *continuous* in the sense

$$lim_{r \to s} O_r(x) = O_s(x) \tag{6.64}$$

then, again, $C_\mu(x,y) \Leftrightarrow (x = y)$, with $EC_{mu}, Tingr_{C_\mu}, NTingr_{C_\mu}$ as above.

An example of continuous μ is μ_ρ: as $\mu_\rho(x,y,r) \Leftrightarrow \rho(x,y) \leq 1 - r$, where ρ a metric, (6.64) is satisfied with μ_ρ.

6.12.2   The Case of Symmetric Non–transitive Rough Inclusions and the General Case

The general case is represented by rough inclusions of the type *indri*, see sect. 6.6 (30), and, as we know, $ingr(x,y)$ means in this case that $|IND(x,s)| \leq$

$|IND(y,s)|$. Thus, if there exists an object z with $IND(z,s) = \emptyset$, then z is an ingredient of each object, hence, any two objects overlap, and the definition (6.62) is not plausible. The only definition that remains is to accept $C(x,y)$ if and only of $x = y$, i.e., when $|IND(X,s)| = |IND(y,s)|$. The case of symmetric non–transitive rough inclusions is represented, e.g., by set–theoretic rough inclusions $\mu^S(x,y,r) \Leftrightarrow \frac{|x \cap y|}{|x|} \geq r$, see sect. 6.4, and geometric rough inclusions $\mu^G(x,y,r) \Leftrightarrow \frac{||x \cap y||}{||x||} \geq r$, see sect. 6.5. We consider a family $\mathcal{F}$ of finite subsets of a certain universe U and a family $\mathcal{C}$ of convex compact regions in a certain space R^k which share the following properties

$$RINC4 \; extr(x,y) \Rightarrow \exists r < 1.extr(O_r(x), O_r(y)) \qquad (6.65)$$

$$RINC5 \; \neg ingr(x,y) \Rightarrow \exists z.ingr(z,x) \wedge extr(z,y) \qquad (6.66)$$

With respect to the property RINC4, one may say that objects in $\mathcal{F}$, respectively, in $\mathcal{C}$ are *well–separated*. We adopt the definition of a connection C in terms of overlapping

$$C(x,y) \Leftrightarrow \forall r < 1.Ov(O_r(x), O_r(y)) \qquad (6.67)$$

Proposition 6.51. *Under RINC4,RINC5, C is a connection.*

Proof. Again, CN1, CN2, are obviously satisfied, and only CN3 needs to be verified. In case $x \neq y$, assume, e.g., that $\neq ingr(x,y)$. Under complementation RINC5, there exists z with $ingr(z,x)$ and $extr(z,y)$. By RNC4, there is $r < 1$ such that $extr(O_r(z), O_r(y))$, whereas, $Ov(O_r(z), O_r(x))$ for each $r < 1$ $\qquad\qquad\qquad\qquad\qquad\qquad\qquad\qquad\qquad\qquad\qquad\qquad\qquad\qquad \square$

Under this definition, external connectedness can be defined as

Proposition 6.52. $EC(x,y)$ *if and only if* $Ov(O_r(x), O_r(y))$ *for each* $r < 1$ *but not* $Ov(x,y)$.

The notion of non–tangential part comes down to

Proposition 6.53. $NTingr_C(x,y) \Leftrightarrow C(z,x) \Rightarrow Ov(z,y)$ *for each* z.

We define the interior $Int_C(x)$ for each object x, as usual as the class of the property of being a non–tangential part

$$NTP(x)(y) \Leftrightarrow NTingr_C(y,x), \qquad (6.68)$$

and

$$Int_C(x) = ClsNTP(x) \qquad (6.69)$$

Hence,

$$ingr(z, Int_C(x)) \Leftrightarrow \exists w. Ov(z, w) \wedge NTingr_C(w, x) \qquad (6.70)$$

Our claim is that

Proposition 6.54. $ingr(z, Int_C(x))$ *implies that* $ingr(O_r(z), x)$ *for some* $r < 1$.

Proof. Assuming to the contrary, that it is not true that $ingr(O_r(z), x)$ for $r < 1$, we obtain that the complement $-x$ is externally connected to x and z, a contradiction   □

A corollary follows

Proposition 6.55. $ingr(Int_C(x), Int(x))$, *where* $Int(x)$ *is the interior of* x *defined in terms of* μ.

The converse also holds

Proposition 6.56. $ingr(Int(x), Int_C(x))$ *holds for each object* x.

Proof. For z with $ingr(z, Int(x))$, we have that $ingr(O_r(z), x)$ for some r; if $C(z, w)$ for any w, then $Ov(O_r(z), O_r(w))$, but this means that $Ov(x, w)$ by RINC4   □

A corollary is

Proposition 6.57. $Int_C(x) = Int(x)$ *for every* x.

Relations of rough mereology to the fuzzy set theory are highlighted by the fact that rough inclusions are higher–order fuzzy equivalences in the sense of Zadeh [25].

6.13   Rough Inclusions as Many–Valued Fuzzy Equivalences

Rough inclusions are defined as relations of a part to a degree and this makes them affine to fuzzy constructs, though they stem form different inspirations and have a different logical structure. Nevertheless, one may try to relate both worlds. We have a result that states, see, e.g., Polkowski [10], [12] that any rough inclusion $\mu(x, y, r)$ does induce on its universe a fuzzy similarity relation in the sense of Zadeh [25], see Ch. 4, sect. 12. First, writing $\mu_y(x) = r$ instead of $\mu(x, y, r)$, we convert the relational notation into the fuzzy–style one. We observe that fuzzy sets of the form μ_y are higher–level sets: values of fuzzy membership degrees here are convex sub–intervals of the unit interval $[0, 1]$ of the form $[0, r]$, i.e., with the left–end point 0, hence the formula $\mu_y(x) = t$ is understood as the statement that the sub–interval $\mu_y(x)$ contains t. Under this proviso, the *fuzzy tolerance relation* $\tau_y^\mu(x)$ is defined by means of

$$\tau_y^\mu(x) = r \Leftrightarrow \mu_y(x) = r \text{ and } \mu_x(y) = r \qquad (6.71)$$

and it does satisfy, clearly,

1. $\tau_x^\mu(x) = 1$;

2. $\tau_x^\mu(y) = \tau_y^\mu(x)$.

We will use now the notation $\tau_r(x, y)$ for $\tau_x(y) = r$, disregarding the rough inclusion μ. Following Zadeh [25], we define *similarity classes* $[x]_\tau$ as fuzzy sets satisfying the condition,

$$\chi_{[x]_\tau}(y) = r \Leftrightarrow \tau_r(x, y) \qquad (6.72)$$

and in this interpretation, τ becomes a fuzzy equivalence in the sense of Zadeh [25], see Polkowski [43], i.e., the family $\{[x]_\tau : x \subseteq U\}$ does satisfy the requirements for a *T–fuzzy partition* in the sense of Zadeh [25], cf., Ch. 4, sect. 12

$$\forall x \exists y. \chi_{[x]_\tau}(y) = 1 \qquad (6.73)$$

and

$$[x]_\tau \neq [z]_\tau \Rightarrow max_y\{min\{\chi_{[x]_\tau}(y), \chi_{[z]_\tau}(y)\}\} < 1, \qquad (6.74)$$

and,

$$\bigcup_x [x]_\tau \times_T [x]_\tau = \tau \qquad (6.75)$$

where $A \times_T B$ denotes the fuzzy set defined via,

$$\chi_{A \times_T B}(u, v) = T(\chi_A(u), \chi_B(v)) \qquad (6.76)$$

and $\bigcup$ denotes the supremum operator.

We include an argument, cf., Polkowski [43], see Ch. 4, sect. 12, for properties FS1–FS3. For (6.73), it is satisfied with $x = y$. To justify (6.74), observe that the existence of y with $\tau(x, y) = 1 = \tau(z, y)$ would imply $\tau(x, z) = 1 = \tau(z, x)$, hence, $x = z$. For (6.75), we have for given x, y, z by FS3 that

$$T(\tau(x, y), \tau(x, z)) = T(\tau(y, x), \tau(x, z)) \leq \tau(y, z)$$

But, for $x = y$, we have by FS1 that

$$[x]_\tau \times_T [x]_\tau(y, z) = [y]_\tau \times_T [y]_\tau(y, z) = T(\tau(y, y), \tau(y, z)) = T(1, \tau(y, z)) =$$

$$\tau(y, z)$$

so (6.75) follows.

References

1. Van Benthem, J.: The Logic of Time. Reidel, Dordrecht (1983)
2. Faucett, W.M.: Compact semigroups irreducibly connected between two idempotents. Proc. Amer. Math.Soc. 6, 741–747 (1955)
3. Hájek, P.: Metamathematics of Fuzzy Logic. Kluwer, Dordrecht (1998)
4. Klösgen, W., Zytkow, J. (eds.): Handbook of Data Mining and Knowledge Discovery. Oxford University Press, Oxford (2002)
5. Kolmogorov, A.N.: On the representation of continuous functions of many variables by superposition of continuous functions of one variable and addition. Amer. Math.Soc. Transl. 28, 55–59 (1963)
6. Ling, C.-H.: Representation of associative functions. Publ. Math. Debrecen 12, 189–212 (1965)
7. Mostert, P.S., Shields, A.L.: On the structure of semigroups on a compact manifold with a boundary. Ann. Math. 65, 117–143 (1957)
8. Ośmiałowski, P.: On path planning for mobile robots: Introducing the mereological potential field method in the framework of mereological spatial reasoning. Journal of Automation, Mobile Robotics and Intelligent Systems (JAMRIS) 3(2), 1–10 (2009)
9. Poincaré, H.: La Science et l'Hypothése. Flammarion, Paris (1902)
10. Polkowski, L.: Rough Sets. Mathematical Foundations. Physica Verlag, Heidelberg (2002)
11. Polkowski, L.: A rough set paradigm for unifying rough set theory and fuzzy set theory. In: Wang, G., Liu, Q., Yao, Y., Skowron, A. (eds.) RSFDGrC 2003. LNCS (LNAI), vol. 2639, pp. 70–78. Springer, Heidelberg (2003)
12. Polkowski, L.: Toward rough set foundations. Mereological approach. In: Tsumoto, S., Słowiński, R., Komorowski, J., Grzymała-Busse, J.W. (eds.) RSCTC 2004. LNCS (LNAI), vol. 3066, pp. 8–25. Springer, Heidelberg (2004)
13. Polkowski, L.: Formal granular calculi based on rough inclusions. In: Proceedings of IEEE 2005 Conference on Granular Computing GrC 2005, pp. 57–62. IEEE Press, Beijing (2005)
14. Polkowski, L.: Rough–fuzzy–neurocomputing based on rough mereological calculus of granules. International Journal of Hybrid Intelligent Systems 2, 91–108 (2005)
15. Polkowski, L.: A model of granular computing with applications. In: Proceedings of IEEE 2006 Conference on Granular Computing GrC 2006, pp. 9–16. IEEE Computer Society Press, Atlanta (2006)
16. Polkowski, L.: Ośmiałowski P. Spatial reasoning with applications to mobile robotics. In: Xing–Jian, J. (ed.) Motion Planning for Mobile Robots: New Advances, InTech, Vienna, pp. 433–453 (2008)

17. Polkowski, L., Ośmiałowski, P.: Navigation for mobile autonomous robots and their formations: An application of spatial reasoning induced from rough mereological geometry. In: Barrera, A. (ed.) Mobile Robots Navigation. I–Tech, Zagreb, pp. 329–354 (2010)
18. Polkowski, L., Skowron, A.: Rough mereology. In: Raś, Z.W., Zemankova, M. (eds.) ISMIS 1994. LNCS, vol. 869, pp. 85–94. Springer, Heidelberg (1994)
19. Polkowski, L., Skowron, A.: Rough mereology: a new paradigm for approximate reasoning. International Journal of Approximate Reasoning 15(4), 333–365 (1997)
20. Polkowski, L., Skowron, A.: Rough mereology in information systems with applications to qualitative spatial reasoning. Fundamenta Informaticae 43, 291–320 (2000)
21. Polkowski, L., Szmigielski, A.: Computing with words via rough mereology in mobile robot navigation. In: Proceedings 2003 IEEE/RSJ Int. Conf. Intell. Robots and Systems IROS 2003, pp. 3498–3503. IEEE Press, Los Alamitos (2003)
22. Szmigielski, A.: A description based on rough mereology of the workspace of a mobile robot by means of a system of ultrasound sensors (in Polish). L. Polkowski (supervisor). PhD Dissertation. Dept. Electronics and Computer Techniques, Warsaw University of Technology (2003)
23. Tarski, A.: What is elementary geometry? In: Henkin, L., Suppes, P., Tarski, A. (eds.) The Axiomatic Method with Special Reference to Geometry and Physics, pp. 16–29. North-Holland, Amsterdam (1959)
24. Zadeh, L.A.: Fuzzy sets. Information and Control 8, 338–353 (1965)
25. Zadeh, L.A.: Similarity relations and fuzzy orderings. Information Sciences 3, 177–200 (1971)

Chapter 7
Reasoning with Rough Inclusions: Granular Computing, Granular Logics, Perception Calculus, Cognitive and MAS Reasoning

Rough mereology allows for a plethora of applications in various reasoning schemes due to universality of its primitive predicate of a part to a degree. We have already stressed that by its nature, rough mereology is especially suited to reasoning with collective concepts like geometric figures or solids, or, concepts learned by machine learning methods, i.e., with collective concepts. Those applications are presented in Ch. 8 and Ch. 9. In this chapter, we begin this discussion with a formal approach to the problem of granulation of knowledge and then we examine rough mereological logics: from our results in Ch. 6 it follows that representing implication with a rough inclusion μ leads to logics which extend and generalize fuzzy logics. As an application, we propose a formal rendering of the idea of perception calculus, due to Zadeh [67]. We apply rough mereological schemes to reasoning by multi–agent (MAS) systems, and finally we present a rough mereological variant of cognitive reasoning in neural–like systems.

7.1 On Granular Reasoning

The creator of Fuzzy Set Theory Lotfi A. Zadeh proposed to compute with granules in Zadeh [66]. The idea was natural, as fuzzy reasonings are carried out in terms of fuzzy membership functions. A fuzzy membership function μ_X maps a universe U of objects into the interval $[0, 1]$ and it represents the membership in a set X as a membership to a degree. The value $\mu_X(x) = r$ is interpreted as the statement that the object x is an element of the set X to the degree of r. The mapping $U \to U/\mu_X$ which sends each object x to its fibre $\mu_X^{-1}(\mu_X(x))$ identifies into the granule $g_\mu(r)$ all objects that belong in X to the degree r.

All fuzzy constructs are then expressed in terms of those granules. In this sense, fuzzy reasoning is in a natural way reasoning with granules. Granules in this reasoning are constructed in a uniform way, i.e., all objects in a granule share the same property of external character: they belong to an "oracle"

L. Polkowski: Approximate Reasoning by Parts, ISRL 20, pp. 259–295.
springerlink.com   © Springer-Verlag Berlin Heidelberg 2011

X to the same degree; changing X produces a variety of granules to reason with. The relation forming any granule is an equivalence R_X: the universe is decomposed into fibres of a fuzzy membership function μ_X in question, and the knowledge base is composed of all relations R_X for subsets $X \subseteq U$ of the universe of objects.

The same conclusion concerns rough set reasoning: elementary objects in reasoning are indiscernibility classes – elementary granules which are elements of a partition of the universe of objects by an indiscernibility relation $IND(B)$ for some set B of attributes, see Ch. 4. These granules are the smallest objects which can be described in terms of attributes and their values, i.e., in terms of descriptors and they are used in forming descriptions of objects, in building decision rules and classifiers as well as control algorithms.

Reasoning by means of aggregating objects, situations, cases, etc., into granules of similar entities is common to all forms of human reasoning. It is therefore important to capture this form of reasoning in its essential and typical facets and render it in mereological environment with help of mereological notions.

In Lin [21], [22], topological character of granules was recognized and the basic notion of a neighborhood system as the meaning of the collection of granules on the universe of objects was brought forth, Lin [22] recognized the import of tolerance relations, see Nieminen [37], cf., Ch. 1, sect. 13, by discussing tolerance induced neighborhoods.

In all hybrid approaches involving fuzzy or rough sets along with neural networks, genetic algorithms, etc., etc., one is therefore bound to compute with granules; this fact testifies to the importance of granular structures.

In search of adequate similarity relations, various forms of granules were proposed and considered as well as experimentally verified as to their effectiveness. In information systems, indiscernibility classes were proposed as granules, or, more generally, *templates* have served that purpose, i.e., meanings of generalized descriptors of the form $(a \in W_a)$ where $W_a \subseteq V_a$ with the meaning $[[(a \in W_a)]] = \{u \in U : a(u) \in W_a\}$, see Nguyen S. H. [36]; clearly, templates are aggregates, in ontological sense, of descriptors, i.e., they form "big" granules. Their usage is motivated by their potentially greater descriptive force than that of descriptors; a judicious choice of sets W_a should allow for constructing of a similarity relation that would reflect satisfactorily well the dependence of decision on conditional attributes.

As means for granule construction, rough inclusions have been considered and applied, in Polkowski [41] – [52]. The idea of granule formation and analysis rests on usage of the mereological class operator in the framework of mereological reasoning.

Granules formed by rough inclusions are used in models of fusion of knowledge, rough–neural computing and in building many–valued logics reflecting the rough set ideology in reasoning, and these forms of reasoning are discussed in further parts of this chapter.

Granulation of knowledge can be considered from a few angles

1. *General purpose of granulation;*

2. *Granules from binary relations;*

3. *Granules in information systems from indiscernibility;*

4. *Granules from generalized descriptors;*

5. *Granules from rough inclusions – mereological approach.*

We briefly examine those facets of granulation.

7.2   On Methods for Granulation of Knowledge

Granulation of knowledge comes into existence for a few reasons; the principal one is founded on the underlying assumption of basically all paradigms, viz., that reality exhibits a fundamental continuity, i.e., objects with identical descriptions in the given paradigm should exhibit the same properties with respect to classification or decision making, in general, in their behavior towards the external world.

For instance, fuzzy set theory assumes that objects with identical membership descriptions should behave identically and rough set theory assumes that objects indiscernible with respect to a group of attributes should behave identically, in particular they should fall into the same decision class.

Hence, granulation is forced by assumptions of the respective paradigm and it is unavoidable once the paradigm is accepted and applied. Granules induced in the given paradigm form the first level of granulation.

In the search for similarities better with respect to applications like classification or decision making, more complex granules are constructed, e.g., as unions of granules of the first level, or more complex functions of them, resulting, e.g., in fusion of various granules from distinct sources.

Among granules of the first two levels, some kinds of them can be exhibited by means of various operators, e.g., the class operator associated with a rough inclusion.

7.2.1   *Granules from Binary Relations*

Granulation on the basis of binary general relations as well as their specializations to, e.g., tolerance relations, has been studied by Lin [23] – [28], in particular as an important notion of a neighborhood system; see also Yao [64], [65]. This approach extends the special case of the approach based on indiscernibility. A general form of this approach according to Yao [64], [65], exploits the classical notion of *Galois connection*, cf., Ch. 1 sect. 4: two mappings $f : X \to Y, g : Y \to X$ form a Galois connection between ordered sets $(X, <)$ and $(Y, \prec)$ if and only if the equivalence $x < g(y) \Leftrightarrow f(x) \prec y$ holds.

In an information system (U, A), for a binary relation R on U, one considers the sets $xR = \{y \in U : xRy\}$ and $Rx = \{y \in U : yRx\}$, called, respectively, the *successor neighborhood* and the *predecessor neighborhood* of x. These sets are considered as granules formed by a specific relation of being affine to x in the sense of the relation R. Other forms of granulation can be obtained by comparing objects with identical neighborhoods, e.g., $x \equiv y \Leftrightarrow xR = yR$.

Saturation of sets of objects X, Y with respect to the relation R leads to sets X^*, Y^* such that $X * R = Y* \Leftrightarrow RY^* = X^*$, forming a Galois connection. This approach is closely related to the Formal Concept Analysis of Wille [63].

7.2.2 Granules in Information Systems from Indiscernibility

Granules based on indiscernibility in information/decision systems, see Ch. 4, sect. 2, are constructed as indiscernibility classes: given an information system (U, A) and the collection $IND = \{IND(B) : B \subseteq A\}$ of indiscernibility relations, each elementary granule is of the form of an indiscernibility relation $[u]_B = \{v \in U : (u, v) \in IND(B)\}$ for some B. Among those granules, there are minimal ones: granules of the form $[u]_A$ induced from the set A of all attributes.

Granules $[u]_A$ form the finest partition of the universe U which can be obtained by means of indiscernibility; given the class $[u]_B$ and the class $[u]_{A \setminus B}$, we have $[u]_A = [u]_B \cap [u]_{A \setminus B}$ hence $[u]_B = \bigcup_{v \in [u]_B \cap DIS_{A \setminus B}(u)} [v]_A$, where $v \in DIS_{A \setminus B}(u)$ if and only if there exists an attribute $a \in A \setminus B$ such that $a(u) \neq a(v)$. It is manifest that granules $[u]_B$ can be arranged into a tree with the root $[u]_\emptyset = U$ and leaves of the form $[u]_A$.

Granules based on indiscernibility form a complete Boolean algebra generated by atoms of the form $[u]_A$: unions of these atomic granules are closed on intersections and complements and these operations induce into granules the structure of a field of sets.

Atomic granules are made into some important unions by approximation operators, see Ch. 4, sect. 2: given a concept $X \subseteq U$, the lower approximation $\underline{B}X$ to X over the set B of attributes is defined as the union $\bigcup\{[u]_B : [u]_B \subseteq X\}$; the operator $L_B : Concepts \to Granules$ sending X to $\underline{B}X$ is monotone increasing and idempotent: $X \subseteq Y$ implies $LX \subseteq LY$ and $L \circ L = L$. Similarly, the upper approximation $\overline{B}X = \bigcup\{[u]_B : [u]_B \cap X \neq \emptyset\}$ to X over B, makes some elementary granules into the union; the operator U^B sending concepts into upper approximations is also monotone increasing and idempotent.

7.2.3   Granules from Generalized Descriptors

Some authors have made use of generalized descriptors called *templates*, see
Nguyen S. H. [36]; a template is a formula $T : (a \in W_a)$, where $W_a \subseteq V_a$,
with the meaning $g(T) : \{u \in U : a(u) \in W_a\}$. The granule $g(T)$ can be
represented as the union $\bigcup_{u \in W_a}[u]_a$; granules of the form $[u]_a$ are also called
blocks, see Grzymala–Busse [15], Grzymala–Busse and Ming Hu [16].

7.3   Granules from Rough Inclusions: The Mereological Approach to Granulation of Knowledge

Assume that a rough inclusion μ is given along with the associated ingredient
relation *ingr*, as in postulate RINC1, in Ch. 6, sect. 1.

The granule $g_\mu(u,r)$ of the radius r about the center u is defined as the
class of property $\Phi^\mu_{u,r}$

$$\Phi^\mu_{u,r}(v) \Leftrightarrow \mu(v,u,r) \tag{7.1}$$

The granule $g_\mu(u,r)$ is defined by means of

$$g_\mu(u,r) = Cls\Phi^\mu_{u,r} \tag{7.2}$$

Properties of granules depend, obviously, on the type of rough inclusion used
in their definitions. We consider separate cases, as some features revealed
by granules differ from a rough inclusion to a rough inclusion. The reader
is asked to refer to Ch. 5 for description of mereological reasoning, which is
going to be used in what follows.

In case of Archimedean t–norm–induced rough inclusions, see Ch. 6, sect.
3, or metric–induced rough inclusions, see Ch. 6, sect. 7, by their transitivity,
and symmetry, the important property holds, see Polkowski [52].

Proposition 7.1. *In case of a symmetric and transitive rough inclusion* μ,
for each pair u,v *of objects , and* $r \in [0,1]$, *ingr*$(v,g_\mu(u,r))$ *if and only if*
$\mu(v,u,r)$ *holds. In effect, the granule* $g_\mu(u,r)$ *can be represented as the set*
$\{v : \mu(v,u,r)\}$.

Proof. Assume that *ingr*$(v,g_\mu(u,r))$ holds. Thus, there exists z such that
$Ov(z,v)$ and $\mu(z,u,r)$. There is x with *ingr*(x,v), *ingr*(x,z), hence, by transitivity of μ, also $\mu(x,u,r)$ holds. By symmetry of μ, *ingr*(v,x), hence, $\mu(v,x,r)$
holds also   □

In case of rough inclusions in information systems, induced by residual implications generated by continuous t–norms, see Ch. 6, sect. 2, $L,P,$ or M,
we have a positive case, for the minimum t–norm M

Proposition 7.2. *For the rough inclusion* μ *induced by the residual implication* $\Rightarrow_M$, *due to the minimum t–norm* M, *and* $r < 1$, *the relation*
ingr$(v,g_\mu(u,r))$ *holds if and only if* $\mu(v,u,r)$ *holds.*

Proof. The rough inclusion μ has the form $\mu(v,u,r)$ if and only if $\frac{|IND(v,s)|}{|A|}$
$\Rightarrow_M \frac{|IND(u,s)|}{|A|} \geq r$. Assume that $ingr(v, g_\mu(u,r))$ holds, so by the class definition, there exists z such that $Ov(v,z)$ and $\mu(z,u,r)$ hold. Thus, we have w
with $ingr(w,v)$ and $\mu(w,u,r)$ by transitivity of μ and the fact that $ingr(w,z)$.
By definition of μ, $ingr(w,v)$ means that $|IND(w,s)| \leq |IND(v,s)|$. As
$\mu(w,u,r)$ with $r < 1$ means that $|IND(u,s)| \geq r$ because of $|IND(w,s)| \geq$
$|IND(u,s)|$, the condition $|IND(w,s)| \leq |IND(v,s)|$ implies that $\mu(v,u,r)$
holds as well   □

The case of the rough inclusion μ induced either by the product t–norm
$P(x,y) = x \cdot y$, or by the Łukasiewicz t–norm L, is a bit more intricate. To
obtain in this case some positive result, we exploit the averaged t–norm $\vartheta(\mu)$
defined for the rough inclusion μ, induced by a t–norm T, by means of the
formula

$$\vartheta(\mu)(v,u,r) \Leftrightarrow \forall z. \exists a,b. \mu(z,v,a), \mu(z,u,b), a \Rightarrow_T b \geq r \qquad (7.3)$$

Our proposition for the case of the t–norm P is

Proposition 7.3. *For* $r < 1$, $ingr(v, g_{\vartheta(\mu)}(u,r))$ *holds if and only if* $\mu(v,u,a \cdot$
$r)$, *where* $\mu(v,t,a)$ *holds for* t *which obeys conditions* $ingr(t,v)$ *and* $\vartheta(\mu)$
(t,u,r).

Proof. $ingr(v, g_{\vartheta(\mu)}(u,r))$ implies that there is w such that $Ov(v,w)$ and
$\vartheta(\mu)(w,u,r)$, so we can find t with properties, $ingr(t,w)$, $ingr(t,v)$, hence,
by transitivity of $\vartheta(\mu)$ also $\vartheta(\mu)(t,u,r)$.
    By definition of $\vartheta(\mu)$, there are a,b such that $\mu(v,t,a)$, $\mu(v,u,b)$, and
$a \Rightarrow_P b \geq r$, i.e., $\frac{b}{a} \geq r$. Thus, $\mu(v,u,b)$ implies $\mu(v,u,a \cdot r)$ □

An analogous reasoning brings forth in case of the rough inclusion μ induced
by residual implication due to the Łukasiewicz implication L, the result that

Proposition 7.4. *For* $r < 1$, $ingr(v, g_{\vartheta(\mu)}(u,r))$ *holds if and only if* $\mu(v,u,r$
$+a-1)$ *holds, where* $\mu(v,t,a)$ *holds for* t *such that* $ingr(t,v)$ *and* $\vartheta(\mu)(t,u,r)$.

The two last propositions can be recorded jointly in the form

Proposition 7.5. *For* $r < 1$, *and* μ *induced by residual implications either*
$\Rightarrow_P$ *or* $\Rightarrow_L$, $ingr(v, g_{\vartheta(\mu)}(u,r))$ *holds if and only if* $\mu(v,u,T(r,a))$ *holds,*
where $\mu(v,t,a)$ *holds for* t *such that* $ingr(t,v)$ *and* $\vartheta(\mu)(t,u,r)$.

Granules as collective concepts can be objects for rough mereological calculi.

7.4 Rough Inclusions on Granules

Due to the feature of mereology that it operates (due to the class operator)
only on level of individuals, one can extend rough inclusions from objects to

granules; the formula for extending a rough inclusion μ to a rough inclusion $\overline{\mu}$ on granules is a modification of mereological axiom M3 of Ch. 5

$$\overline{\mu}(g, h, r) \Leftrightarrow \forall z.ingr(z, g) \Rightarrow \exists w.ingr(w, h), \mu(z, w, r). \tag{7.4}$$

Proposition 7.6. *The predicate $\overline{\mu}(g, h, r)$ is a rough inclusion on granules.*

Proof. $\mu(g, h, 1)$ means that for each object z with $ingr(z, g)$ there exists an object w with $ingr(w, h)$ such that $\mu(z, w, 1)$, i.e., $ingr(z, w)$, which, by the inference rule implies that $ingr(g, h)$. This proves RINC1. For RINC2, assume that $\mu(g, h, 1)$ and $\mu(k, g, r)$ so for each $ingr(x, k)$ there is $ingr(y, g)$ with $\mu(x, y, r)$. For y there is z such that $ingr(z, h)$ and $\mu(y, z, 1)$, hence, $\mu(x, z, r)$ by property RINC2 of μ. Thus, $\mu(k, h, r)$. RINC2 follows and RINC3 is obviously satisfied □

We now examine rough mereological granules with respect to their properties.

7.5 General Properties of Rough Mereological Granules

They are collected below in

Proposition 7.7. *The following constitute a set of basic properties of rough mereological granules*

1. *If $ingr(y, x)$ then $ingr(y, g_\mu(x, r))$;*

2. *If $ingr(y, g_\mu(x, r))$ and $ingr(z, y)$ then $ingr(z, g_\mu(x, r))$;*

3. *If $\mu(y, x, r)$ then $ingr(y, g_\mu(x, r))$;*

4. *If $s < r$ then $ingr(g_\mu(x, r), g_\mu(x, s))$,*

which follow straightforwardly from properties RINC1–RINC3 of rough inclusions and the fact that $ingr$ is a partial order, in particular it is transitive, regardless of the type of the rough inclusion μ.

For T–transitive rough inclusions, we can be more specific, and prove

Proposition 7.8. *For each T–transitive rough inclusion μ,*

1. *If $ingr(y, g_\mu(x, r)$ then $ingr(g_\mu(y, s), g_\mu(x, T(r, s))$;*

2. *If $\mu(y, x, s)$ with $1 > s > r$, then there exists $\alpha < 1$ with the property that $ingr(g_\mu(y, \alpha), g_\mu(x, r))$.*

Proof. Property 1 follows by transitivity of μ with the t–norm T. Property 2 results from the fact that the inequality $T(s,\alpha) \geq r$ has a solution in α, e.g., for $T = P$, $\alpha \geq \frac{r}{s}$, and, for $T = L$, $\alpha \geq 1 - s + r$ $\qquad\qquad$ □

It is natural to regard granule system $\{g_r^{\mu_t}(x) : x \in U; r \in (0,1)\}$ as a neighborhood system for a topology on U that may be called the *granular topology*.

In order to make this idea explicit, we define classes of the form

$$N^T(x,r) = Cls(\psi_{r,x}^{\mu_T}) \qquad (7.5)$$

where

$$\psi_{r,x}^{\mu_T}(y) \Leftrightarrow \exists s > r.\mu_T(y,x,s) \qquad (7.6)$$

We declare the system $\{N^T(x,r) : x \in U; r \in (0,1)\}$ to be a neighborhood basis for a topology θ_μ. This is justified by the following

Proposition 7.9. *Properties of the system $\{N^T(x,r) : x \in U; r \in (0,1)\}$ are as follows*

1. $y \; ingr \; N^T(x,r) \Rightarrow \exists \delta > 0.N^T(y,\delta) \; ingr \; N^T(x,r)$;

2. $s > r \Rightarrow N^T(x,s) \; ingr \; N^T(x,r)$;

3. $z \; ingr \; N^T(x,r) \wedge z \; ingr \; N^T(y,s) \Rightarrow \exists \delta > 0 \; N^T(z,\delta) \; ingr \; N^T(x,r) \wedge N^T(z,\delta) \; ingr \; N^T(y,s)$.

Proof. For Property 1, $y \; ingr \; N^t(x,r)$ implies that there exists an $s > r$ such that $\mu_t(y,x,s)$. Let $\delta < 1$ be such that $t(u,s) > r$ whenever $u > \delta$; δ exists by continuity of t and the identity $t(1,s) = s$. Thus, if $z \; ingr \; N^t(y,\delta)$, then $\mu_t(z,y,\eta)$ with $\eta > \delta$ and $\mu_t(z,x,t(\eta,s))$ hence $z \; ingr \; N^t(x,r)$.

Property 2 follows by RINC3 and Property 3 is a corollary to properties 1 and 2. This concludes the argument $\qquad\qquad$ □

Granule systems defined above form a basis for applications, where approximate reasoning is a crucial ingredient.

We begin with a basic application in which approximate reasoning itself is codified as a many–world (intensional) logic, where granules serve as possible worlds.

7.6 Reasoning by Granular Rough Mereological Logics

The idea of a granular rough mereological logic, see Polkowski [42], Polkowski and Semeniuk–Polkowska [53], consists in measuring the meaning of a unary predicate in the model which is a universe of an information system against a granule defined by means of a rough inclusion. The result can be regarded as the degree of truth (the logical value) of the predicate with respect to the given granule. The obtained logics are intensional as they can be regarded as mappings from the set of granules (possible worlds) to the set of logical values in the interval $[0, 1]$, the value at a given granule regarded as the extension at that granule of the generally defined intension.

The problem of meaning in intensional contexts, pointed to by usage of, e.g., linguistic contexts like 'he believes that ...', 'he knows that...', 'he thinks that....' for long has attracted the attention of philosophers and logicians, it suffices to mention Immanuel Kant and John Stuart Mill.

Problems that arise here are distinctly illustrated with the well–known example of two phrases: 'the morning star' and 'the evening star'. Both describe the planet Venus in two distinct appearances; once, as the star appearing in the sky at the sunrise, and secondly, as the star that appears in the sky at the sunset. One says that both phrases have the same *denotation*, i.e., the planet Venus, but distinct *meanings*. Terms 'denotation', 'meaning', call for a precise definition and a formal mechanism of inference with them.

The problem was addressed by Gottlob Frege [12] in 'Über Sinn und Bedeutung', terms usually rendered as *sense* (Sinn) and *reference* (Bedeutung). Thus, phrases 'the morning star' and 'the evening star' have different senses but the common reference – the planet Venus.

Among later approaches to the problem, we would like to point to that proposed by Rudolph Carnap [6]. Carnap introduces an important notion of a *state description* understood as a logical status of a state and determined by assignment of truth or falsity to each atomic object of logic. Carnap calls senses *intensions* and formalizes them as functions on state descriptions whereas references, or, denotations, are defined as values of intensions at particular state descriptions and are called *extensions*.

In this way, intensions are given a functional character, e.g., 'the morning star' and 'the evening star' as intensions describe some various astronomic objects at various solar systems, but at the Sun solar system, at the planet Earth, their extension is the same: the planet Venus.

Functional character of intensions was exploited ingeniously in Montague [34], [35] most notably culminating in His formal model of natural language grammar, the *Montague Grammar*, in which hierarchies of functional intensions model the structure of a sentence.

The idea of functionality of intensions has been exploited most famously in Kripke [19] who produced a semantics for modal logics, cf., Ch. 3, sect. 9: state descriptions have been replaced with *possible worlds* collected in the set of possible worlds W and intensions are valuations at possible worlds on logical variables, the extension at a possible world being the value of a given variable at that world. Imposing on possible worlds an *accessibility relation* R allows for formal definitions of modal notions of *necessity* and *possibility* in the *frame* (W, R).

For a given world w, and a formula ϕ, one declares ϕ *necessarily true* at w if and only if ϕ is true at each world w' accessible from w, i.e., such that $(w, w') \in R$; similarly, ϕ is *possibly true* at w if and only if there exists w' such that ϕ is true at w' and $(w, w') \in R$.

Gallin [13] proposed an axiomatics for intensional logic along with completeness proof; expositions of intensional and modal logics can be found in Van Benthem [3], Hughes and Creswell [18] and Fitting [11].

Our approach to rough mereological logics has an intentional tint, as we assume that possible worlds are granules obtained by means of a rough inclusion μ, and, against these worlds, unary predicates interpreted in the collection of objects, on which μ is defined, are evaluated as to their degree of truth. Thus, intensions are functions on granule collection, and extensions are values of truth at particular granules for particular predicates.

Any attempt at assigning various degrees of truth to logical statements places one in the realm of many–valued logic. These logics describe formally logical functors as mappings on the set of truth values/states into itself hence they operate a fortiori with values of statements typically as fractions or reals in the unit interval $[0, 1]$, see in this respect, e.g., Łukasiewicz [29], [30], Łukasiewicz and Tarski [32], and Hájek [17], see also Ch. 3.

In logics based on implication given by residua of t–norms, negation is defined usually as $\neg x = x \Rightarrow_T 0$. Thus, the Łukasiewicz negation is $\neg_L x = 1 - x$ whereas Goguen's as well as Gödel's negation is $\neg_G x = 1$ for x=0 and is 0 for $x > 0$, see Ch. 3. Other connectives are defined with usage of the t–norm itself as semantics for the strong conjunction and ordinary conjunction and disjunction are interpreted semantically as, respectively, *min, max*.

In this approach a rule $\alpha \Rightarrow \beta$ is evaluated by evaluating the truth state $[[\alpha]]$ as well as the truth state $[[\beta]]$ and then computing the values of $[[\alpha]] \Rightarrow_T [[\beta]]$ for a chosen t–norm T. Similarly other connectives are evaluated.

In the rough set context, this approach would pose the problem of evaluating the truth state of a conjunct α of descriptors; to this end, one can invoke the idea of Łukasiewicz [29] and assign to α a value $[[\alpha]]_L = \frac{|\{u \in U : u \models \alpha\}|}{|U|}$.

Clearly, this approach does not take into account the logical containment or its lack between α and β, and this fact makes the many–valued approach of a small use when data mining tasks are involved.

For this reason, we propose an approach to logic of decision rules which is based on the idea of measuring the state of truth of a formula against a concept constructed as a granule of knowledge; concepts can be regarded as "worlds" and our logic becomes intensional: logical evaluations at a given world are extensions of the intension which is the mapping on worlds valued in the set of logical values of truth.

For our purpose it is essential to extend rough inclusions to sets, cf., Ch. 6 sect. 5; we use the t–norm L along with the representation $L(r, s) = g(f(r) + f(s))$ already introduced in Ch. 4, sect. 9. We denote rough inclusions on sets with the generic symbol ν.

For finite sets X, Y, we let,

$$\nu_L(X, Y, r) \Leftrightarrow g(\frac{|X \setminus Y|}{|X|}) \geq r \tag{7.7}$$

As $g(x) = 1 - x$, we have that $\nu_L(X, Y, r)$ holds if and only if $\frac{|X \cap Y|}{|X|} \geq r$.

Let us observe that ν_L is *regular*, i.e., $\nu_L(X, Y, 1)$ if and only if $X \subseteq Y$ and $\nu_L(X, Y, r)$ only with $r = 0$ if and only if $X \cap Y = \emptyset$.

Thus, the ingredient relation associated with a regular rough inclusion is the improper containment $\subseteq$ whereas the underlying part relation is the strict containment $\subset$.

Other rough inclusion on sets which we will use is the 3–valued rough inclusion ν_3 defined via the formula,

$$\nu_3(X, Y, r) \Leftrightarrow \begin{cases} X \subseteq Y \text{ and } r = 1 \\ X \cap Y = \emptyset \text{ and } r = 0 \\ r = \frac{1}{2} \text{ otherwise} \end{cases} \tag{7.8}$$

We now proceed with a construction of the rough mereological logic.

7.7 A Logic for Information Systems

We assume that an information/decision system (U, A, d) is given, along with a rough inclusion ν on the subsets of the universe U; for a collection of unary predicates Pr, interpreted in the universe U (meaning that for each predicate $\phi \in Pr$ the meaning $[[\phi]]$ is a subset of U), we define the intensional logic GRM_ν by assigning to each predicate ϕ in Pr its intension $I_\nu(\phi)$ defined by its extension $I_\nu^\vee(g)$ at each particular granule g, as

$$I_\nu^\vee(g)(\phi) \geq r \Leftrightarrow \nu(g, [[\phi]], r) \tag{7.9}$$

With respect to the rough inclusion ν_L, the formula (7.9) becomes

$$I_{\nu_L}^\vee(g)(\phi) \geq r \Leftrightarrow \frac{|g \cap [[\phi]]|}{|g|} \geq r \tag{7.10}$$

The counterpart for ν_3 is specified by definition (7.8), and it comes down to the following

$$I_{\nu_3}^\vee(g)(\phi) \geq r \Leftrightarrow \begin{cases} g \subseteq [[\phi]] \text{ and } r = 1 \\ g \cap [[\phi]] \neq \emptyset \text{ and } r \geq \frac{1}{2} \\ g \cap [[\phi]] = \emptyset \text{ and } r = 0 \end{cases} \tag{7.11}$$

We say that a formula ϕ interpreted in the universe U of an information system (U, A) is *true* at a granule g with respect to a rough inclusion ν if and only if $I_\nu^\vee(g)(\phi) = 1$.

Hence, for every regular rough inclusion ν, a formula ϕ interpreted in the universe U, with the meaning $[[\phi]] = \{u \in U : u \models \phi\}$, is true at a granule g with respect to ν if and only if $g \subseteq [[\phi]]$.

In particular, for a decision rule $r : p \Rightarrow q$ in the descriptor logic, see Ch. 4, the rule r is true at a granule g with respect to a regular rough inclusion ν if and only if $g \cap [[p]] \subseteq [[q]]$. We state these facts in the following

Proposition 7.10. *For every regular rough inclusion ν, a formula ϕ interpreted in the universe U, with the meaning $[[\phi]]$, is true at a granule g with respect to ν if and only if $g \subseteq [[\phi]]$. In particular, for a decision rule $r : p \Rightarrow q$ in the descriptor logic, the rule r is true at a granule g with respect to a regular rough inclusion ν if and only if $g \cap [[p]] \subseteq [[q]]$.*

Proof. Indeed, truth of ϕ at g means that $\nu(g, [[\phi]], 1)$ which in turn, by regularity of ν is equivalent to the inclusion $g \subseteq [[\phi]]$ □

We will say that a formula ϕ is a *tautology* of our intensional logic if and only if ϕ is true at every world g.

The preceding proposition implies that,

Proposition 7.11. *For every regular rough inclusion ν, a formula ϕ is a tautology if and only if $Cls(G) \subseteq [\phi]$, where G is the property of being a granule; in the case when granules considered cover the universe U this condition simplifies to $[[\phi]] = U$. This means for a decision rule $p \Rightarrow q$ that it is a tautology if and only if $[[p]] \subseteq [[q]]$.*

Hence, the condition for truth of decision rules in the logic GRM_ν is the same as the truth of an implication in descriptor logic, see Ch. 4, under caveat that granules considered cover the universe U of objects.

Our rough mereological intensional logic depends obviously on the chosen rough inclusion μ on sets – concepts in the universe of objects. We are going to examine relationships of this logic, in case of two rough inclusions, viz., the Łukasiewicz t–norm induced ν_L and the 3–valued rough inclusion ν_3, to many–valued logics based on the Łukasiewicz residual implication, studied in Ch. 3. It turns out that in a sense, rough mereological intensional logics are embedded in, respectively, [0,1]–valued and 3–valued Łukasiewicz logics.

We apply to this end the idea of *collapse*. Collapse in this case consists in omitting the variable symbols while preserving connectives and in this way transforming open formulas of unary predicate calculus into propositional formulas. We show that theorems of rough mereological intensional logic are after collapsing them theorems of the respective Łukasiewicz many valued logic. We also apply a kind of inverse collapse by regarding in the coming sections formulas of many valued propositional calculus as formulas of unary predicate calculus with variables omitted.

7.7.1 Relations to Many–Valued Logics

We examine some axiomatic schemes for many–valued logics with respect to their meanings under the stated in introductory section assumption that $[[p \Rightarrow q]] = (U \setminus [[p]]) \cup [[q]]$, $[[\neg p]] = U \setminus [[p]]$.

We examine first axiom schemes for 3–valued Łukasiewicz logic; we recall, see Ch. 3, axiom schemes given in Wajsberg [62].

W1 $q \Rightarrow (p \Rightarrow q)$

W2 $(p \Rightarrow q) \Rightarrow ((q \Rightarrow r) \Rightarrow (p \Rightarrow r))$

W3 $((p \Rightarrow \neg p) \Rightarrow p) \Rightarrow p$

W4 $(\neg q \Rightarrow \neg p) \Rightarrow (p \Rightarrow q)$

We have as meanings of those formulas

$$[[W1]] = (U \setminus [[q]]) \cup (U \setminus [[p]]) \cup [[q]] = U \qquad (7.12)$$

$$[[W2]] = ([[p]] \setminus [[q]]) \cup ([[q]] \setminus [[r]]) \cup (U \setminus [[p]]) \cup [[r]] = U \qquad (7.13)$$

$$[[W3]] = (U \setminus [[p]]) \cup [[p]] = U \qquad (7.14)$$

$$[[W4]] = ([[p]] \setminus [[q]]) \cup [[q]] = U \qquad (7.15)$$

It follows that

Proposition 7.12. *All instances of Wajsberg axiom schemes for 3–valued Łukasiewicz logic are tautologies of our intensional logic in case of regular rough inclusions on sets.*

The deduction rule in 3–valued Łukasiewicz logic is Modus Ponens: $\frac{p, p \Rightarrow q}{q}$.

In our setting this is a valid deduction rule: if $p, p \Rightarrow q$ are tautologies than q is a tautology. Indeed, if $[[p]] = U = [[p \Rightarrow q]]$, then $[[q]] = U$.

We have obtained

Proposition 7.13. *Each tautology of 3–valued Łukasiewicz logic is a tautology of rough mereological granular logic in case of a regular rough inclusion on sets.*

In an analogous manner, we examine axiom schemes for infinite valued Łukasiewicz logic, proposed by Łukasiewicz, see Łukasiewicz and Tarski [32] in a form modified due to Meredith [33] and Chang [7], see Ch. 3, sect. 4

L1 $q \Rightarrow (p \Rightarrow q)$

L2 $(p \Rightarrow q) \Rightarrow ((q \Rightarrow r) \Rightarrow (p \Rightarrow r))$

L3 $((q \Rightarrow p) \Rightarrow p) \Rightarrow ((p \Rightarrow q) \Rightarrow q)$

L4 $(\neg q \Rightarrow \neg p) \Rightarrow (p \Rightarrow q)$

As L1 is W1, L2 is W2 and L4 is W4, it remains to examine L3.
In this case, we have

$$[[(q \Rightarrow p) \Rightarrow p]] = (U \setminus [[q \Rightarrow p]]) \cup [[p]]) =$$

$$(U \setminus ((U \setminus [[q]]) \cup [[p]])) \cup [[p]] = ([[q]] \setminus [[p]]) \cup [[p]] =$$

$$[[q]] \cup [[p]]$$

Similarly,

$$[[(p \Rightarrow q) \Rightarrow q]]$$

is

$$[[p]] \cup [[q]]$$

by symmetry, and finally, the meaning $[[L3]]$ is

$$(U \setminus ([[q]] \cup [[p]])) \cup [[p]] \cup [[q]] = U$$

It follows that

Proposition 7.14. *All instances of axiom schemes for infinite–valued Łukasiewicz logic are tautologies of rough mereological granular logic.*

As Modus Ponens remains a valid deduction rule in infinite–valued case, we obtain, analogous to Prop. 7.13,

Proposition 7.15. *Each tautology of infinite–valued Łukasiewicz logic is a tautology of rough mereological granular logic in case of a regular rough inclusion on sets.*

It follows from Prop.7.15 that all tautologies of *Basic logic*, see Hájek [17], cf., Ch. 3, sect. 6, i.e., logic which is intersection of all many–valued logics with implications evaluated semantically by residual implications of continuous t–norms are tautologies of rough mereological granular logic for each regular rough inclusion ν.

The assumption of regularity of a rough inclusion ν is essential: considering the drastic rough inclusion ν_1, we find that an implication $p \Rightarrow q$ is true only at the world $(U \setminus [[p]]) \cup [[q]]$, so it is not any tautology; this concerns all schemes systems W and L above as they are true only at the global world U.

7.8   A Graded Notion of Truth

As already stated, the usual interpretation of functors $\vee, \wedge$ in many–valued logics is $[[p \vee q]] = max\{[[p]], [[q]]\}$ and $[[p \wedge q]] = min\{[[p]], [[q]]\}$, where

$[[p]]$ is the state of truth of p. In case of concept–valued meanings, we admit the interpretation which conforms to accepted in many valued logics (especially in the context of fuzzy set theory) interpretation of min as $\cap$ and max as $\cup$.

The formula $\nu(g, [\phi], 1)$ stating the truth of ϕ at g, ν with ν regular can be regarded as a condition of orthogonality type, with the usual consequences.

1. *If ϕ is true at granules g, h, then it is true at $g \cup h$;*

2. *If ϕ is true at granules g, h then it is true at $g \cap h$;*

3. *If ϕ, ψ are true at a granule g, then $\phi \vee \psi$ is true at g;*

4. *If ϕ, ψ are true at a granule g, then $\phi \wedge \psi$ is true at g;*

5. *If ψ is true at a granule g, then $\phi \Rightarrow \psi$ is true at g for every formula ϕ;*

6. *If ϕ is true at a granule g then $\phi \Rightarrow \psi$ is true at g if and only if ψ is true at g.*

The graded relaxation of truth is given obviously by the condition that a formula ϕ is *true to a degree at least r at g, ν* if and only if

$$I_\nu^\vee(g)(\phi) \geq r \qquad (7.16)$$

i.e., $\nu(g, [[\phi]], r)$ holds.

In particular, ϕ is *false* at g, ν if and only if $I_\nu^\vee(g)(\phi) \geq r$ implies $r = 0$, i.e., $\nu(g, [\phi], r)$ implies $r = 0$.

Proposition 7.16. *The following properties hold with the Łukasiewicz residual implication $\Rightarrow_L$*

1. For each regular ν, a formula α is true at g, ν if and only if $\neg\alpha$ is false at g, ν;

2. For $\nu = \nu_L, \nu_3$, $I_\nu^\vee(g)(\neg\alpha) \geq r$ if and only if $I_\nu^\vee(g)(\alpha) \geq s$ implies $s \leq 1-r$;

3. For $\nu = \nu_L, \nu_3$, the implication

$$\alpha \Rightarrow_L \beta$$

is true at g if and only if

$$g \cap [\alpha] \subseteq [\beta]$$

and $\alpha \Rightarrow_L \beta$ is false at g if and only if

$$g \subseteq [\alpha] \setminus [\beta];$$

4. For $\nu = \nu_L$, if

$$I_\nu^\vee(g)(\alpha \Rightarrow_L \beta) \geq r$$

then $\Rightarrow_L (t, s) \geq r$, where $I_\nu^\vee(g)(\alpha) \geq t$ and $I_\nu^\vee(g)(\beta) \geq s$.

Further analysis should be split into the case of ν_L and the case of ν_3 as the two differ essentially with respect to the form of reasoning they imply.

Property 4 in Proposition 7.16 shows in principle that the value of $I_\nu^\vee(g)(\alpha \Rightarrow \beta)$ is bounded from above by the value of $I_\nu^\vee(g)(\alpha) \Rightarrow_{t_L} I_\nu^\vee(g)(\beta))$.

This suggests that the idea of collapse attributed to Leśniewski can be applied to formulas of rough mereological logic in the following form: for a formula $q(x)$ we denote by the symbol q^* the formula q regarded as a sentential formula (i.e., with variable symbols removed) subject to relations

1. $(\neg q(x))^*$ *is* $\neg(q(x)^*)$;

2. $(p(x) \Rightarrow q(x))^*$ *is* $p(x)^* \Rightarrow q(x)^*$.

As the value $[[q^*]]_g$ of the formula $q(x)^*$ at a granule g, we take the value of $\frac{|g \cap [[q(x)]]|}{|g|}$, i.e, $argmax_r\{\nu_L(g, [[q^*]]_g, r)\}$. Thus, Property 4 in Proposition 7.16 can be rewritten in the form

$$I_\nu^\vee(g)(\alpha \Rightarrow \beta) \leq [\alpha^*]_g \Rightarrow_L [\beta^*]_g \qquad (7.17)$$

The statement follows

Proposition 7.17. *If $\alpha \Rightarrow \beta$ is true at g, then the collapsed formula has the value 1 of truth at the granule g in the Łukasiewicz logic.*

This gives a necessity condition for verification of implications of rough mereological logics

Proposition 7.18. *If $[\alpha^*]_g \Rightarrow_L [\beta^*]_g < 1$, then the implication $\alpha \Rightarrow \beta$ is not true at g.*

This concerns in particular decision rules

Proposition 7.19. *A decision rule $p(v) \Rightarrow q(v)$, is true on a granule g if and only if $[[p^*]]_g \leq [[q^*]]_g$.*

In case of ν_3, one can check on the basis of definitions that $I_\nu^\vee(g)(\neg\alpha) \geq r$ if and only if $I_\nu^\vee(g)(\alpha) \leq 1 - r$; thus the negation functor in rough mereological logic based on ν_3 is the same as the negation functor in the 3–valued Łukasiewicz logic. For implication, the relations between granular

$$\begin{array}{c|ccc} \Rightarrow & 0 & 1 & \frac{1}{2} \\ \hline 0 & 1 & 1 & 1 \\ 1 & 0 & 1 & \frac{1}{2} \\ \frac{1}{2} & \frac{1}{2} & 1 & 1 \end{array}$$

Fig. 7.1 Truth values for implication in L_3

$\Rightarrow$	$I_{\nu_3}^{\vee}(g)(q) = 0$	$I_{\nu_3}^{\vee}(g)(q) = 1$	$I_{\nu_3}^{\vee}(g)(q) = \frac{1}{2}$
$I_{\nu_3}^{\vee}(g)(p) = 0$	1	1	1
$I_{\nu_3}^{\vee}(g)(p) = 1$	0	1	$\frac{1}{2}$
$I_{\nu_3}^{\vee}(g)(p) = \frac{1}{2}$	$\frac{1}{2}$	1	1 when $g \cap [\alpha] \subseteq [\beta]$; $\frac{1}{2}$ otherwise

Fig. 7.2 Truth values for implication $p \Rightarrow q$ in logic based on ν_3

rough mereological logic and 3–valued logic of Łukasiewicz follow from truth tables for respective functors of negation and implication.

Table 7.1 shows truth values for implication in 3–valued logic of Łukasiewicz. We recall that these values obey the implication $x \Rightarrow_L y = min\{1, 1 - x + y\}$. Values of x correspond to rows and values of y correspond to columns in the table of Fig. 7.1.

The table of Fig. 7.2 shows values of implication for rough mereological logic based on ν_3. Values are shown for the extension $I_{\nu}^{\vee}(g)(p \Rightarrow q)$ of the implication $p \Rightarrow q$. Rows correspond to p, columns correspond to q.

We verify values shown in Fig. 7.2.

First, we consider the case when $I_{\nu_3}^{\vee}(g)(p) = 0$, i.e., the case when $g \cap [[p]] = \emptyset$. As

$$g \subseteq (U \setminus [[p]]) \cup [[q]]$$

for every value of $[[q]]$, we have only values of 1 in the first row of Table 2.

Assume now that $I_{\nu_3}^{\vee}(g)(p) = 1$, i.e., $g \subseteq [[p]]$. As

$$g \cap (U \setminus [[p]]) = \emptyset$$

the value of $I_{\nu}^{\vee}(g)(p \Rightarrow q)$ depends only on a relation between g and $[[q]]$. In case

$$g \cap [[q]] = \emptyset$$

the value in Table 2 is 0, in case $g \subseteq [[q]]$ the value in Table 2 is 1, and in case

$$I_{\nu_3}^{\vee}(g)(q) = \frac{1}{2}$$

the value in Table 2 is $\frac{1}{2}$.

Finally, we consider the case when

$$I_{\nu_3}^{\vee}(g)(p) = \frac{1}{2}$$

i.e.,

$$g \cap [[p]] \neq \emptyset \neq g \setminus [[p]]$$

In case $g \cap [[q]] = \emptyset$, we have

$$g \cap ((U \setminus [[p]]) \cup [[q]]) \neq \emptyset$$

and it is not true that

$$g \subseteq ((U \setminus [[p]]) \cup [[q]])$$

so the value in table is $\frac{1}{2}$.

In case $g \subseteq [[q]]$, the value in Table is clearly 1. The case when

$$I_{\nu_3}^{\vee}(g)(q) = \frac{1}{2}$$

remains. Clearly, when $g \cap [[p]] \subseteq [[q]]$, we have

$$g \subseteq (U \setminus [[p]]) \cup [[q]]$$

so the value in Table is 1; otherwise, the value is $\frac{1}{2}$.

Thus, negation in both logic is semantically treated in the same way, whereas treatment of implication differs only in case of implication $p \Rightarrow q$ from the value $\frac{1}{2}$ to $\frac{1}{2}$, when $g \cap [[p]]$ is not any subset of $[[q]]$.

It follows from these facts that given a formula α and its collapse α^*, we have,

$$I_{\nu_3}^{\vee}(g)(\neg\alpha) = [[(\neg\alpha)^*]]_{L_3}, I_{\nu_3}^{\vee}(g)(\alpha \Rightarrow \beta) \leq [[(\alpha \Rightarrow \beta)^*]]_{L_3} \qquad (7.18)$$

A more exact description of implication in both logics is as follows.

Proposition 7.20. *The following statements are true*

1. *If* $I_{nu_3}^{\vee}(g)(\alpha \Rightarrow \beta) = 1$ *then* $[(\alpha \Rightarrow \beta)^*]_{L_3} = 1$;

2. *If* $I_{nu_3}^{\vee}(g)(\alpha \Rightarrow \beta) = 0$ *then* $[(\alpha \Rightarrow \beta)^*]_{L_3} = 0$;

3. *If* $I_{nu_3}^{\vee}(g)(\alpha \Rightarrow \beta) = \frac{1}{2}$ *then* $[(\alpha \Rightarrow \beta)^*]_{L_3} \geq \frac{1}{2}$ *and this last value may be 1.*

Proof. We offer a simple check–up on Proposition 7.20. In Case 1, we have

$$g \subseteq ((U \setminus [[\alpha]]) \cup [[\beta]])$$

For the value of $[(\alpha \Rightarrow \beta)^*]$, consider some subcases.

Subcase 1.1 $g \subseteq U \setminus [[\alpha]]$. Then $[[\alpha^*]] = 0$ and

$$[[(\alpha \Rightarrow \beta)^*]] = [[\alpha^*]] \Rightarrow [[\beta^*]]$$

is always 1 regardless of a value of $[\beta^*]$.

Subcase 1.2 $g \cap [[\alpha]] \neq \emptyset \neq g \setminus [[\alpha]]$ so $[\alpha^*] = \frac{1}{2}$. Then $g \cap [\beta] = \emptyset$ is impossible, i.e., $[\beta^*]$ is at least $\frac{1}{2}$ and $[(\alpha \Rightarrow \beta)^*]=1$.

Subcase 1.3 $g \subseteq [[\alpha]]$ so $[[\alpha^*]] = 1$; then $g \subseteq [[\beta]]$ must hold, i.e., $[[\beta^*]] = 1$ which means that $[[(\alpha \Rightarrow \beta)^*]]=1$.

For case 2, we have

$$g \cap ((U \setminus [[\alpha]]) \cup [[\beta]]) = \emptyset$$

hence, $g \cap [[\beta]] = \emptyset$ and $g \subseteq [[\alpha]]$, i.e., $[\alpha^*] = 1, [\beta^*] = 0$ so $[\alpha^*] \Rightarrow [\beta^*]=0$.

In case 3, we have

$$g \cap ((U \setminus [[\alpha]]) \cup [[\beta]]) \neq \emptyset$$

and

$$g \cap [[\alpha]] \setminus [[\beta]] \neq \emptyset$$

Can $[\alpha^*] \Rightarrow [\beta^*]$ be necessarily 0? This would mean that $[\alpha^*] = 1$ and $[\beta^*] = 0$, i.e., $g \subseteq [[\alpha]]$ and $g \cap [[\beta]] = \emptyset$ but then

$$g \cap ((U \setminus [[\alpha]]) \cup [[\beta]]) = \emptyset$$

a contradiction. Thus the value $[[\alpha^*]] \Rightarrow [[\beta^*]]$ is at least $\frac{1}{2}$. In the subcase: $g \subseteq [[\alpha]], g \cap [[\beta]] \neq \emptyset \neq g \setminus [[\beta]]$, the value of $[[\alpha^*]] \Rightarrow [[\beta^*]]$ is $0 \Rightarrow_L \frac{1}{2} = 1$, and the subcase is consistent with case 3 □

Let us mention that a 3–valued intensional variant of Montague Grammar is considered in Alves and Guerzoni[1].

7.9 Dependencies and Decision Rules

It is an important feature of rough set theory that it allows for an elegant formulation of the problem of dependency between two sets of attributes, see Ch. 4, sect. 4, cf., Pawlak [39], [40], in terms of indiscernibility relations.

We recall, see Ch. 4, sect. 4, that for two sets $C, D \subseteq A$ of attributes, one says that D *depends functionally on* C when $IND(C) \subseteq IND(D)$, symbolically denoted $C \mapsto D$. Functional dependence can be represented locally by means of functional dependency rules of the form

$$\phi_C(\{v_a : a \in C\}) \Rightarrow \phi_D(\{w_a : a \in D\}) \tag{7.19}$$

where $\phi_C(\{v_a : a \in C\})$ is the formula $\bigwedge_{a \in C}(a = v_a)$, and $[\phi_C] \subseteq [\phi_D]$.

A proposition holds

Proposition 7.21. *If* $\alpha : \phi_C \Rightarrow \phi_D$ *is a functional dependency rule, then* α *is a tautology of logic induced by* ν_3.

Proof. For each granule g, we have $g \cap [[\phi_C]] \subseteq [[\phi_D]]$ $\square$

Let us observe that the converse statement is also true, i.e., if a formula $\alpha : \phi_C \Rightarrow \phi_D$ is a tautology of logic induced by ν_3, then this formula is a functional dependency rule in the sense of (7.19). Indeed, assume that α is not any functional dependency rule, i.e., $[[\phi_C]] \setminus [[\phi_D]] \neq \emptyset$. Taking $[[\phi_C]]$ as the witness granule g, we have that g is not any subset of $[[\alpha]]$, i.e, $I_{\nu_3}^{\vee}(g)(\alpha) \leq \frac{1}{2}$, so α is not true at g, a fortiori it is not any tautology.

Let us observe that these characterizations are valid for each regular rough inclusion on sets ν.

A more general and also important notion is that of a local proper dependency: a formula $\phi_C \Rightarrow \phi_D$ where $\phi_C(\{v_a : a \in C\})$ is the formula $\bigwedge_{a \in C}(a = v_a)$, similarly for ϕ_D, is a local proper dependency when $[\phi_C] \cap [\phi_D] \neq \emptyset$.

We will say that a formula α is *acceptable with respect to a collection M of worlds* when

$$I_{\nu_3}^{\vee}(g)(\alpha) \geq \frac{1}{2} \tag{7.20}$$

for each world $g \in M$, i.e, when α is false at no world $g \in M$. Then,

Proposition 7.22. *If a formula* $\alpha : \phi_C \Rightarrow \phi_D$ *is a local proper dependency rule, then it is acceptable with respect to all C–exact worlds.*

Proof. Indeed, for a C–exact granule g, the case that $I_{\nu_3}^{\vee}(g)(\alpha) = 0$ means that $g \subseteq [[\phi_C]]$ and $g \cap [[\phi_D]] = \emptyset$.

As g is C–exact and $[[\phi_C]]$ is a C–indiscernibility class, either $[[\phi_C]] \subseteq g$ or $[[\phi_C]] \cap g = \emptyset$. When $[[\phi_C]] \subseteq g$, then $[[\phi_C]] = g$ which makes $g \cap [[\phi_D]] = \emptyset$ impossible.

When $[[\phi_C]] \cap g = \emptyset$, then $g \cap [[\phi_D]] = \emptyset$ is impossible. In either case, $I_{\nu_3}^{\vee}(g)(\alpha) = 0$ cannot be satisfied with any C–exact granule g $\square$

Again, the converse is true: when α is not local proper, i.e., $[[\phi_C]] \cap [[\phi_D]] = \emptyset$, then $g = [[\phi_C]]$ does satisfy $I_{\nu_3}^{\vee}(g)(\alpha) = 0$.

A corollary of the same forms follows for *decision rules* in a given decision system (U, A, d), i.e., dependencies of the form $\phi_C \Rightarrow (d = w)$.

7.10   An Application to Calculus of Perceptions

Calculus of perceptions, posed as an idea by Zadeh [67], should render formally vague statements and queries, and answer vague questions by giving a semantic interpretation to vague statements . Here, we apply granular mereological logics toward this problem. To this end, we would like to borrow a

Fig. 7.3 Decision system *Age*

object n	age_1	age_2	age_3	Age
1	3	15	22	30	58
2	3	10	12	16	42
3	2	6	10	--	30
4	2	24	33	--	56
5	2	28	35	--	62
6	3	22	33	40	67
7	2	18	25	--	60
8	2	26	35	--	63
9	2	22	38	--	70
10	3	8	12	16	38
11	2	22	32	--	58
12	3	24	36	40	63
13	2	28	34	--	60
14	3	26	30	35	65
15	3	18	25	35	60
16	3	6	12	16	40
17	3	22	30	35	65
18	2	24	34	--	60
19	3	22	30	34	58
20	2	24	35	--	62

part of a complex percept in Zadeh [67] and interpret it in terms of granular logic.

The percept is: (i) *Carol has two children: Robert who is in mid–twenties and Helen who is in mid–thirties* with a query (ii) *how old is Carol*. To interpret it, we begin with Table 7.3 in which a decision system *Age* is given with attributes n – the number of children, a_i – the age of the i–th child for $i \leq 3$, and with the decision *age* – the age of the mother.

We define for a fuzzy concept X represented by the fuzzy membership function μ_X on the domain D_X, the *c–cut* where $c \in [0,1]$ as the concept $X_C = \{x \in D_X : \mu_X(x) \geq c\}$. Concepts *in mid–twenties, in mid–thirties* are represented by fuzzy membership functions, μ_{20}, μ_{30}, respectively

$$\mu_{20}(x) = \begin{cases} 0.25(x - 20), & x \in [20, 24] \\ 1, & x \in [24, 26] \\ 1 - 0.25(x - 26), & x \in [26, 30] \end{cases} \quad (7.21)$$

and

$$\mu_{30}(x) = \begin{cases} 0.25(x - 30), & x \in [30, 24] \\ 1, & x \in [34, 36] \\ 1 - 0.25(x - 36), & x \in [36, 40] \end{cases} \quad (7.22)$$

The concept *old* is interpreted as

$$\mu_{old}(x) = \begin{cases} 0.02(x - 30), & x \in [30, 60] \\ 0.04(x - 60) + 0.6, & x \in [60, 70] \\ 0, \ else \end{cases} \tag{7.23}$$

We interpret our query by a function $q : [0, 1]^3 \rightarrow [0, 1]$, where $f(u, v, w) = t$ would mean that for cut levels u, v, t, respectively for *old, in mid–twenties, in mid thirties*, the truth value of the statement *Carol is at least $supp_u$ old to the degree of t with respect to v, w*.

In our example, letting $u = 0.6$, we obtain $old_{0.6} = [60, 70]$; letting $v = 0.5 = w$, we obtain *in mid twenties*$_{0.5} = [23, 27]$ and *in mid thirties*$_{0.5} = [33, 37]$. In order to evaluate the truth degree t, we refer to the world knowledge of Table 7.3 and we find the set of objects $\Lambda_{v,w}$ with two children of ages respectively in the intervals $[23, 27], [33, 37]$ corresponding to values of v, w as well as the set of objects Γ_u having the value of *Age* in old_u. In our case we have $\Lambda_{0.5,0.5} = \{4, 8, 12, 18, 20\}$ and $\Gamma_{0.6} = \{5, 6, 7, 8, 9, 12, 13, 14, 15, 17, 18, 20\}$.

Finally, we evaluate the truth degree of the predicate *in old*$_{0.6}(x)$ represented in the universe of Table 7.3 by the set $\Gamma_{0.6}$ with respect to the granule $\Lambda_{0.5,0.5}$. We obtain by applying the Lukasiewicz rough inclusion μ_L,

$$(I^{\mu_L}_{\Lambda_{0.5,0.5}})^{\vee}(old_{0.6}(x)) = \frac{|\Lambda_{0.5,0.5} \cap \Gamma_{0.6}|}{|\Lambda_{0.5,0.5}|} = 0.8 \tag{7.24}$$

The result is the statement: *Carol is over 60 years old to degree of 0.8 under the assumed interpretation of in mid twenties*$_{0.5}$, *in mid thirties*$_{0.5}$ *with respect to knowledge in* Table 7.3.

7.11 Modal Aspects of Rough Mereological Logics

Modal logics are concerned with formal rendering of *modalities* of *necessity* and *possibility*. Possible world semantics (Kripke semantics) is set in terms of a set W of *possible worlds* endowed with an accessibility relation R: in case $(w, w') \in R$, one says that the world w' is *accessible* from the world w. In case of propositional calculus, intensions are valuations on propositional variables: for each world w and a variable p, a valuation *val* assigns the truth value $val(p, g)$ (True or False). This assignment extends by standard logical calculations to formulas, giving for each formula ϕ and a world w, the value $val(\phi, w)$, see Ch. 3, sect. 9.

Necessity is introduced by the requirement that a formula ϕ is *necessarily true* at a world w if and only if ϕ is true at every world w' accessible from w via R. Possibility is defined as the requirement that ϕ is *possibly true* at w if and only if there is w' accessible from w via R with ϕ true at w'. A formula

ϕ is *true at the frame* (W, R) if and only if it is true at each world $w \in W$, and ϕ is *true* if and only if it is true at each frame.

A hierarchy of modal logics is built by demanding that they satisfy some postulates about the nature of necessity and possibility, see Hughes and Creswell [18], or, Lemmon and Scott [20].

Necessity is denoted with L and $L\phi$ reads *necessarily* ϕ, possibility is denoted with M and $M\phi$ reads *possibly* ϕ.

A basic postulate going back to Aristotle, cf., Łukasiewicz [31] is that necessity of an implication along with the necessity of the premise implies necessity of the conclusion, expressed by means of a formula (K),

$$(K) \ L(\phi \Rightarrow \psi) \Rightarrow (L\phi \Rightarrow L\psi) \tag{7.25}$$

It is well–known, see Ch. 3, sect. 9, that (K) is true at every frame regardless of the relation R.

The system in which (K) is satisfied is denoted as the system K.

The next postulate concerns the nature of necessity L and is expressed with a formula (T)

$$(T) \ L\phi \Rightarrow \phi \tag{7.26}$$

The formula (T) is satisfied in every frame in which the relation R is reflexive; the system K with (T) added is denoted T.

As a next step usually the formula $(S4)$ is added to the system T

$$(S4) \ L\phi \Rightarrow LL\phi \tag{7.27}$$

(S4) is true in every frame in which the relation R is transitive; the system $S4$ in which $K, T, S4$ hold is satisfied in every frame with the relation R reflexive and transitive.

The postulate $(S5)$ establishes a relation between L and M

$$(S5) \ M\phi \Rightarrow LM\phi \tag{7.28}$$

The system $S5$ resulting from adding $(S5)$ to the system $S4$ is valid in all frames in which the relation R is an equivalence.

7.11.1 *A Modal Logic with Ingredient Accessibility*

In an attempt at expressing modalities afforded by the rough mereological context, we reach to the relation of an ingredient as an accessibility relation. Thus, we adopt the collection of granules induced by a rough inclusion μ as the set W of possible worlds, and the relation $R(g, h) \Leftrightarrow ingr(h, g)$, i.e., a world (granule) h is accessible via R from a world (granule) g if and only if h is an ingredient of g, where the relation of an ingredient on granules is induced from the relation of ingredient on objects by means of (7.4).

For a predicate α, and a granule g, it follows that α is necessarily true at g if and only if α is true at every ingredient h of g, which means that, for a regular rough inclusion ν on sets, that $[[h]] \subseteq [[\alpha]]$ for each ingredient h of g. Possibility of α at g means that there exists a granule h with $ingr(h, g)$ such that α is true at h, i.e., $[[h]] \subseteq [[g]]$.

As ingredient relation is reflexive and transitive, we have

Proposition 7.23. *The modal logic obtained by taking the ingredient relation on granules as accessibility relation r and granules of knowledge as possible worlds W satisfies requirements for the system $S4$ in the frame (W, R).*

7.11.2 Modal Logic of Rough Set Approximations

Possibility and necessity are introduced in rough set theory by means of approximations: the upper and the lower, respectively. A logical rendering of these modalities in rough mereological logics exploits the approximations. We define two modal operators: M (possibility) and L (necessity).

To this end, we let

$$\begin{cases} I_\nu^\vee(g)(M\alpha) \geq r \Leftrightarrow \nu_L(g, \overline{[\alpha]}, r) \\ I_\nu^\vee(g)(L\alpha) \geq r \Leftrightarrow \nu_L(g, \underline{[\alpha]}, r) \end{cases} \tag{7.29}$$

Then we have the following criteria for necessarily or possibly true formulas.

A formula α is *necessarily true at a granule* g if and only if $g \subseteq \underline{[\alpha]}$; α is *possibly true at* g if and only if $g \subseteq \overline{[[\alpha]]}$.

This semantics of modal operators M, L can be applied to show that rough set structures carry the semantics of S5 modal logics, i.e., the following relations hold at each granule g.

1. $L(\alpha \Rightarrow \beta) \Rightarrow [(L\alpha) \Rightarrow L(\beta)]$;

2. $L\alpha \Rightarrow \alpha$;

3. $L\alpha \Rightarrow LL\alpha$;

4. $M\alpha \Rightarrow LM\alpha$.

We need to show that the meaning of each of formulas 1–4 is U.

Concerning formula 1 (modal formula (K)), we have $[[L(\alpha \Rightarrow \beta) \Rightarrow (L\alpha) \Rightarrow L(\beta)]] = (U \setminus (U \setminus [[\alpha]]) \cup [[\beta]]) \cup (U \setminus [[\alpha]]) \cup [[\beta]]$.

Assume that $u \notin (U \setminus [[\alpha]]) \cup [[\beta]]$; thus, (i) $[u]_A \subseteq [[\alpha]]$ and (ii) $[u]_A \cap [[\beta]] = \emptyset$. If it were $u \in (U \setminus [[\alpha]]) \cup [[\beta]]$ then we would have $[u]_A \subseteq (U \setminus [[\alpha]]) \cup [[\beta]]$, a contradiction with (i), (ii). Thus, the meaning of (K) is U.

For formula 2, modal formula (T), we have $[[L\alpha \Rightarrow \alpha]] = ((U \setminus \underline{[[\alpha]]}) \cup [[\alpha]])$; as $\underline{[[\alpha]]} \subseteq [[\alpha]]$, it follows that the meaning of (T) is U.

In case of formula 3, modal formula (S4), the meaning is $(U \setminus \underline{[[\alpha]]}) \cup \underline{[[\alpha]]} = (U \setminus \underline{[[\alpha]]}) \cup \underline{[[\alpha]]} = U$.

The meaning of formula 4, modal formula (S5), is $(U \setminus \overline{[[\alpha]]}) \cup \overline{\underline{[[\alpha]]}} = (U \setminus \overline{[[\alpha]]}) \cup \overline{[[\alpha]]} = U$.

It follows that the logic S5 is satisfied within logic induced by ν_L and more generally in logic induced by any regular rough inclusion ν.

7.12 Reasoning in Multi–agent and Distributed Systems

Approximate reasoning is often concerned with 'complex cases' like, e.g., robotic soccer, in which performing successfully tasks requires participation of a number of 'agents' bound to cooperate, and in which a task is performed with a number of steps, see, e.g., Stone [60]; other areas where such approach seems necessary concern assembling and design, see Amarel [2], fusion of knowledge, e.g., in robotics, fusion of information from sensors, see, e.g., Canny [5], Choset et al. [8], or, Stone [60], as well as in machine learning and fusion of classifiers, see, e.g., Dietterich [10].

Rough mereological approach to these problems was initiated with Polkowski and Skowron [54] – [57] and here we give these topics a logical touch.

Rough inclusions and granular intensional logics based on them can be applied in describing workings of a collection of intelligent agents which are called here *granular agents*.

A granular agent a will be represented as a tuple

$$(U_a, \mu_a, L_a, prop_a, synt_a, aggr_a)$$

where

1. U_a *is a collection of objects available to the agent a;*

2. μ_a *is a rough inclusion on objects in U_a;*

3. L_a *is a set of unary predicates in first–order open calculus, interpretable in U_a;*

4. $prop_a$ *is the propagation function that describes how uncertainty expressed by rough inclusions at agents connected to a propagates to a;*

5. $synt_a$ is the logic propagation functor which expresses how formulas of logics at agents connected to the agent a are made into a formula at a;

6. $aggr_a$ is the synthesis function which describes how objects at agents connected to a are made into an object at a.

We assume for simplicity that agents are arranged into a rooted tree; for
each agent a distinct from any leaf agent, we denote by B_a the children of
a, understood as agents connected to a and directly sending to a objects,
logical formulas describing them, and uncertainty coefficients like values of
rough inclusions.
    For $b \in B_a$, the symbol x_b will denote an object in U_b; similarly, ϕ_b will
denote a formula of L_b, and μ_b will be a rough inclusion at b with values r_b.
The same convention will be obeyed by objects at a.
Our scheme should obey some natural postulates stemming from an assumption of regularity of reasoning.

MA1 If $ingr_b(x'_b, x_b)$ for each $b \in B_a$, then $ingr_a(aggr(\{x'_b\}), aggr(\{x_b\})$

This postulate does assure that ingredient relations are in agreement with
aggregate operator of forming complex objects: ingredients of composed objects form an ingredient of a complex object. We can say that $aggr \circ ingr =
ingr \circ aggr$, i.e, the resulting diagram commutes.

MA2 If $x_b \models \phi_b$, then $aggr(\{x_b\}) \models synt(\{\phi_b\})$

This postulate is about agreement between aggregation of objects and their
logical descriptions: descriptions of composed objects merge into a description of the resulting complex object.

MA3 If $\mu_b(x_b, y_b, r_b)$ for $b \in B_a$, then $\mu_a(aggr(\{x_b\}), aggr(\{y_b\}), prop\{r_b\})$

This postulate introduces the propagation function, which does express how
uncertainty at connected agents is propagated to the agent a. One may observe the uniformity of $prop$, which in the setting of MA3 depends only on
values of r_b's; this is undoubtedly a simplifying assumption, but we want to
avoid unnecessary and obscuring the general view complications, which of
course can be multiplied at will.
    For Archimedean or metric induced rough inclusions μ, in whose cases
$g_\mu(u, r) = \{v : \mu(v, u, r)\}$, see Proposition 7.1, in this chapter, MA3 induces
an aggregation operator on granules

MA4 For $b \in B_a$, $ingr_b(x_b, g_{mu_r}(u_b, r_b))$ implies

$$ingr_a(aggr(\{x_b\}), g_{\mu_a}(aggr(\{u_b\}), prop(\{r_b\})))$$

Admitting MA4, we may also postulate that in case agents have at their disposal variants of rough mereological granular logics, intensions propagate according to the *prop* functor

MA5 *If* $I_{\nu_b}^\vee(g_b)(\phi_b) \geq r_b$ *for each* $b \in B_a$, *then*

$$I_{\nu_a}^\vee(aggr(\{g_b\}))(synt(\{\phi_b\})) \geq prop(\{r_b\}))$$

Here, we abuse language a bit, as we write *prop* in the same form as in MA3, again, it is done for simplicity of exposition.

We examine, for example's sake, a simple case of knowledge fusion, cf., Polkowski [52].

Proposition 7.24. *We consider an agent* $a \in Ag$ *and – for simplicity rea-sons – we assume that* a *has two incoming connections from agents* b, c; *the number of outgoing connections is of no importance as* a *sends along each of them the same information. Thus,* $B_a = \{b, c\}$.

We assume that each agent is applying the rough inclusion $\mu = \mu_L$ *induced by the Łukasiewicz t–norm* L *in the frame of the respective information system* $(U_a, A_a), (U_b, A_b), (U_c, A_c)$. *Each agent is also applying the rough inclusion on sets of the form (7.7) in evaluations related to extensions of formulae in-tensions.*

We consider a simple fusion scheme in which information systems at b, c *are combined object–wise to make the information system at* a; *thus,* $aggr_a(x, y) = (x, y)$. *Such case may happen, e.g., when an object is described with help of a camera image by some features and at the same time it is perceived and recog-nized with range sensors like infrared or laser sensors and some localization means like GPS.*

Then: uncertainty propagation and granule propagation are described by the Łukasiewicz t–norm L *and extensions of logical intensions propagate ac-cording to the product t–norm* P.

Proof. The set A_a of attributes at a equals then $A_b \times A_c$, i.e., attributes in A_a are pairs (a_1, a_2) with $a_1 \in A_b$ and $a_2 \in A_c$, and a fortiori, the value of this attribute is defined as

$$(a_1, a_2)(x, y) = (a_1(x), a_2(y))$$

It follows that the condition holds

$$IND_a(aggr_a(x, y), aggr_a(x', y')) \Leftrightarrow IND_b(x, x') \text{ and } IND_c(y, y') \quad (7.30)$$

Concerning the function $prop_a$, we consider objects x, x', y, y'; clearly,

$$DIS_a(aggr_a(x, y), aggr_a(x', y')) \quad (7.31)$$

is contained (as a subset) in

$$DIS_b(x, x') \times A_c \cup A_b \times DIS_c(y, y') \tag{7.32}$$

It follows by (7.31), (7.32) that

$$|DIS_a(aggr_a(x, y), aggr_a(x', y'))| \tag{7.33}$$

is less or at most equal to

$$|DIS_b(x, x')| \cdot |A_c| + |A_b| \cdot |DIS_c(y, y')| \tag{7.34}$$

As we know

$$\mu_a(aggr_a(x, y), aggr_a(x', y'), t) \tag{7.35}$$

is satisfied with the maximal value of t equal to

$$1 - \frac{|DIS_a(aggr_a(x, y), aggr_a(x', y'))|}{|A_b| \cdot |A_c|} \tag{7.36}$$

As the value of

$$\frac{|DIS_a(aggr_a(x, y), aggr_a(x', y'))|}{|A_b| \cdot |A_c|} \tag{7.37}$$

in the left–side of (7.36) is not less than

$$\frac{|DIS_b(x, x')| \cdot |A_c| + |A_b| \cdot |DIS_c(y, y')|}{|A_b| \cdot |A_c|} \tag{7.38}$$

which in turn is not less than

$$(\frac{|DIS_b(x, x')|}{|A_b|} + \frac{|DIS_c(y, y')|}{|A_c|} + 1 - 1 \tag{7.39}$$

it follows that

$$1 - \frac{|DIS_a(aggr_a(x, y), aggr_a(x', y'))|}{|A_b| \cdot |A_c|} \tag{7.40}$$

is greater or equal to

$$(1 - \frac{|DIS_b(x, x')|}{|A_b|}) + (1 - \frac{|DIS_c(y, y')|}{|A_c|}) - 1 \tag{7.41}$$

which is equivalent to

$$L(max\{r : \mu_b(x, x', r\}, max\{s : \mu_c(y, y', s\}) \tag{7.42}$$

We have
Claim
    If $\mu_b(x, x', r)$ and $\mu_c(y, y', s)$, then

$$\mu_a(aggr_a(x,y), aggr_a(x',y'), L(r,s))$$

It follows that the propagation function prop is defined by the Lukasiewicz *t–norm:* $prop(r,s) = L(r,s) = max\{0, r+s-1\}$.

In consequence, the granule propagation functor $prop_a^g$ can be defined in our example as indicated by the following equation

$$prop_a(g_{\mu_b}(x_b, r_b), g_{\mu_c}(y_c, s_c)) = (g_{\mu_a}(aggr_a(x_b, y_c), L(r_b, s_c))) \qquad (7.43)$$

The logic synthesizer $synt_a$ is defined by our assumptions as

$$synt_a(\phi_b, \phi_c) = \phi_b \wedge \phi_c \qquad (7.44)$$

Finally, we consider extensions of our logical operators of intensional logic.

We have for the extension

$$I(\mu_a)^{\vee}_{prop_a^g(g_b, g_c)}(synt_a(\phi_b, \phi_c)) \qquad (7.45)$$

that it is equal to

$$I(\mu_b)^{\vee}_{g_b}(\phi_b) \cdot I(\mu_c)^{\vee}_{g_c}(\phi_c) \qquad (7.46)$$

i.e., it is the defined by the product t–norm P $\qquad\qquad\qquad\qquad$ □

In such schemes one can discuss *synthesis* processes: when agents are arranged in a tree T, synthesis over T consists in outputting at the root agent $root(T)$ an object x_T. A characterization of x_T can be given by a postulate that it falls into a granule about a specified *standard* s_T within a radius of r_T.

Aggregation operators of agents $aggr_a$ for non–leaf agents a,compose into a global aggregate operator $aggr_T$, which sends objects chosen at leaf agents, at most one for each agent, into an object at the root of T. In a similar vein, propagation operators $prop_a$ compose into the global propagation operator $prop_T$.

The synthesis procedure could be presented as

Input: $g_{\mu_T}(s_t, r_T)$

1. *Find recursively going down the tree to leaf agents: in each leaf agent b_l, an object s_l such that*

$$aggr_T(\{s_l\}) = s_T \qquad (7.47)$$

2. *Find recursively radii r_l at leaf agents which satisfy the condition that,*

$$prop_T(\{r_l\}) \geq r_T \qquad (7.48)$$

3. *For each leaf agent b_l, find an object x_l with the property that,*

$$\mu_l(x_l, s_l, r_l) \; for \; each \; leaf \; agent \; b_l \qquad (7.49)$$

4. *Output the object $aggr_T(\{x_l\})$.*

This reasoning scheme seems fairly universal also for networking schemes of reasoning like in neural networks. However, as learning in neural networks is usually based on the idea of gradient search, and in consequence it requires differentiable rough inclusions, we devote to neural network reasoning a separate paragraph.

7.13   Reasoning in Cognitive Schemes

Neural networks are motivated by the functioning of neural system. Their inception was possible due to achievements of neurophysiology, notably, discovery of the structure of the biological neuron by Ramón y Cajal [58].

On this discovery, McCulloch and Pitts [9] built their model of a computing machine called now *McCulloch–Pitts neuron*. In its simplest form, it consisted of a cell endowed with a threshold Θ, and inputs $x_1, x_2, ..., x_n$ through which binary signals of 0 or 1 could be sent. The one output y could issue a binary signal of either 0 or 1. Inputs are counterparts to the biological *synapses*, and the output models the biological *axon*.

Computation by the neuron is governed by the rule

$$y = \begin{cases} 1 \; in \; case \; \sum_i x_i \geq \Theta \\ 0 \; otherwise \end{cases} \qquad (7.50)$$

Thus, the neuron *fires* when the summary input exceeds the threshold, otherwise the neuron remains inactive.

The neuron of McCulloch–Pitts can produce a binary *linear classifier*: the separating hyper–plane, which is to divide the training set into positive and negative examples, is of the form

$$\sum_i x_i = \Theta \qquad (7.51)$$

The idea of networks of neurons was advocated by Alan Turing [61] who proposed a learning scheme for networks of neurons connected through *modifiers*.

Increasing classification possibilities were given a neuron with the idea of a *perceptron* due to Rosenblatt [59]. A simplified perceptron adds to McCulloch–Pitts neuron *weights* on inputs, and an additional input with constant value of 1 and a weight b, called *bias*. Thus, the classifying hyper–plane has the form

$$\sum_i w_i \cdot x_i + b = \Theta \qquad (7.52)$$

Parameters w_i, b, Θ which can be varied, allow for *learning*. As proved by Novikoff [38], if a binary concept can at all be classified with the hyperplane of the form (7.52), then it is possible to find in a finite number of steps of the *learning procedure* a proper set of parameters starting from, e.g., a randomly chosen set.

The idea of a network of perceptrons returned with Grossberg [14]; it was shown that a network of perceptrons can classify, hence learn, any Boolean function.

A further step in improving efficiency of computation and learning of perceptron networks was due to Bryson and Ho [4]: replacement of the Heaviside–type threshold function with a sigmoid–type differentiable function allowed for *supervised learning* by means of the gradient search, called the *backpropagation*, based on computing the gradient of the *error function* $E(w) = \sum_{x_j} (y_j - t_j)^2$, where w is the vector of weights on all connections in the network, x_j are consecutive input examples, y_j are corresponding responses of the network, and t_j are desired outputs to inputs x_j.

We would like to endow neurons with intelligence based on rough mereological reasoning, see Polkowski [44]. In neural models of computation, an essential feature of neurons is differentiability of transfer functions; hence, we introduced a special type of rough inclusions, called *gaussian* in Polkowski [44] because of their form, by letting,

$$\mu_G(x, y, r) \text{ iff } e^{-|\sum_{a \in DIS(x,y)} w_a|^2} \geq r \qquad (7.53)$$

where $w_a \in (0, +\infty)$ is a weight associated with the attribute a for each attribute $a \in A$.

Let us observe in passing that μ_G can be factored through the indiscernibility relation $IND(A)$, and thus its arguments can be objects as well as indiscernibility classes; we will freely use this fact.

Properties of gaussian rough inclusions are, see Polkowski [44]

Proposition 7.25. *The following hold for each gaussian rough inclusion μ_G, where ingr is the ingredient relation associated with μ_G by postulate RINC1, Ch. 5, sect. 2,*

1. *$ingr(x, y)$ if and only if $DIS(x, y) = \emptyset$;*

2. *There exists a function $\eta(r, s)$ such that $\mu_G(x, y, r), \mu_G(y, z, s)$ imply $\mu_G(x, z, t)$ with $t \leq \eta(r, s))$;*

3. *If $ingr(x, g_{\mu_G}(y, r))$ and $ingr(x, g_{\mu_G}(z, s))$, then $ingr(g_{\mu_G}(x, t), g_{\mu_G}(y, r))$, $ingr(g_{\mu_G}(x, t), g_{\mu_G}(z, s))$ for $t \geq max\{r^4, s^4\}$.*

Proof. Property 1 follows by definition 7.25.

To verify Property 2, we observe that by definition (7.25), we have

$$\sum_{a \in DIS(x,y)} w_a \leq (-logr)^{\frac{1}{2}} \tag{7.54}$$

and

$$\sum_{a \in DIS(y,z)} w_a \leq (-logs)^{\frac{1}{2}} \tag{7.55}$$

As $DIS(x,z) \subseteq DIS(x,y) \cup DIS(y,z)$, we denote by t the maximum value of q such that $\mu_G(x,z,q)$, and we obtain that

$$(-logt)^{\frac{1}{2}} \leq (-logr)^{\frac{1}{2}} + (-logs)^{\frac{1}{2}} \tag{7.56}$$

from which by taking squares of both sides, we get at

$$logt \leq log(r \cdot s) - 2(logr \cdot logs)^{\frac{1}{2}} \tag{7.57}$$

so finally

$$t \leq r \cdot s \cdot e^{-2(logr \cdot logs)^{\frac{1}{2}}} \tag{7.58}$$

Thus, it suffices to take $\eta(r,s) = r \cdot s \cdot e^{-2(logr \cdot logs)^{\frac{1}{2}}}$.

In case of Property 3, we observe that as μ_G is symmetric by definition and transitive by Property 2, we have $ingr(x, g_{mu_G}(y,r)) \Leftrightarrow \mu_G(x,y,r)$. Hence, in order to verify Property 3, we have to find t such that $\eta(r,t) \geq r, \eta(s,t) \geq s$. As $\eta(r,t) \geq r$ implies $t \geq r^4$, it suffices to have $t \geq max\{r^4, s^4\}$ to satisfy both conditions   □

7.13.1 *Rough Mereological Perceptron*

The rough mereological perceptron is modeled on the perceptron, and it consists of an intelligent agent a, endowed with a gaussian rough inclusion μ_a on the information system $I_a = (U_a, A_a)$ of the agent a.

The input to a is in the form of a finite tuple $\overline{x} = (x_1, ..., x_k)$ of objects, and the input $\overline{x}$ is converted at a into an object $x = aggr_a(\overline{x}) \in U_a$ by means of an operator $aggr_a$.

The rough mereological perceptron is endowed with a set of *target concepts* $T_a \subseteq U_a/IND(A_a)$, each target concept a class of the indiscernibility IND_a.

Formally, a rough mereological perceptron is thus a tuple

$$RMP = (a, I_a, \mu_a, aggr_a, T_a)$$

Computing by a network of RMP's, is directed by the gradient of the error function, which in this case has the form of the gradient of the function

$$f(x,y) = e^{-|\sum_{a \in DIS(x,y)} w_a|^2} \qquad (7.59)$$

which is

$$\frac{\partial f}{\partial w} = f \cdot (-2 \cdot \sum w_a) \qquad (7.60)$$

It follows from the last equation that gradient search would go in direction of minimizing the value of $\sum_a w_a$.

The result of computation with a target $g_{\mu_G}(t,r)$ for a a sample $x_1, ..., x_k$ is a granule $g = g_{\mu_G}(aggr_a(x_1, .., x_k), r(res))$ such that $ingr(g, g_{\mu_G}(t,r))$.

During computation, weights are incremented according to the ideology of backpropagation, by the recipe,

$$w_a \leftarrow w_a + \Delta \cdot \frac{\partial E}{\partial w_a} \qquad (7.61)$$

where Δ is the *learning rate*.

At a stage *current* of computing, where $\gamma = r_{current} - r$, for a natural number k, the value of $\Delta_{current}$ which should be taken at the step *current* in order to achieve the target in at most k steps should be taken as, see Polkowski [44]

$$\Delta_{current} \simeq \frac{\gamma}{2 \cdot k \cdot f^2 \cdot (\sum_a w_a)^2} \qquad (7.62)$$

References

1. Alves, E.H., Guerzoni, J.A.D.: Extending Montague's system: A three–valued intensional logic. Studia Logica 49, 127–132 (1990)
2. Amarel, S.: Panel on AI and Design. In: Proceedings of 12th Intern. Conf. on AI, Sydney, pp. 563–565 (1991)
3. Van Benthem, J.: A Manual of Intensional Logic. CSLI Stanford University, Stanford (1988)
4. Bryson, A.E., Yu–Chi, H.: Applied Optimal Control: Optimization, Estimation and Control. Blaisdell Publishing Company, Waltham (1969)
5. Canny, J.F.: The Complexity of Robot Motion Planning. MIT Press, Cambridge (1988)
6. Carnap, R.: Necessity and Meaning. Chicago University Press, Chicago (1947)
7. Chang, C.C.: Proof of an axiom of Łukasiewicz. Trans. Amer. Math. Soc. 87, 55–56 (1958)
8. Choset, H., Lynch, K.M., Hutchinson, S., Kantor, G., Burgard, W., Kavraki, L.E., Thrun, S.: Principles of Robot Motion. Theory, Algorithms, and Implementations. MIT Press, Cambridge (2005)
9. McCulloch, W., Pitts, W.: A logical calculus of the ideas immanent in nervous activity. Bulletin of Mathematical Biophysics 7, 115–133 (1943)
10. Dietterich, T.G.: Ensemble Methods in Machine Learning. In: Kittler, J., Roli, F. (eds.) MCS 2000. LNCS, vol. 1857, pp. 1–15. Springer, Heidelberg (2000)
11. Fitting, M.C.: First–order intensional logic. Annals of Pure and Applied Logic 127, 171–193 (2004)
12. Frege, G.: Über Sinn und Bedeutung. Zeitschrift für Philosophie und Philosophische Kritik NF 100, 25–50 (1892)
13. Gallin, D.: Intensional and higher–order modal logic. North Holland, Amsterdam (1975)
14. Grossberg, S.: Contour enhancement, short–term memory, and constancies in reverberating neural networks. Studies in Applied Mathematics 52, 213–257 (1973)
15. Grzymała-Busse, J.W.: Data with missing attribute values: Generalization of indiscernibility relation and rule induction. In: Peters, J.F., Skowron, A., Grzymała-Busse, J.W., Kostek, B.z., Świniarski, R.W., Szczuka, M.S. (eds.) Transactions on Rough Sets I. LNCS, vol. 3100, pp. 78–95. Springer, Heidelberg (2004)
16. Grzymała-Busse, J.W., Hu, M.: A Comparison of Several Approaches to Missing Attribute Values in Data Mining. In: Ziarko, W.P., Yao, Y. (eds.) RSCTC 2000. LNCS (LNAI), vol. 2005, pp. 378–385. Springer, Heidelberg (2001)

17. Hájek, P.: Metamathematics of Fuzzy Logic. Kluwer, Dordrecht (2001)
18. Hughes, G.E., Creswell, M.J.: A New Introduction to Modal Logic. Routledge, London (1996)
19. Kripke, S.: Semantical considerations on modal logics. Acta Philosophica Fennica. Modal and Many–Valued Logics, pp. 83–94 (1963)
20. Lemmon, E.J., Scott, D.S., Sederberg, K. (eds.): The Lemmon Notes. An Introduction to Modal Logic. Basil Blackwell, Oxford (1963)
21. Lin, T.Y.: Neighborhood systems and relational database. Abstract. In: Proceedings of CSC 1988, p. 725 (1988)
22. Lin, T.Y.: Neighborhood systems and approximation in Database and Knowledge Based Systems. In: Proceedings of the 4th International Symposium on Methodologies for Intelligent Systems (ISMIS), pp. 75–86 (1989) .
23. Lin, T.Y.: Topological and fuzzy rough sets. In: Słowiński, R. (ed.) Intelligent Decision Support. Handbook of Applications and Advances of the Rough Sets Theory, pp. 287–304. Kluwer, Dordrecht (1992)
24. Lin, T.Y.: From rough sets and neighborhood systems to information granulation and computing with words. In: Proceedings of the European Congress on Intelligent Techniques and Soft Computing, pp. 1602–1606 (1994)
25. Lin, T.Y.: Granular Computing: Fuzzy logic and rough sets. In: Zadeh, L.A., Kacprzyk, J. (eds.) Computing with Words in Information/Intelligent Systems, vol. 1, pp. 183–200. Physica Verlag, Heidelberg (1999)
26. Lin, T.Y.: Granular computing. In: Wang, G., Liu, Q., Yao, Y., Skowron, A. (eds.) RSFDGrC 2003. LNCS (LNAI), vol. 2639, pp. 16–24. Springer, Heidelberg (2003)
27. Lin, T.Y.: Granular computing: Examples, intuitions, and modeling. In: Proceedings of IEEE 2005 Conference on Granular Computing GrC 2005, pp. 40–44. IEEE Press, Beijing (2005)
28. Lin, T.Y.: A roadmap from rough set theory to granular computing. In: Wang, G.-Y., Peters, J.F., Skowron, A., Yao, Y. (eds.) RSKT 2006. LNCS (LNAI), vol. 4062, pp. 33–41. Springer, Heidelberg (2006)
29. Łukasiewicz, J.: Die Logischen grundlagen der Wahrscheinlichtkeitsrechnung. Cracow (1913)
30. Łukasiewicz, J.: O logice trójwartościowej (On three–valued logic, in Polish). Ruch Filozoficzny 5, 170–171 (1920)
31. Łukasiewicz, J.: Aristotle's Syllogistic from the Standpoint of Modern Formal Logic, 2nd edn. Oxford University Press, Oxford (1957)
32. Łukasiewicz, J., Tarski, A.: Untersuchungen ueber den Aussagenkalkuels. C.R. Soc. Sci. Lettr. Varsovie 23, 39–50 (1930)
33. Meredith, C.A.: The dependence of an axiom of Łukasiewicz. Trans. Amer. Math. Soc. 87, 54 (1958)
34. Montague, R.: Pragmatics and intensional logic. Synthese 22, 68–94 (1970)
35. Montague, R., Thomason, R. (eds.): Formal Philosophy. Yale University Press, New Haven (1974)
36. Nguyen, S.H.: Regularity analysis and its applications in Data Mining. In: Polkowski, L., Tsumoto, S., Lin, T.Y. (eds.) Rough Set Methods and Applications. New Developments in Knowledge Discovery in Information Systems, pp. 289–378. Physica Verlag, Heidelberg (2000)
37. Nieminen, J.: Rough tolerance equality and tolerance black boxes. Fundamenta Informaticae 11, 289–296 (1988)

38. Novikoff, A.B.: On convergence proofs on perceptrons. In: Symposium on the Mathematical Theory of Automata, vol. 12, pp. 615–622. Brooklyn Polytechnic Institute, New York (1962)
39. Pawlak, Z.: Rough sets. Intern. J. Comp. Inform. Sci. 11, 341–366 (1982)
40. Pawlak, Z.: Rough Sets: Theoretical Aspects of Reasoning about Data. Kluwer, Dordrecht (1991)
41. Polkowski, L.: A rough set paradigm for unifying rough set theory and fuzzy set theory. In: Wang, G., Liu, Q., Yao, Y., Skowron, A. (eds.) RSFDGrC 2003. LNCS (LNAI), vol. 2639, pp. 70–78. Springer, Heidelberg (2003)
42. Polkowski, L.: A note on 3–valued rough logic accepting decision rules. Fundamenta Informaticae 61, 37–45 (2004)
43. Polkowski, L.: Toward rough set foundations. Mereological approach. In: Tsumoto, S., Słowiński, R., Komorowski, J., Grzymała-Busse, J.W. (eds.) RSCTC 2004. LNCS (LNAI), vol. 3066, pp. 8–25. Springer, Heidelberg (2004)
44. Polkowski, L.: A rough–neural computation model based on rough mereology. In: Pal, S.K., Polkowski, L., Skowron, A. (eds.) Rough–Neural Computing. Techniques for Computing with Words, pp. 85–108. Springer, Heidelberg (2004)
45. Polkowski, L.: Formal granular calculi based on rough inclusions. In: Proceedings of IEEE 2005 Conference on Granular Computing GrC 2005, pp. 57–62. IEEE Computer Society Press, Beijing (2005)
46. Polkowski, L.: Rough–fuzzy–neurocomputing based on rough mereological calculus of granules. International Journal of Hybrid Intelligent Systems 2, 91–108 (2005)
47. Polkowski, L.: A model of granular computing with applications. In: Proceedings of IEEE 2006 Conference on Granular Computing GrC 2006, pp. 9–16. IEEE Computer Society Press, Atlanta (2006)
48. Polkowski, L.: Granulation of knowledge in decision systems: The approach based on rough inclusions. The method and its applications. In: Kryszkiewicz, M., Peters, J.F., Rybiński, H., Skowron, A. (eds.) RSEISP 2007. LNCS (LNAI), vol. 4585, pp. 69–79. Springer, Heidelberg (2007)
49. Polkowski, L.: The paradigm of granular rough computing. In: Proceedings of 6th IEEE Intern. Conf. on Cognitive Informatics, ICCI 2007, pp. 145–163. IEEE Computer Society, Los Alamitos (2007)
50. Polkowski, L.: Rough mereology in analysis of vagueness. In: Wang, G., Li, T., Grzymala-Busse, J.W., Miao, D., Skowron, A., Yao, Y. (eds.) RSKT 2008. LNCS (LNAI), vol. 5009, pp. 197–204. Springer, Heidelberg (2008)
51. Polkowski, L.: A unified approach to granulation of knowledge and granular computing based on rough mereology: A survey. In: Pedrycz, W., Skowron, A., Kreinovich, V. (eds.) Handbook of Granular Computing, pp. 375–400. John Wiley and Sons Ltd., Chichester (2008)
52. Polkowski, L.: Granulation of Knowledge: Similarity Based Approach in Information and Decision Systems. In: Meyers, R.A. (ed.) Springer Encyclopedia of Complexity and System Sciences, Springer, Berlin (2009) article 00 788
53. Polkowski, L., Semeniuk–Polkowska, M.: On rough set logics based on similarity relations. Fundamenta Informaticae 64, 379–390 (2005)
54. Polkowski, L., Skowron, A.: Rough mereological foundations for design, analysis, synthesis and control in distributed systems. Information Sciences. An International Journal 104(1-2), 129–156 (1998)

55. Polkowski, L., Skowron, A.: Grammar systems for distributed synthesis of of approximate solutions extracted from experience. In: Paun, G., Salomaa, A. (eds.) Grammatical models of Multi-Agent Systems, pp. 316–333. Gordon and Breach, Amsterdam (1999)

56. Polkowski, L., Skowron, A.: Towards adaptive calculus of granules. In: Zadeh, L.A., Kacprzyk, J. (eds.) Computing with Words in Information/Intelligent Systems 1. Foundations, Physica Verlag, Heidelberg (1999)

57. Polkowski, L., Skowron, A.: Rough mereological calculi of granules: A rough set approach to computation. Computational Intelligence. An Intern. Journal 17(3), 472–492 (2001)

58. Ramóny Cajal, S.: Sur la morphologie et les connexions des elements de la retine des oiseaux. Anatomisches Anzeiger 4, 111–121 (1889)

59. Rosenblatt, F.: The perceptron: A probabilistic model for information storage and organization in the brain. Psychological Review 65(6), 386–408 (1958)

60. Stone, P.: Layered Learning in Multiagent Systems: A Winning Approach to Robotic Soccer. MIT Press, Cambridge (2000)

61. Turing, A.M.: Intelligent Machinery. A report. National Physical Laboratory. Mathematical Division (1948)

62. Wajsberg, M.: Axiomatization of the 3–valued sentential calculus (in Polish, a summary in German). C. R. Soc. Sci. Lettr. Varsovie 24, 126–148 (1931)

63. Wille, R.: Restructuring lattice theory: An approach based on hierarchies of concepts. In: Rival, I. (ed.) Ordered Sets, pp. 445–470. D. Reidel, Dordrecht (1982)

64. Yao, Y.Y.: Granular computing: Basic issues and possible solutions. In: Proceedings of the 5th Joint Conference on Information Sciences I. Assoc. Intell. Machinery, pp. 186–189. Assoc. Intell. Machinery, Atlantic (2000)

65. Yao, Y.Y.: Perspectives of granular computing. In: Proceedings of IEEE 2005 Conference on Granular Computing GrC 2005, pp. 85–90. IEEE Press, Beijing (2005)

66. Zadeh, L.A.: Fuzzy sets and information granularity. In: Gupta, M., Ragade, R., Yager, R.R. (eds.) Advances in Fuzzy Set Theory and Applications, pp. 3–18. North–Holland, Amsterdam (1979)

67. Zadeh, L.A.: Toward a unified theory of uncertainty. In: Proceedings of IPMU 2004, Perugia, Italy, vol. 1, pp. 3–4 (2004)

Chapter 8
Reasoning by Rough Mereology in Problems of Behavioral Robotics

In Ch. 6, we have developed basic notions and propositions of rough mere-ogeometry and rough mereotopology. We have stressed that by its nature, rough mereology does address collective concepts, relations among which are expressed by partial containment rendered as the predicate of a part to a degree. Behavioral robotics falls into this province, as usually robots as well as obstacles and other environmental objects are modeled as figures or solids. In this chapter, we discuss planning and navigation problems for mobile autonomous robots and their formations. In particular, we give a formal definition of a robot formation based on the betweenness relation, cf., Ch. 6., sect. 10. First, we introduce the subject of planning in robotics.

8.1 Planning of Robot Motion

Planning is concerned with setting a trajectory for a robot endowed with some sensing devices which allow it to perceive the environment in order to reach by the robot a goal in the environment at the same time bypassing obstacles.

Planning methods, cf., e.g., Choset et al. [17], vary depending on the robot abilities, features of the environment and chosen methodology. Among them are simple geometric methods designed for a robot endowed with sensors detecting obstacles, e.g., touch sensors or range sensors and able to detect distance between any pair of points. These methods are called 'contour following', as for such a robot, the idea can be implemented of moving to goal in a straight line segment and in case of meeting with an obstacle to bypass it by circumnavigating its boundary until the straight line to goal is encountered anew.

In this class belong so called 'bug algorithms' like BUG 1 algorithm due to Lumelsky and Stepanov [35], and its modification, the 'tangent bug' planner, Kamon et al. [27], cf., Choset et al. [17], in which the robot performs a heuristic search of A^* type, see, e.g., Russell and Norvig [52] or Choset et al.

L. Polkowski: Approximate Reasoning by Parts, ISRL 20, pp. 297–318.
springerlink.com   © Springer-Verlag Berlin Heidelberg 2011

[17] with the heuristic function $h(x) = \rho(x, O) + \rho(O, goal)$ where x is the current position of the robot, and the point O is selected as an end–point of the continuity interval of ρ – the distance function, whose values are bound by a constant R. When the distance measured by range sensors exceeds R the value of ρ is set to infinity. The graph of ρ against the position x exhibits then discontinuities and continuity intervals clearly outline boundaries of obstacles, hence, the idea of selecting O as a boundary continuity point. Minimization of h leads to optimization of the chosen safe trajectory.

To stay with geometric methods, we proceed with *Voronoï diagrams*. Most often, obstacles are modeled as two–dimensional polygons, and then the Voronoï diagram V is a 1–dimensional set consisting of points which are at equal distance from two closest to them obstacles, i.e., $x \in V$ if and only if $\rho(x, W_i) = \rho(x, W_j) \leq \rho(x, W_k)$ for $k \neq i, j$. Clearly, navigating the robot along V keeps it at safe distance from obstacles, hence, the planner based on the idea of the Voronoï diagram transports the robot from the starting point to the nearest point on V and then along V to the point in V nearest to the goal, see Choset et al. [17].

Another geometric idea is implemented in *visibility graphs*, see Latombe [32] and Li and Canny [34]. Vertices of obstacles, again, modeled as 2–dimensional polygons, constitute nodes of the graph. Two distinct nodes are joined by an edge if and only if two corresponding vertices can be connected by a straight line segment avoiding any contact with any of obstacles except for these two points. For particular start and goal points *start, goal*, these points are added to the graph as nodes and also connected by edges with other nodes when the straight line visibility condition is satisfied.

A further development is provided by *silhouette methods*, see Canny [15]. It exploits the *sweeping algorithm* which consists in moving the *sweeping line* of the form, e.g., $\{x_1\} \times R$ by increasing x_1; at each position, the end points of segments in the line obtained by intersection with obstacles constitute the *silhouette*, or, the *Canny roadmap*.

In some cases, not only the aim is to transport the robot from start to goal, but to cover all free space, e.g., when one wants the robot to paint the floor. One says in such cases of the *coverage problem*. Then, a method exploited is the *cell decomposition*. One represents the free space as a union of *cells* and builds on cells as nodes of a graph, the *adjacency graph* joining two nodes–cells with an edge if and only if the cells share a boundary.

There are various implementations of the idea, e.g., *trapezoidal decomposition*, see de Berg et al. [10]. This implementation looks at each vertex v of an obstacle for half–lines going 'up' and 'down' and registering points on them of intersection with obstacles or boundaries of working space. This defines a decomposition into cells.

As 'exact' planning algorithms are infeasible in some environments, see Schwartz and Sharir [54], Kavraki [28], or Canny [15] approximate planners based on probabilistic sampling were developed, see, e.g., Barraquand et al. [6], Kavraki et al. [29].

When many queries about paths are intended, then it is useful to build a probabilistic roadmap planner, see Kavraki et al. [29]. It is built on a standard planner, say P, which explores the possibility of a path between points in the working space and builds a metric on it. Nodes of the graph are configurations sampled from possible ones and roadmaps are built incrementally; any time a configuration, say c, is sampled, k nearest neighbors already sampled are selected and the planner P checks whether there is a safe path between c and any neighbor, adding an edge to the graph for each pair with a safe path.

Planning is necessarily coupled with *localization*, in order to plan one should know the robot position. To this effect, bayesian filtering, in particular Kalman filtering is applied, see Kalman [26], cf. Choset et al. [17].

A method referring to physical inspiration is the *potential field* method, see Khatib [30]. A potential field is composed of attractive potentials for goals and repulsive potentials for obstacles.

An example may be taken as the quadratic potential function

$$U_{attractive}(x) = \frac{1}{2} \cdot ||x - x_{goal}||^2 \tag{8.1}$$

which induces the gradient

$$\nabla U_{attractive}(x) = x - x_{goal} \tag{8.2}$$

which assures that the force (the gradient) exerted on the robot is greater when the robot is far from the goal and diminishes to zero as the robot is approaching the goal.

A repulsive potential should have opposite properties: it should exert a force tending to ∞ with the distance to the obstacle reaching 0. Denoting the distance from a point x to the closest obstacle with $s(x)$, the repulsive potential can be defined as in

$$U_{repulsive}(x) = \frac{1}{2} \cdot \left[\frac{1}{s(x)}\right] \tag{8.3}$$

with the gradient

$$\nabla U_{repulsive}(x) = -\frac{1}{s(x)^2} \cdot \nabla s(x) \tag{8.4}$$

The global potential function U is the sum of the attractive and repulsive parts:

$$U(x) = U_{attractive}(x) + U_{repulsive}(x)$$

Given U, the robot performs a well–known *gradient descent* : it does follow the direction of the gradient in small steps : the $(i + 1)$–th position is given from the i-th position and the gradient therein as

$$x_{i+1} = x_i + \xi_i \cdot \nabla U(x_i) \tag{8.5}$$

Potential fields method suffers from the local minima problem immanent to gradient descent method: though the absolute minimum of 0 is achieved by the potential function at the goal, yet superposition of many fields from obstacles and goals induces local minima and saddle points typical to many–dimensional landscape. A way out of local minima can be found by means of small random perturbations, see Barraquand et al. [6].

This closes our very brief glimpse at planning methods showing their variety. For a discussion of an underlying architecture, see Brooks [12]. One can ask, what rough mereology can add to this repertoire? We have proposed a new variant of potential field method, based on rough inclusion technique, see Ośsmiałowski [41] and Polkowski and Ośmiałowski [48], [49].

8.2   Potential Fields from Rough Inclusions

Classical methodology of potential fields works with integrable force field given by formulas of Coulomb or Newton which prescribe force at a given point as inversely proportional to the squared distance from the target; in consequence, the potential is inversely proportional to the distance from the target. The basic property of the potential is that its density (=force) increases in the direction toward the target. We observe this property in our construction.

We refer to mereogeometry of Ch. 6, sect. 10, and we recall the rough inclusion

$$\mu(x, y, r) \Leftrightarrow \frac{||x \cap y||}{||x||} \tag{8.6}$$

where $||x||$ is the area of the region x. In our construction of the potential field, region will be squares: this corresponds with the robots used which are disc–shaped Roomba (a trademark of iRobot, Inc.) robots, so they can be represented by squares circumscribed on them.

Geometry induced by means of a rough inclusion can be used to define a generalized potential field: the force field in this construction can be interpreted as the density of squares that fill the workspace and the potential is the integral of the density. We present now the details of this construction, see Ośmiałowski [41], Polkowski and Ośmiałowski [49].

We construct the potential field by a discrete construction. The idea is to fill the free workspace of a robot with squares of fixed size in such a way that the density of the square field (measured, e.g., as the number of squares intersecting the disc of a given radius r centered at the target) increases toward the target.

To ensure this property, we fix a real number – the **field growth step** in the interval (0, square edge length); in our exemplary case the parameter **field growth step** is set to 0.01.

The collection of squares grows recursively with the distance from the target by adding to a given square in the $(k + 1) - -th$ step all squares obtained from it by translating it by $k \times$ **field growth step** (with respect to Euclidean distance) in basic eight directions: N, S, W, E, NE, NW, SE, SW (in the implementation of this idea, the *floodfill algorithm* with a queue has been used, see Ośmiałowski [41]. Once the square field is constructed, the path for a robot from a given starting point toward the target is searched for.

The idea of this search consists in finding a sequence of *way–points* which delineate the path to the target. Way–points are found recursively as centroids of unions of squares mereologically closest to the square of the recently found way–point. We recall, see Ch. 6, that the mereological distance between squares x, y is defined by means of

$$k(x, y) = min\{\max\{r, s\} : \mu(x, y, r), \mu(y, x, s)\} \qquad (8.7)$$

We also remind that the mereological distance $k(x, y)$ takes on the value 1 when $x = y$ and the minimal value of 0 means that $x \cap y \subseteq Bd(x) \cap Bd(y)$. In order do define a "potential" of the rough mereological field, let us consider how many generations of squares will be centered within the distance r from the target. Clearly, we have

$$d + 2d + ... + kd \leq r \qquad (8.8)$$

where d is the field growth step, k is the number of generations. Hence,

$$k^2 d \leq \frac{k(k+1)}{2} d \leq r \qquad (8.9)$$

and thus

$$k \leq (\frac{r}{d})^{\frac{1}{2}} \qquad (8.10)$$

The potential $V(r)$ can be taken as $\sim r^{\frac{1}{2}}$. The force field $F(r)$ is the negative gradient of $V(r)$,

$$F(r) = -\frac{d}{dr} V(r) \sim -\frac{1}{r^{\frac{1}{2}}} \qquad (8.11)$$

Hence, the force decreases with the distance r from the target slower than traditional Coulomb force. It has advantages of slowing the robot down when it is closing on the target. Parameters of this procedure are: the **field growth step** set to 0.01, and the size of squares which in our case is 1.5 times the diameter of the Roomba robot.

The path planner designed in this way, accepts target point coordinates and provides list of way–points from given robot position to the goal. To do its job, it needs a map of static obstacles that a robot should avoid while approaching target point. A robot and a target should both lay within the area delimited by surrounding static obstacles that form borders of robot workspace. There can be other static obstacles within the area, all marked on the provided map. After the path is proposed a robot is lead through the path until it reaches given target. If a robot cannot move towards goal position for some longer time (e.g., it keeps on hitting other robot reaching its target or any other unknown non–static obstacle), new path is proposed. We tested our planner device running simulations in which we have had a model of Roomba robot traveling inside artificial workspace. Real Roomba robots are round and therefore easy to model, however they do not provide many useful sensor devices (except bumpers which we were using to implement lower–level reaction for hitting unexpected obstacles). Also odometry of Roomba robots is unreliable, Tribelhorn and Dodds [59], hence, we assume that simulated robots are equipped with a global positioning system. A map of an environment as used in simulations along with a potential field generated for a given goal is is shown in Fig. 8.1.

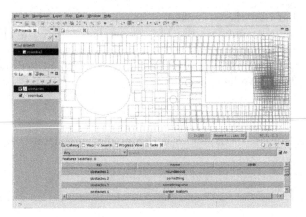

Fig. 8.1 Obstacles and potential field layer

A robot should follow the path proposed by planner by going from one area centroid to another until the goal is reached. The proposed path is marked on the map, see Fig. 8.2, Fig. 8.3.

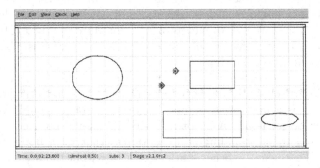

Fig. 8.2 Stage simulator: iRobot Roomba robots starting to a goal

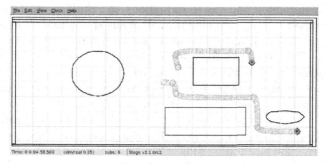

Fig. 8.3 Planned paths of Roomba robots to their targets

To perform re–planning of the path, usually the planner should repeat the planning routine. Using our method, only second stage of planning routine is done during replanning as potential field is computed only once (unless the database is updated with new obstacles). Searching for a path within already computed potential field is computationally cheap as it is limited to database lookup operations (therefore speed of database communication is critical if this method is intended to be working fast).

8.3 Planning for Teams of Robots

Both theoretical interests, see Walter [61], [62] as well as practical motivations, have driven the attention of researchers in robotics toward problems related to teams of robots. From purely intellectual point of view, this opens a new venue for solving problems of cooperation, communication, task–sharing and division, and planning non–collision paths for robots.

According to Cao et al. [16], robot teams can provide a useful playground for studies of cognitive theories, biology, ethology, organization and management. They can also lead to new solutions to problems of artificial intelligence. Passing from a single robot to teams of robots can be motivated also by pragmatic reasons, Cao et al. [16], as tasks for robots can be too complex for a single robot, or many robots can do the task easier at a lesser cost, or many robots can perform the task more reliably. Practical studies along these lines were concerned with moving large objects of irregular shapes by groups of robots, see Kube and Zhang [31], search and rescue, see Jennings et al. [25], formations of planetary outposts of mobile robots, see Huntsberger et al. [23]. Simulations of systems a few robots were studied, e.g., in CEBOT, see Fukuda and Nakagawa [21], ACTRESS, see Asama et al. [3], GOFER, see Caloud et al. [14], cf., the ALLIANCE architecture in Parker [46].

In Cao et al. [16], main research directions in this area were systematized, among them, Geometric Problems, involving multiple–robot path planning, formation maintaining, moving to formations, marching, pattern generation. To be more specific, path planning involves many specialized strategies; some of them propose initial individual paths for each robot, often straight lines to the goal, with strategies for obstacle negotiating and conflict resolution by either negotiations between robots or by a supervising agent. The choice here is between prioritized planning which takes one robot at a time according to some priority scheme, see Erdmann and Lozano–Perez [19] and path coordination method which plans paths by scheduling configuration space resources. Problems of cooperation and negotiations are discussed in Naffin and Suthname [39] and Parker [45].

The Formation Problem as well as Marching Problem require of robots to move into a prescribed formation and march to the goal maintaining the formation. Various have been solutions adopted for these problems. A study of the concept of a robot team was initially based on a perception of animal behavior like herding, swarming, flocking or schooling. Work by Reynolds [51] brought forth an approach to flocking in groups of birds called 'boids' based on simple behaviors like: collision–avoidance, velocity matching, flock centering, where each bird senses its neighbors only, and geometric positions of birds are not specified. A similar position was adopted in Mataric [36], [37] who studied flocking in wheeled robots induced by simple behaviors: wandering, homing, following, avoidance, aggregation, dispersion, cf., [20], [38]. See Wilson [63] for a deep study in sociobiology and Agah [1], Agah and Bekey [2], and Bekey [8] for robotic counterparts.

These observations brought forth some basic principles of *behavioral approach*, see, e.g., Balch and Arkin [5]: it is vital to keep all robots within a certain distance from one another (e.g., to ensure mutual visibility), to move away when the distance becomes too close (to avoid congestion, collision, or resource conflict), to adapt own movement to movement of neighbors (e.g., by adjusting velocity of motion), to orient oneself on a leader, or a specific location, e.g., the gravity center of the group.

Balch and Arkin, op.cit., implement these principles by imposing geometric constraints on individual robots: each of them is to keep a precisely specified geometric position. They investigate four types of formations: *line* where robots move line–abreast, *column* where robots travel in an *Indian file*, *diamond*, *wedge*. Each robot in a formation is given its ID number. Maintaining position in a formation is provided by two processes: *detect–formation–position* detects the current position of a robot on basis of sensory data, *maintain–formation* produces motor commands to move robot to the proper position. They evaluate three *referencing techniques* for computing a proper position of a robot in the formation: *unit–center–reference*, *leader–reference*, *neighbor–reference*.

In *unit–center–reference* each robot computes the coordinates of the center of the formation and keeps position with respect to it.

In *leader–referencing*, one robot is designated as the leader whose position is irrelevant and other robots are keeping the formation.

In *neighbor–reference* each robot keeps position with respect to one specified neighbor. The local coordinate system in which positions are computed is at each step given by the unit center and the line from it through the next navigation way–point. The parameter *spacing* of the procedure *detect-formation–position* is used to keep distances between robots.

This scheme sets demands on sensory and motor capabilities of robots. Authors propose using dead reckoning, or GPS, or direct perception of neighbors as means of sensory determination of a position in the formation. One can say that Balch and Arkin, op. cit., are proponents of the *geometric approach* to formations: it uses *referencing* techniques; reference is made either to the team center or to the team leader, or to a specified neighbor in a coordinate system given by the position of the team center or the leader along with the orientation given by the nearest navigation point; positions are determined, e.g., with the help of GPS or dead reckoning. Spacing between robots is imposed and robots keep their geometric positions. In this approach, complex behaviors of a team of robots result from an interaction of primitive behaviors organized by sensory and motor components of robots. Dorigo et al. [18] situate themselves with their approach in the realm of self–organization and swarm intelligence, see Bonabeau et al. [11], Payton et al. [47], and evolutionary computing, see Baeck et al. [4]. Authors study s–bots: robots of limited capabilities, with respect to their coordinated behaviors. S–bots have ability to connect one to another by physical links therefore making rigid formations and the objective of the authors is to study coordinated movement of robots in a formation. The method of evaluating performance of the complex controller is by a genetic process. Coordinated behavior of a team results as a complex product of individual behaviors.The important issue raised in this work is stressing the importance of ability of a formation to change shape or to adapt to changing environment and pointing to the problem of dependence of complex behaviors of a team on the set of individual behaviors of robots in that team. Current status of the field of self–configurable robots, a variation on the theme of coupled formations can be found in Stoy et al. [57].

The potential field methodology has also been extended to teams of robots. A good example of this approach is provided in Leonard and Fiorelli [33] which combines potential field approach with the virtual leader–reference approach. Potentials define interaction forces among robots forcing them to keep at desired distance one from another. Virtual leaders are moving reference points for robots to control the group movement and maintain group geometry. In this approach, the already mentioned by us biological behaviors of swarms like avoidance of close neighbors, keeping distance to the group, velocity matching are encoded by means of artificial local potentials defined as functions of relative distances between pairs of neighbors; control forces are then defined as negative gradients of the sum of potentials affecting a given robot. By their action, robots are driven to the absolute minimum of the global potential function; local potentials can be designed as to correspond to a given geometry of the group.

In addition to local neighbor–pair potentials, virtual leaders, i.e., moving reference beacons are added, each of them generating its own potential field with the aim of manipulating the group, directing it or herding robots into a group. Authors discuss motions like schooling and flocking. Schooling is a maneuver in which a steady group translation occurs, and flocking takes place when robots circle a stationary point. For instance, authors demonstrate a stationary movement of a group of 6 robots forming vertices of a hexagonal lattice.

An approach to controlling of a group of robots using the leader idea is presented in Shao et al. [55]. The leader–follower paradigm means that each robot in a group has a neighbor assigned as its leader whom it follows with prescribed distance and eventual other parameter values. For a group on N robots, authors propose to express the group structure in the form of a tree in which pairs of the form parent–child are pairs the leader, the follower, encoded in the usual form of an adjacency matrix; another matrix is the parameter matrix: authors discuss four parameters for each robot: distance, orientation error, angle and the Boolean attribute Presence meaning visibility of the leader. Authors show some patterns: a *hexagon*, a *diamond* of twelve robots, a *column* (line), a *wedge*. An interesting fact is that authors raise the problem of changing patterns.

Another method for forming a geometric formation relies on a direct usage of a metric, see, e.g., Sugihara and Suzuki [58]: given a threshold δ, and a parameter D – the circle diameter, for each robot M in a team, its farthest neighbor M_1 and the nearest neighbor M_2

1. If $\rho(M, M_1) > D$, then M moves toward M_1;

2. If $\rho(M, M_1) < D - \delta$, then M moves away from M_1.

   These two steps assure that the diameter of the set of robots in each cross-section is about D. Finally

3. If $D - \delta < \rho(M, M_1) < D$, then M moves away from M_2.

By this method, robots are arranged on an approximation to a circle of diameter D. This procedure is performed iteratively and in each iteration robots move sequentially.

In Schneider et al. [53] authors attempt a systematic discussion of metrics for formation navigation. They use the term 'metric' in order to denote some criterion of performance evaluation for a group of robots. Authors assume a group of identical robots, communicating among themselves, with ability to sense the environment and one another. Among some metrics of that type authors mention

1. *Path length ratio: the ratio of the average path length by robots in a group to the straight–line distance to the goal;*

2. *Average position error: average displacement from the correct position in the group during the run;*

3. *Percentage of time out of formation;*

4. *As an additional measure, time to convergence, i.e., time needed for the group to assume a given pattern is added.*

Authors define a formation of robots in strict geometric terms, characterizing a formation by means of a finite set of segments and angles between them, such that

1. *Uniform dispersion is secured: all neighboring robots keep the same distance d with maximum error ε;*

2. *Shape is proper: each robot keeps its position within an error ε;*

3. *Orientation: the angles are kept within the error ε_a.*

In order to keep a formation, potential fields are used. In addition to the goal potential and the attractive potentials of obstacles, each robot in a group exerts attractive and repulsive forces on other robots. All these potentials sum up for each robot inducing the directing forces.

Planning paths for multiple robots adapts and modifies planing methodology for a single robot, see Hwang and Ahuja [24] or Latombe [32] for surveys. The *centralized* approach, finding a path in a complex configuration space describing the system, provides *complete* planners which always find a path for the system if it exists; however, this comes at the cost of exponential complexity: the problem of planning for rectangular robots in the rectangular workspace is **P**–space complete, see Hopcroft et al. [22]. Schwartz and Sharir [54] describe planners of polynomial complexity based on cell decomposition for disc shaped robots in polygonal obstacle world. Variants of the method of potential field, were applied as well in centralized planning, see Barraquand and Latombe [7]; they proposed a randomized path planner based on a a potential field induced by goals with random fluctuations for escaping local minima. In the area of *decoupled planning*, the problem is to merge plans for individual robots into one general plan for the whole system; here, the idea of *prioritization* was put forth in Erdmann and Lozano–Perez [19].

High complexity of centralized planning and incompleteness of decoupled planning prompted research in the area between the two and the idea of separate *roadmaps* for robots emerged in which separate roadmaps are combined into a global roadmap, see La Valle and Hutchinson [60]. A general problem of mission planning for multiple vehicles and concurrent goals is addressed in Brumitt et al. [13] where a distributed planner is introduced in the context of a dynamic allocation of goals to autonomous vehicles. Planning is effected in the environment of a *mission grammar MG*:

1. $m \rightarrow M(r,g)$—$m \wedge m$—$m \vee m$—$m \Rightarrow m$—(m);

2. $r \rightarrow R_i$—$r \wedge r$—$r \vee r$—(r);

3. $g \rightarrow G_j$—$g \wedge g$—$g \vee g$—$g \Rightarrow g$—(g),

where $a \Rightarrow b$ means "a followed by b", $a \wedge b$ means "a and b", $a \vee b$ means "a or b", R_i means "robot i", G_i means "goal i", $M(r,g)$ means "move robot r to goal g". For instance, the expression $M((R_1 \wedge R_2), G_1 \Rightarrow G_2)$ means that robots 1 and 2 are to go to goal 1 and then to goal 2. These simple grammar expressions are examined by the mission planner and parsed into sequences of executable commands and planning of paths for them uses the D^* search algorithm, see Choset et al. [17].

So now again the recurrent question: what rough mereology can do for robot teams in terms of planning and navigation? We may observe that one can hardly find in literature a formal definition of what a robot formation is, independent of the context, e.g., a metric. Rather, formations are defined by setting constraints on individual robot either absolute or relative to a leader, or a neighbor. A definition absolute in a sense, abstracted from metric context, can be useful. Also, rigidity of constraints, e.g., necessity of keeping distances and angles, rids the team of flexibility, necessary when, e.g., obstacles force the team to change the formation in order to pass, e.g., a bottleneck. Thus, we strive for a definition and conditions for a formations which on one hand would secure its maintenance through manoeuvering to goal, and, on the other hand would permit a flexible behavior, e.g., bypassing an obstacle whereas keeping the formation. We propose a solution based on mereogeometry of Ch. 6.

8.4   Rough Mereological Approach to Robot Formations

We again resort to mereogeometry, see Ch. 6, sect. 10, Ośmiałowski [41], and Polkowski and Ośmiałowski [48], [49]. We recall that on the basis of the rough inclusion μ, and mereological distance κ defined as

$$\kappa(X,Y) = min\{max\ r, max\ s:\ \mu(X,Y,r), \mu(Y,X,s)\} \qquad (8.12)$$

geometric predicates of *nearness* and *betweenness*, see Ch. 6, sect. 10, are redefined in the mereological frame.

The relation N of *nearness* proposed by Van Benthem [9] is defined in mereological context as

$$N(X,U,V) \text{ if and only if } \kappa(X,U) > \kappa(V,U) \qquad (8.13)$$

Here, $N(X,U,V)$ means that X is closer to U than V is to U.

The *betweenness* relation T_B, see Van Benthem [9], is defined as

$$T_B(Z,U,V) \text{ if and only if [for each } W\ (Z=W) \text{ or } N(Z,U,W) \text{ or } N(Z,V,W)]$$
$$(8.14)$$

The principal example bearing on our approach to robot control deals with rectangles in 2D space regularly positioned, i.e., having edges parallel to coordinate axes. We model robots (which are represented in the plane as discs of the same radius in 2D space) by means of their safety regions about robots; those regions are modeled as rectangles circumscribed on robots. One of advantages of this representation is that safety regions can be always implemented as regularly positioned rectangles circumscribed on discs representing robots.

Given two robots a, b as discs of same radii, and their safety regions as circumscribed regularly positioned rectangles A, B, we search for a proper choice of a region X containing A, and B with the property that a robot C contained in X can be said to be between A and B. For two (possibly but not necessarily) disjoint rectangles A, B, we define the *extent*, $ext(A,B)$ of A and B as the smallest rectangle containing the union $A \cup B$. Then we have the claim, obviously true by definition of T_B, see Ch. 6, sect. 10

Proposition 8.1. *We consider a context in which objects are rectangles positioned regularly, i.e., having edges parallel to axes in* R^2. *The measure* μ *is* μ^G, *see Ch. 6, sect. 6. In this setting, given two disjoint rectangles* C, D, *the only object between* C *and* D *in the sense of the predicate* T_B *is the extent* $ext(C,D)$ *of* C, D, *, i.e., the minimal rectangle containing the union* $C \cup D$.

For details of the exposition which we give now, please consult Ośmiałowski [42], [43], Polkowski and Ośmiałowski [49], Ośmiałowski and Polkowski [44]. The notion of betweenness along with Proposition 8.1 permits to define the notion of betweenness for robots. Recall that we represent the disc–shaped Roomba robots by means of safety squares around them, regularly placed, i.e., with sides parallel to coordinate axes.

For robots a, b, c, we say that a robot b is *between robots a and c*, in symbols

$$(between\ b\ a\ c) \qquad (8.15)$$

in case the rectangle $ext(b)$ is contained in the extent of rectangles $ext(a)$, $ext(c)$, i.e.

$$\mu_0(ext(b), ext(ext(a), ext(c)), 1) \tag{8.16}$$

i.e., see Ch. 6, sect. 10, $ext(b) \subseteq ext(ext(a), ext(b))$.

This allows as well for a generalization of the notion of betweenness to the notion of *partial betweenness* which models in a more realistic manner spatial relations among a, b, c; we say in this case that robot b is *between robots a and c to a degree of at least r*, in symbols,

$$(\text{between–degr } b \ a \ c \ ) \tag{8.17}$$

if and only if

$$\mu_0(ext(b), ext[ext(a), ext(c)], r) \tag{8.18}$$

i.e., $\frac{||ext(b) \cap ext(ext(a), ext(c))||}{||ext(b)||} \geq r$.

For a team of robots, $T(r_1, r_2, ..., r_n) = \{r_1, r_2, ..., r_n\}$, an *ideal formation IF* on $T(r_1, r_2, ..., r_n)$ is a betweenness relation (between...) on the set $T(r_1, r_2, ..., r_n)$ of robots.

In implementations, ideal formations are represented as lists of expressions of the form

$$(\text{between } r_0 \ r_1 \ r_2) \tag{8.19}$$

indicating that the object r_0 is between r_1, r_2, for all such triples, along with a list of expressions of the form

$$(\text{not–between } r_0 \ r_1 \ r_2) \tag{8.20}$$

indicating triples which are not in the given betweenness relation.

To account for dynamic nature of the real world, in which due to sensory perception inadequacies, dynamic nature of the environment etc., we allow for some deviations from ideal formations by allowing that the robot which is between two neighbors can be between them to a degree in the sense of (8.17). This leads to the notion of a real formation.

For a team of robots, $T(r_1, r_2, ..., r_n) = \{r_1, r_2, ..., r_n\}$, a *real formation RF* on $T(r_1, r_2, ..., r_n)$ is a betweenness to degree relation (between–deg ....) on the set $T(r_1, r_2, ..., r_n)$ of robots.

In practice, real formations will be given as a list of expressions of the form,

$$(\text{between–deg } \delta \ r_0 \ r_1 \ r_2), \tag{8.21}$$

indicating that the object r_0 is to degree of δ in the extent of r_1, r_2, for all triples in the relation (between–deg ....), along with a list of expressions of the form,

$$(\text{not–between } r_0 \ r_1 \ r_2), \tag{8.22}$$

indicating triples which are not in the given betweenness relation.

Description of formations, as proposed above, can be a list of relation instances of large cardinality, cf., examples below. The problem can be posed of finding a minimal set of instances sufficient for describing a given formation, i.e., implying the full list of instances of the relation (between...). This problem turns out to be NP–hard, see Ośmiałowski and Polkowski [44].

Proposition 8.2. *The problem of finding a minimal description of a formation is NP–hard.*

Proof. (Ośmiałowski and Polkowski [44]) We construct an *information system*, see Ch. 4, *Formations* as a triple (U, A, f) where U is a set of objects, A is a set of attributes and f is a value assignment, i.e., a mapping $f : A \times U \rightarrow V$, where V is a set of possible values of attributes in A on objects in U. For a formation F, with robots $r_1, ..., r_n$ we let $U = T(r_1, ..., r_n)$, a team of robots; $A = \{[r_k, r_l, r_m] : r_k, r_l, r_m$ pairwise distinct robots$\}$. For a given formation F of robots $r_1, ..., r_n$, the value assignment f is defined as follows,

$$f([r_k, r_l, r_m], r_i) = \begin{cases} 1 \text{ in case } r_i = r_l \text{ and (between } r_l \ r_k \ r_l) \\ \frac{1}{2} \text{ in case } r_i = r_l \text{ or } r_i = r_m \text{ and (between } r_l \ r_k \ r_m) \\ 0 \text{ in case } r_i \neq r_l r_k r_m \end{cases}$$

$$(8.23)$$

The system *Formations* describes the formation F.

Clearly, reducts of the system *Formations* provide a complete description of the formation F and correspond to minimal descriptions of the formation. As shown by Skowron and Rauszer [56] the problem of finding a minimum size reduct of a given information system is NP–hard $\qquad \square$

To describe formations we have proposed a language derived from LISP–like s–expressions: a formation is a list in LISP meaning with some restrictions that formulates our language. We will call elements of the list the objects. Typically, LISP lists are hierarchical structures that can be traversed using recursive algorithms. We restrict that top–level list (a root of whole structure) contains only two elements where the first element is always a formation identifier (a name). For instance

Example 8.1. (formation1 (some_predicate *param1 ... paramN*))

For each object on a list (and for a formation as a whole) an extent can be derived and in facts, in most cases only extents of those objects are considered. We have defined two possible types of objects

1. *Identifier: robot or formation name (where formation name can only occur at top–level list as the first element)*;

2. *Predicate: a list in LISP meaning where first element is the name of given predicate and other elements are parameters; number and types of parameters depend on given predicate.*

Minimal formation should contain at least one robot. For example

Example 8.2. (formation2 roomba0)

To help understand how predicates are evaluated, we need to explain how extents are used for computing relations between objects. Suppose we have three robots (*roomba0, roomba1, roomba2*) with *roomba0* between *roomba1* and *roomba2* (so the *between* predicate is fulfilled). We can draw an extent of this situation as the smallest rectangle containing the union *roomba1* ∪ *roomba2* oriented as a regular rectangle, i.e., with edges parallel to coordinate axes. This extent can be embedded into bigger structure: it can be treated as an object that can be given as a parameter to predicate of higher level in the list hierarchy. For example:

Example 8.3. (formation3 (between (between roomba0 roomba1 roomba2) roomba3 roomba4))

We can easily find more than one situation of robots that fulfill this example description. That is one of the features of our approach: one s–expression can describe many situations. This however makes very hard to find minimal s–expression that would describe already given arrangement of robots formation (as stated earlier in this chapter , the problem is NP–hard). An exemplary s–description is shown in Fig. 8.4.

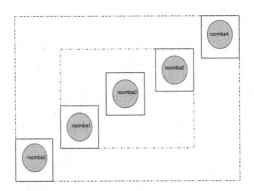

Fig. 8.4 Formation described by an s–expression: *(formation3 (between (between roomba0 roomba1 roomba2) roomba3 roomba4))*

Typical formation description may look like below, see Ośmiałowski [42], [43], Polkowski and Ośmiałowski [49], [50]

Example 8.4. (cross
    (set
        (max–dist 0.25 roomba0 (between roomba0 roomba1 roomba2))

        (max–dist 0.25 roomba0 (between roomba0 roomba3 roomba4))
        (not–between roomba1 roomba3 roomba4)
        (not–between roomba2 roomba3 roomba4)
        (not–between roomba3 roomba1 roomba2)
        (not–between roomba4 roomba1 roomba2)
)
)

This is a description of a formation of five Roomba robots arranged in a cross shape. The *max–dist* relation is used to bound formation in space by keeping all robots close one to another.

The final stage of planning is in checking its soundness by navigating robots in an environment with obstacles. We show results of navigating with a team of robots in the initial formation of cross–shape in a crowded environment, see Fig. 8.5. In order to bypass a narrow avenue between an obstacle and the border of the environment, the formation changes to a line, see Fig. 8.6, see Ośmiałowski [40], [43].

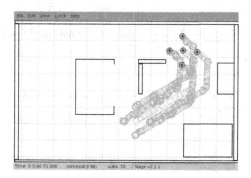

Fig. 8.5 Trails of robots arranged in cross formation following the leader

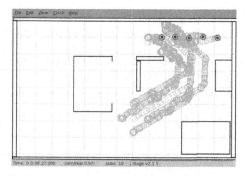

Fig. 8.6 Trails of robots moving to their positions in the line formation

After the line was formed and robots passed through the passage, the line formation can be restored to the initial cross–shaped formation, see Figs. 8.7, 8.8. These behaviors witness the flexibility of our definition of a robot formation: first, robots can change formation, next, as the definition of a formation is relational, without metric constraints on robots,the formation can manage an obstacle without losing the prescribed formation (though, this feature is not illustrated in figures in this chapter).

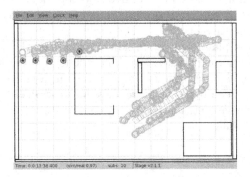

Fig. 8.7 Trails of robots moving in the line formation through the passage

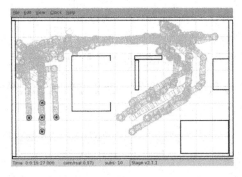

Fig. 8.8 Trails of robots in the restored cross formation in the free workspace after passing through the passage

References

1. Agah, A.: Robot teams, human workgroups and animal sociobiology. A review of research on natural and artificial multi–agent autonomous systems. Advanced Robotics 10, 523–545 (1996)
2. Agah, A., Bekey, G.A.: Tropism based cognition: A novel software architecture for agents in colonies. Journal of Experimental and Theoretical Artificial Intelligence 9(2-3), 393–404 (1997)
3. Asama, H., Matsumoto, A., Ishida, Y.: Design of an autonomous and distributed robot system: ACTRESS. In: Proceedings IEEE/RSJ IROS, pp. 283–290 (1989)
4. Baeck, T., Fogel, D., Michalewicz, Z.: Evolutionary Computation 1: Basic algorithms and operators. Taylor and Francis, New York (1997)
5. Balch, T., Arkin, R.C.: Behavior–based formation control for multiagent robot teams. IEEE Transactions on Robotics and Automation 14(12) (1998)
6. Barraquand, J., Kavraki, L.E., Latombe, J.-C., Li, T.-Y., Motvani, R., Raghavan, P.: A random sampling scheme for robot path planning. Intern. J. Robotic Research 16(6), 759–774 (1997)
7. Barraquand, J., Latombe, J.C.: A Monte–Carlo algorithm for path planning with many degrees of freedom. In: Proceedings of the IEEE International Conference on Robotics and Automation, pp. 1712–1717 (1990)
8. Bekey, G.A.: Autonomous Robots. From Biological Inspiration to Implementation and Control. MIT Press, Cambridge (2005)
9. Van Benthem, J.: The Logic of Time. D. Reidel, Dordrecht (1983)
10. de Berg, M., van Kreveld, M., Overmars, M.: Computational Geometry: Algorithms and Applications. Springer, Berlin (1997)
11. Bonabeau, E., Dorigo, M., Theraulaz, G.: Swarm Intelligence. From Natural to Artificial Systems. Oxford University Press, Oxford (1999)
12. Brooks, R.: Cambrian Intelligence: The Early History of the New AI. MIT Press, Cambridge (1999)
13. Brumitt, B., Stentz, A., Hebert, M.: The CMU UGV Group. Autonomous driving with concurrent goals and multiple vehicles: Mission planning and architecture. Autonomous Robots 11(2), 103–115 (2001)
14. Caloud, P., Choi, W., Latombe, J.-C., Pape Le, C., Yin, M.: Indoor automation with many mobile robots. In: Proceedings IEEE/RSJ IROS, pp. 67–72 (1990)
15. Canny, J.F.: The Complexity of Robot Motion Planning. MIT Press, Cambridge (1988)
16. Uny, C.Y., Fukunaga, A.S., Kahng, A.B.: Cooperative mobile robotics: Antecedents and directions. Autonomous Robots 4, 7–27 (1997)

17. Choset, H., Lynch, K.M., Hutchinson, S., Kantor, G., Burgard, W., Kavraki, L.E., Thrun, S.: Principles of Robot Motion. Theory, Algorithms, and Implementations. MIT Press, Cambridge (2005)
18. Dorigo, M., Trianni, V., Sahin, E., Gross, R., Labella, T., Baldassare, G., Nolfi, S., Deneubourg, J.–L., Mondada, F., Floreano, D., Gambardella, L.: Evolving self–organizing behaviors for a Swarm–Bot. Autonomous Robots 17, 223–245 (2004)
19. Erdmann, M., Lozano–Perez, T.: On multiple moving objects. In: Proceedings IEEE ICRA, pp. 1419–1424 (1986)
20. Fredslund, J., Matarić, M.: A general algorithm for robot formation using local sensing and minimal communication. IEEE Transactions on Robotics and Automation 18(5), 837–846 (2002)
21. Fukuda, T., Nakagawa, S.: A dynamically reconfigurable robotic system (concept of a system and optimal configurations). In: International Conference on Industrial Electronics, Control, and Instrumentation, pp. 588–595 (1987)
22. Hopcroft, J., Schwartz, J.T., Sharir, M.: On the complexity of motion planning for multiple independent objects: P–space hardness of the warehouseman's problem. International Journal of Robotic Research 3(4), 76–88 (1984)
23. Huntsberger, T., Stroupe, A., Aghazarian, H., Garrett, M., Younse, P., Powell, M.: TRESSA: Teamed robots for exploration and science of steep areas. Journal of Field Robotics 24(11/12), 1015–1031 (2007)
24. Hwang, Y., Ahuja, N.: Gross motion planning - A survey. ACM Computing Surv. 24(3), 219–291 (1992)
25. Jennings, J.S., Whelan, G., Evans, W.F.: Cooperative search and rescue with a team of mobile robots. In: SPIE, Aerosense, Orlando, FL, vol. 4364 (2001)
26. Kalman, R.: A new approach to linear filtering and prediction problem. Transactions of the ASME. Journal of Basic Engineering 82, 35–45 (1960)
27. Kamon, I., Rimon, E., Rivlin, E.: A range–sensor based navigation algorithm. Intern. J. Robotic Research 17(9), 934–953 (1998)
28. Kavraki, L.E.: Random Networks in Configuration Space for Fast Planning. PhD Thesis. Stanford University (1995)
29. Kavraki, L.E., Švestka, P., Latombe, J.–C., Overmars, M.: Probabilistic roadmaps for path planning in high–dimensional configuration spaces. IEEE Transactions on Robotics and Automation 12(4), 566–580 (1996)
30. Khatib, O.: Real–time obstacle avoidance for manipulators and mobile robots. Intern. J. Robotic Research 5, 90–98 (1986)
31. Kube, C.R., Zhang, H.: The use of perceptual cues in multi–robot box–pushing. In: Proceedings IEEE Intern. Conference on Robotics and Automation, pp. 2085–2090 (1996)
32. Latombe, J.C.: Robot Motion Planning. Kluwer Academic Publishers, Boston (1991)
33. Leonard, N.E., Fiorelli, E.: Virtual leaders, artificial potentials and coordinated control of groups. In: Proceedings of the 40th IEEE Conference on Decision and Control, pp. 2968–2973 (2001)
34. Li, Z., Canny, J.F.: Nonholonomic Motion Planning. Kluwer Academic Publishers, Boston (1993)
35. Lumelsky, V., Stepanov, A.: Path planning strategies for point mobile automaton moving amidst unknown obstacles of arbitrary shape. Algorithmica 2, 403–430 (1987)

36. Matarić, M.: Kin recognition, similarity and group behavior. In: Proceedings of 15th Annual Cognitive Society Conference, pp. 705–710. Lawrence Erlbaum, Mahwah (1993)
37. Matarić, M.: Interaction and Intelligent Behavior. PhD Dissertation. MIT EECS Dept (1994)
38. Matarić, M.: Behavior–based control: Examples from navigation, learning, and group behavior. Journal of Experimental and Theoretical Artificial Intelligence 9(2,3), 323–336 (1997)
39. Naffin, D., Sukhatme, G.: Negotiated formations. In: Proceedings of the Intern. Conference on Intelligent Autonomous Systems, pp. 181–190 (2004)
40. Ośmiałowski, P.: Player and Stage at PJIIT Robotics Laboratory. Journal of Automation, Mobile Robotics and Intelligent Systems (JAMRIS) 2, 21–28 (2007)
41. Ośmiałowski, P.: On path planning for mobile robots: Introducing the mereological potential field method in the framework of mereological spatial reasoning. Journal of Automation, Mobile Robotics and Intelligent Systems (JAMRIS) 3(2), 24–33 (2009)
42. Ośmiałowski, P.: A Case of Planning and Programming of a Concurrent Behavior: Planning and Navigating with Formations of Robots. In: Proceedings of CSP 2009. Concurrency, Specification, Programming. Warsaw University Press, Kraków (2009)
43. Ośmiałowski, P., Polkowski, L.: (supervisor). Spatial reasoning based on rough mereology in planning and navigation problems of autonomous mobile robots. PhD dissertation. Polish–Japanese Institute of IT, Warszawa
44. O'smiaıowski, P., Polkowski, L.: Spatial reasoning based on rough mereology: A notion of a robot formation and path planning problem for formations of mobile autonomous robots. In: Peters, J.F., Skowron, A., Słowiński, R., Lingras, P., Miao, D., Tsumoto, S. (eds.) Transactions on Rough Sets XII. LNCS, vol. 6190, pp. 143–169. Springer, Heidelberg (2010)
45. Parker, L.E.: Cooperative motion control for multi–target observations. In: IEEE/RSJ International Conference on Intelligent Robots and Systems, pp. 1591–1598 (1997)
46. Parker, L.E.: ALLIANCE: An architecture for fault tolerant multirobot cooperation. IEEE Transactions on Robotics and Automation 14(2), 220–240 (1998)
47. Payton, D., Estkowski, R., Howard, M.: Progress in pheromone robotics. In: Proceedings of the 7th Intern. Conference on Intelligent Autonomous Systems, Marina del Rey, CA (2002)
48. Polkowski, L.: Ośmiałowski P. Spatial reasoning with applications to mobile robotics. In: Xing–Jian, J. (ed.) Mobile Robots Motion Planning. New Challenges. I-Tech, Vienna, pp. 433–453 (2008)
49. Polkowski, L., Ośmiałowski, P.: A framework for multiagent mobile robotics: Spatial reasoning based on rough mereology in player/Stage system. In: Chan, C.-C., Grzymala-Busse, J.W., Ziarko, W.P. (eds.) RSCTC 2008. LNCS (LNAI), vol. 5306, pp. 142–149. Springer, Heidelberg (2008)
50. Polkowski, L., Ośmiałowski, P.: Navigation for mobile autonomous robots and their formations: An application of spatial reasoning induced from rough mereological geometry. In: Barrera, A. (ed.) Mobile Robots Navigation. InTech, Zagreb, pp. 329–354 (2010)
51. Reynolds, C.: Flocks, herds and schools. A distributed behavioral model. Comput. Graph. 21(4), 25–34 (1987)

52. Russell, S.J., Norvig, P.: Artificial Intelligence: Modern Approach, 3rd edn. Prentice Hall, Upper Saddle River (2009)
53. Schneider, F., Wildermuth, D., Kräusling, A.: Discussion of exemplary metrics for multi–robot systems for formation navigation. In: Advanced Robotic Systems, pp. 345–353 (2005)
54. Schwartz, J.T., Sharir, M.: On the piano movers' problem II. General techniques for computing topological properties of real algebraic manifolds. Advances in Applied Mathematics 4, 298–351 (1983)
55. Shao, J., Xie, G., Yu, J., Wang, L.: Leader–following formation control of multiple mobile robots. In: Proceedings of the 2005 IEEE Intern. Symposium on Intelligent Control, pp. 808–813 (2005)
56. Skowron, A., Rauszer, C.: The discernibility matrices and functions in decision systems. In: Słowiński, R. (ed.) Intelligent Decision Support. Handbook of Applications and Advances of the Rough Sets Theory, pp. 311–362. Kluwer, Dordrecht (1992)
57. Stoy, K., Brandt, D., Christensen, D.J.: Self–Reconfigurable Robots. The MIT Press, Cambridge (2010)
58. Sugihara, K., Suzuki, I.: Distributed motion coordination of multiple mobile robots. In: Proceedings 5th IEEE Intern. Symposium on Intelligent Control, pp. 138–143 (1990)
59. Tribelhorn, B., Dodds, Z.: Evaluating the Roomba: A low-cost, ubiquitous platform for robotics research and education. In: IEEE International Conference on Robotics and Automation ICRA 2007, Roma, Italy, pp. 1393–1399 (2007)
60. La Valle, S.M., Hutchinson, S.A.: Optimal motion planning for multiple robots having independent goals. In: Proceedings IEEE International Conference on Robotics and Automation, pp. 2847–2852 (1996)
61. Walter, W.G.: An imitation of life. Scientific American 182, 42–45 (1950)
62. Walter, W.G.: The Living Brain. Norton, New York (1953)
63. Wilson, E.O.: Sociobiology: The New Synthesis, 25th Anniversary edn. Belknap Press of Harvard University Press, Cambridge (2000)

Chapter 9
Rough Mereological Calculus of Granules in Decision and Classification Problems

The idea of mereological granulation of knowledge, proposed and presented in detail in Ch. 7, sect. 3, finds an effective application in problems of synthesis of classifiers from data tables. This application consists in granulation of data at preprocessing stage in the process of synthesis: after granulation, a new data set is constructed, called a granular reflection, to which various strategies for rule synthesis can be applied. This application can be regarded as a process of *filtration* of data, aimed at reducing noise immanent to data. This chapter presents this application.

9.1   On Decision Rules

In Ch. 4, we have given an introduction to the problem of decision rule synthesis from decision systems, so basic notions and results are known to the reader. Now, we comment on a more specific problem of quality of decision rules and classifiers. We recall that decision rules are formed in the frame of a *decision system* (U, A, d), as implications of the form

$$\bigwedge_i (a_i = v_i) \Rightarrow (d = v) \tag{9.1}$$

where a_i are *conditional attributes*, v_i are their values, and, v is a value of the decision.

A *decision algorithm, classifier* is a judiciously chosen set of decision rules, approximating possibly most closely the real decision function, which, by necessity, is not known to us. This comes down to a search in the space of possible descriptors in order to find their successful combinations. In order to judge the quality, or, degree of approximation, decision rules are learned on a part of the decision system, the *training set* and then the decision algorithm is *tested* on the remaining part of the decision system, called the *test set*. Degree of approximation is measured by some coefficients of varied character. Simple measures of statistical character are found from the *contingency table*,

L. Polkowski: Approximate Reasoning by Parts, ISRL 20, pp. 319–334.
springerlink.com   © Springer-Verlag Berlin Heidelberg 2011

see Arkin and Colton [5]. This table is built for each decision rule r and a decision value v, by counting the number n_t of training objects, the number n_r of objects satisfying the premise of the rule r (caught by the rule), $n_r(v)$ is the number of objects counted in n_r and with the decision v, and $n_r(\neg v)$ is the number of objects counted in n_r but with decision value distinct from v. To these factors, we add n_v, the number of training objects with decision v and $n_{\neg v}$, the number of remaining objects, i.e, $n_{\neg v} = n_t - n_v$.

For these values, *accuracy of the rule r relative to v* is the quotient

$$acc(r, v) = \frac{n_r(v)}{n_r} \qquad (9.2)$$

and *coverage of the rule r relative to v* is

$$cov(r, v) = \frac{n_r(v)}{n_v} \qquad (9.3)$$

These values are useful as indicators of a *rule strength* which is taken into account when classification of a test object is under way: to assign the value of decision, a rule pointing to a decision with a maximal value of accuracy, or coverage, or combination of both can be taken; methods for combining accuracy and coverage into a single criterion are discussed, e.g., in Michalski [10]. Accuracy and coverage can, however, be defined in other ways; for a decision algorithm D, trained on a training set Tr, and a test set Tst, the *accuracy* of D is measured by its efficiency on the test set and it is defined as the quotient

$$accuracy(D) = \frac{n_{corr}}{n_{caught}} \qquad (9.4)$$

where n_{corr} is the number of test objects correctly classified by D and n_{caught} is the number of test objects classified.

Similarly, *coverage* of D is defined as

$$coverage(D) = \frac{n_{caught}}{n_{test}} \qquad (9.5)$$

where n_{test} is the number of test objects. Thus, the product $accuracy(D) \cdot coverage(D)$ gives the measure of the fraction of test objects correctly classified by D.

We have already mentioned that accuracy and coverage are often advised to be combined in order to better express the trade–off between the two: one may have a high accuracy on a relatively small set of caught objects, or a lesser accuracy on a larger set of caught by the classifier objects. Michalski [10] proposes a combination rule of the form

$$MI = \frac{1}{2} \cdot A + \frac{1}{4} \cdot A^2 + \frac{1}{2} \cdot C - \frac{1}{4} \cdot A \cdot C \qquad (9.6)$$

where A stands for accuracy and C for coverage. With the symbol MI, we denote the *Michalski index* as defined in (9.6).

Statistical measures of correlation between the rule r and a decision class v are expressed, e.g., by χ^2 statistic

$$\chi^2 = \frac{n_t \cdot (n_r(v) \cdot n_{\neg r}(\neg v) - n_r(\neg v) \cdot n_{\neg r}(v))^2}{n(v) \cdot n(\neg v) \cdot n_r \cdot n_{\neg r}} \tag{9.7}$$

where $n_{\neg r}$ is the number of objects not caught by the rule r, see Bruning and Kintz [8].

We now restrict ourselves to rough set framework of decision systems and we denote a rule r by a convenient shortcut

$$r : \phi \Rightarrow (d = v) \tag{9.8}$$

where ϕ is a conjunct of descriptors, see Ch. 4, sect. 2. We recall that $[[\phi]]$ denotes the meaning of a rule ϕ, i. e., the set of objects satisfying ϕ. An object $u \in U$, is caught by r (or, *matches* r) in case $u \in \phi$; $match(r)$ is the number of objects matching r.

Support, $supp(r)$, of r is the number of objects in $[[\phi]] \cap [[(d = v)]]$; the fraction

$$cons(r) = \frac{supp(r)}{match(r)} \tag{9.9}$$

is the *consistency degree* of r; $cons(r) = 1$ means that the rule is *certain*, or *true*.

Strength, $strength(r)$, of the rule r is defined, see, e. g., Bazan [6], and Grzymala–Busse and Ming Hu [9], as the number of objects correctly classified by the rule in the training phase; *relative strength* is defined as the fraction

$$rel - strength(r) = \frac{supp(r)}{|[[(d = v)]]|} \tag{9.10}$$

Specificity of the rule r, $spec(r)$, is the number of descriptors in the premise ϕ of the rule r, Grzymala–Busse and Ming Hu [9].

In the testing phase, rules vie among themselves for object classification when they point to distinct decision classes; in such case, negotiations among rules or their sets are necessary. In these negotiations rules with better characteristics are privileged.

For a given decision class $c : d = v$, and an object u in the test set, the set $Rule(c, u)$ of all rules matched by u and pointing to the decision v, is characterized globally by

$$Support(Rule(c, u)) = \sum_{r \in Rule(c,u)} strength(r) \cdot spec(r) \tag{9.11}$$

The class c for which $Support(Rule(c, u))$ is the largest wins the competition and the object u is classified into the class $c : d = v$, see, e.g., Grzymala–Busse and Ming Hu [9].

It may happen that no rule in the available set of rules is matched by the test object u and *partial matching* is necessary, i.e., for a rule r, the *matching factor match* $- fact(r, u)$ is defined as the fraction of descriptors in the premise ϕ of r matched by u to the number $spec(r)$ of descriptors in ϕ. The rule for which the *partial support*

$$Part - Support(Rule(c, u)) = \sum_{r \in Rule(c,u)} match - fact(r, u) \cdot strength(r) \cdot spec(r)$$

(9.12)

is the largest wins the competition and it does assign the value of decision to u, see Grzymala–Busse and Ming Hu [9].

In a similar way, notions based on relative strength can be defined for sets of rules and applied in negotiations among them, see Bazan et al. [7].

A combination of rough set methods with k-nearest neighbor idea, is a further refinement of the classification based on similarity or analogy, cf., an implementation in RSES [21]. In this approach, training set objects are endowed with a metric, and the test objects are classified by voting by k nearest training objects for some k that is subject to optimization, cf., Polkowski [18].

Our idea of augmenting existing strategies for rule induction consists in using granules of knowledge. The principal assumption we can make is that the nature acts in a continuous way: if objects are similar with respect to judiciously and correctly chosen attributes, then decisions on them should also be similar. A granule collecting similar objects should then expose the most typical decision value for objects in it while suppressing outlying values of decision, reducing noise in data, hence, leading to a better classifier.

These ideas were developed and proposed in Polkowski [12] – [15] see also surveys Polkowski [16] – [18]. In Polkowski and Artiemjew [19], [20] and in Artiemjew [1] – [4] the theoretical analysis was confirmed as to its application merits. We proceed with a summary of methods and results of these verification.

9.2   The Idea of Granular Rough Mereological Classifiers

We assume that we are given a decision system (U, A, d) from which a classifier is to be constructed; on the universe U, a rough inclusion μ is given, and a radius $r \in [0, 1]$ is chosen, see Polkowski [12] – [15].

We can find granules $g_\mu(u, r)$ defined as in Ch. 7, sect. 6, for all $u \in U$, and make them into the set $G(\mu, r)$.

From this set, a covering $Cov(\mu, r)$ of the universe U can be selected by means of a chosen strategy $\mathcal{G}$, i.e.,

$$Cov(\mu, r) = \mathcal{G}(G(\mu, r)) \tag{9.13}$$

We intend that $Cov(\mu, r)$ becomes a new universe of the decision system whose name will be the *granular reflection* of the original decision system. It remains to define new attributes for this decision system.

Each granule g in $Cov(\mu, r)$ is a collection of objects; attributes in the set $A \cup \{d\}$ can be factored through the granule g by means of a chosen strategy $\mathcal{S}$, i.e., for each attribute $q \in A \cup \{d\}$, the new factored attribute $\bar{q}$ is defined by means of the formula

$$\bar{q}(g) = \mathcal{S}(\{a(v) : ingr(v, g_\mu(u, r))\}) \tag{9.14}$$

In effect, a new decision system $(Cov(\mu, r), \{\bar{a} : a \in A\}, \bar{d})$ is defined. The object v with

$$Inf(v) = \{(\bar{a} = \bar{a}(g)) : a \in A\} \tag{9.15}$$

is called the *granular reflection of g*. Granular reflections of granules need not be objects found in data set; yet, the results show that they mediate very well between the training and test sets.

source	method	accuracy	coverage	MI
Bazan [6])	$SNAPM(0.9)$	error = 0.130	--	--
Nguyen SH [11]	simple.templates	0.929	0.623	0.847
Nguyen SH [11]	general.templates	0.886	0.905	0.891
Nguyen SH [11]	tolerance.gen.templ.	0.875	1.0	0.891
Wroblewski [25]	adaptive.classifier	0.863	–	--

Fig. 9.1 Best results for Australian credit by some rough set based algorithms

paradigm	system/method	Austr.credit
Stat.Methods	Logdisc	0.141
Stat.Methods	SMART	0.158
Neural Nets	Backpropagation2	0.154
Neural Networks	RBF	0.145
Decision Trees	CART	0.145
Decision Trees	C4.5	0.155
Decision Trees	ITrule	0.137
Decision Rules	CN2	0.204

Fig. 9.2 A comparison of errors in classification by rough set and other paradigms

algorytm	accuracy	coverage	rule number	MI
covering($p = 0.1$)	0.670	0.783	589	0.707
covering($p = 0.5$)	0.670	0.783	589	0.707
covering($p = 1.0$)	0.670	0.783	589	0.707
LEM2($p = 0.1$)	0.810	0.061	6	0.587
LEM2($p = 0.5$)	0.906	0.368	39	0.759
LEM2($p = 1.0$)	0.869	0.643	126	0.804

Fig. 9.3 Train and Test (trn=345 objects, tst=345 objects) ; Australian Credit; Comparison of RSES implemented algorithms exhaustive, covering and LEM

The procedure just described for forming a granular reflection of a decision system can be modified as proposed in Artiemjew [1] with help of the procedure of *concept dependent granulation*. In this procedure, the granule $g_{mu}(u, r)$ is modified to the granule

$$g_{\mu}^{c}(u, r) = g_{\mu}(u, r) \cap [u]_d$$

i.e., it is computed relative to the decision class of u.

We collect best results for an exemplary data set, the Australian credit data set, see [23], by various rough set based methods in the table of Fig. 9.1. For a comparison we include in the table of Fig. 9.2 results obtained by some other methods, as given in Statlog. In the table of Fig. 9.3, we give a comparison of performance of rough set classifiers, exhaustive, covering and LEM implemented in RSES [21] system. We begin in the next section with granular classifiers in which granules are induced from the training set.

9.3 Classification by Granules of Training Objects

We begin with a classifier in which granules computed by means of the rough inclusion μ_L form a granular reflection of the data set and then to this new data set the exhaustive classifier, see [21], is applied.

Procedure of the test

1. *The data set (U, A, d) is input;*
2. *The training set is chosen at random. On the training set, decision rules are induced by means of exhaustive, covering and LEM algorithms implemented in the RSES system;*
3. *Classification is performed on the test set by means of classifiers of pt. 2;*
4. *For consecutive granulation radii r, granule sets $G(\mu, r)$ are found;*
5. *Coverings $Cov(\mu, r)$ are found by a random irreducible choice;*
6. *For granules in $Cov(\mu, r)$, for each r, we determine the granular reflection by factoring attributes on granules by means of majority voting with random resolution of ties;*

7. *For found granular reflections, classifiers are induced by means of algorithms in pt. 2;*
8. *Classifiers found in pt. 7, are applied to the test set;*
9. *Quality measures: accuracy and coverage for classifiers are applied in order to compare results obtained, respectively, in pts. 3 and 8.*

In the table of Fig. 9.4, the results are collected of results obtained after the procedure described above is applied. The classifier applied was exhaustive one; the method was train–and–test. The rough inclusion applied was the Łukasiewicz t–norm induced μ_L of Ch. 6, sect. 4.

r	tst	trn	rulex	aex	cex	MI
nil	345	345	5597	0.872	0.994	0.907
0.0	345	1	0	0.0	0.0	0.0
0.0714286	345	1	0	0.0	0.0	0.0
0.142857	345	2	0	0.0	0.0	0.0
0.214286	345	3	7	0.641	1.0	0.762
0.285714	345	4	10	0.812	1.0	0.867
0.357143	345	8	23	0.786	1.0	0.849
0.428571	345	20	96	0.791	1.0	0.850
0.5	345	51	293	0.838	1.0	0.915
0.571429	345	105	933	0.855	1.0	0.896
0.642857	345	205	3157	0.867	1.0	0.904
0.714286	345	309	5271	0.875	1.0	0.891
0.785714	345	340	5563	0.870	1.0	0.890
0.857143	345	340	5574	0.864	1.0	0.902
0.928571	345	342	5595	0.867	1.0	0.904

Fig. 9.4 Train–and–test; Australian Credit; Granulation for radii r; RSES exhaustive classifier; r=granule radius, tst=test set size, trn=train set size, rulex=rule number, aex=accuracy, cex=coverage

We can compare results expressed in terms of the Michalski index MI as a measure of the trade–off between accuracy and coverage; for template based methods, the best MI is 0.891, for covering or LEM algorithms the best value of MI is 0.804, for exhaustive classifier (r=nil) MI is equal to 0.907 and for granular reflections, the best MI value is 0.915 with few other values exceeding 0.900.

What seems worthy of a moment's reflection is the number of rules in the classifier. Whereas for the exhaustive classifier (r=nil) in non–granular case, the number of rules is equal to 5597, in granular case the number of rules can be surprisingly small with a good MI value, e.g., at $r = 0.5$, the number of rules is 293, i.e., 5 percent of the exhaustive classifier size, with the best MI at all of 0.915. This compression of classifier seems to be the most impressive feature of granular classifiers.

It is an obvious idea that this procedure can be repeated until a stable system is obtained to which further granulation causes no change; it is the procedure of *layered granulation*, see Artiemjew [1]. The table of Fig. 9.5 shows some best results of this procedure for selected granulation radii. As coverage in all reported cases is equal to 1.0, the Michalski index MI is equal to accuracy.

r	acc	cov
0.500000	0.436	1.000
0.571429	0.783	1.000
0.642857	0.894	1.000
0.714286	0.957	1.000

Fig. 9.5 Train–and–test; Australian Credit;(layered–granulation)

This initial, simple granulation, suggests further ramifications. For instance, one can consider, for a chosen value of $\varepsilon \in [0,1]$, granules of the form

$$g_\mu(u,r,\varepsilon) = \{v \in U : \forall a \in A.|a(u) - a(v)| \leq \varepsilon\} \qquad (9.16)$$

and repeat with these granules the procedure of creating a granular reflection and building from it a classifier.

Another yet variation consists in mimicking the performance of the Lukasiewicz based rough inclusion and introducing a counterpart of the granulation radius in the form of the *catch radius*, r_{catch}. The granule is then dependent on two parameters: ε and r_{catch}, and its form is

$$g_\mu(u,\varepsilon,r_{catch}) = \{v \in U : \frac{|\{a \in A : |a(u) - a(v)| \leq \varepsilon}{|A|} \geq r_{catch}\} \qquad (9.17)$$

Results of classification by granular classifier induced from the granular reflection obtained by means of granules (9.17) are shown in the table of Fig. 9.6.

9.4 A Treatment of Missing Values

A particular but important problem in data analysis is the treatment of missing values. In many data, some values of some attributes are not recorded due to many factors, like omissions, inability to take them, loss due to some events etc.

Analysis of systems with missing values requires a decision on how to treat missing values; Grzymala–Busse and Ming Hu [9] analyze nine such methods, among them, *1. most common attribute value, 2. concept restricted most common attribute value, 3. assigning all possible values to the missing*

r_catch	optimal eps	acc	cov
nil	nil	0.845	1.0
0	0	0.555073	1.0
0.071428	0	0.83913	1.0
0.142857	0.35	0.868116	1.0
0.214286	0.5	0.863768	1.0
0.285714	0.52	0.831884	1.0
0.357143	0.93	0.801449	1.0
0.428571	1.0	0.514493	1.0
0.500000	1.0	0.465217	1.0
0.571429	1.0	0.115942	1.0

Fig. 9.6 ε_{opt}=optimal value of ε, acc=accuracy, cov=coverage. Best r_{catch} = 0.1428, $\varepsilon_{opt} = 0.35$: accuracy= 0.8681, coverage=1.0

location, 4. treating the unknown value as a new valid value, etc. etc. Their results indicate that methods *3,4* perform very well and in a sense stand out among all nine methods.

We adopt and consider two methods, i.e., *3, 4* from the above mentioned. As usual, the question on how to use granular structures in analysis of incomplete systems, should be answered first.

The idea is to embed the missing value into a granule: by averaging the attribute value over the granule in the way already explained, it is hoped the the average value would fit in a satisfactory way into the position of the missing value.

We will use the symbol $*$, commonly used for denoting the missing value; we will use two methods 3, 4 for treating $*$, i.e, either $*$ is a *don't care* symbol meaning that any value of the respective attribute can be substituted for $*$, hence, $* = v$ for each value v of the attribute, or $*$ is a new value on its own, i.e., if $* = v$ then v can only be $*$.

Our procedure for treating missing values is based on the granular structure $(G(\mu, r), \mathcal{G}, \mathcal{S}, \{a* : a \in A\})$; the strategy $\mathcal{S}$ is the majority voting, i.e., for each attribute a, the value $a^*(g)$ is the most frequent of values in $\{a(u) : u \in g\}$. The strategy $\mathcal{G}$ consists in random selection of granules for a covering.

For an object u with the value of $*$ at an attribute a,, and a granule $g = g(v, r) \in G(\mu, r)$, the question whether u is included in g is resolved according to the adopted strategy of treating $*$: in case $* = don't\ care$, the value of $*$ is regarded as identical with any value of a hence $|IND(u, v)|$ is automatically increased by 1, which increases the granule; in case $* = *$, the granule size is decreased. Assuming that $*$ is sparse in data, majority voting on g would produce values of a^* distinct from $*$ in most cases; nevertheless the value of $*$ may appear in new objects g^*, and then in the process of classification, such value is repaired by means of the granule closest to g^* with respect to the rough inclusion μ_L, in accordance with the chosen method for treating $*$.

In plain words, objects with missing values are in a sense absorbed by close to them granules and missing values are replaced with most frequent values in objects collected in the granule; in this way the method *3* or *4* in [9] is combined with the idea of a frequent value, in a novel way.

We have thus four possible strategies:

1. *Strategy A: in building granules* $*=don't$ *care, in repairing values of* $*$, $*=don't$ *care;*
2. *Strategy B: in building granules* $*=don't$ *care, in repairing values of* $*$, $* = *$;
3. *Strategy C: in building granules* $* = *$, *in repairing values of* $*$, $*=don't$ *care;*
4. *Strategy D: in building granules* $* = *$, *in repairing values of* $*$, $* = *$.

We show how effective are these strategies, see Polkowski and Artiemjew [20] by perturbing the data set Pima Indians Diabetes, from UCI Repository [23]. First, in the table of Fig. 9.7 we show results of granular classifier on the non–perturbed (i.e., without missing values) Pima Indians Diabetes data set.

r	$macc$	$mcov$
0.0	0.0	0.0
0.125	0.0	0.0
0.250	0.6835	0.9956
0.375	0.7953	0.9997
0.500	0.9265	1.0
0.625	0.9940	1.0
0.750	1.0	1.0
0.875	1.0	1.0

Fig. 9.7 10-fold CV; Pima; exhaustive algorithm, r=radius,macc=mean accuracy,mcov=mean coverage

We now perturb this data set by randomly replacing 10 percent of attribute values in the data set with missing $*$ values. Results of granular treatment in case of Strategies A,B,C,D in terms of accuracy are reported in the table of Fig. 9.8. As algorithm for rule induction, the exhaustive algorithm of the RSES system has been selected. 10–fold cross validation (CV–10) has been applied.

Strategy A reaches the accuracy value for data with missing values within 94 percent of the value of accuracy without missing values (0.9407 to 1.0) at the radius of .875. With Strategy B, accuracy is within 94 percent from the radius of .875 on. Strategy C is much better: accuracy with missing values

r	$maccA$	$maccB$	$maccC$	$maccD$
0.250	0.0	0.0	0.0	0.645
0.375	0.0	0.0	0.0	0.7779
0.500	0.0	0.0	0.0	0.9215
0.625	0.5211	0.5831	0.5211	0.9444
0.750	0.7705	0.7769	0.7705	0.9994
0.875	0.9407	0.9407	0.9407	0.9987

Fig. 9.8 Accuracies of strategies A, B, C, D. 10-fold CV; Pima Indians; exhaustive algorithm; r=radius,maccA=mean accuracy of A,maccB=mean accuracy of B, maccC=mean accuracy of C, maccD=mean accuracy of D

reaches 99 percent of accuracy in no missing values case from the radius of .625 on. Strategy D gives results slightly better than C with the same radii.

We conclude that the essential for results of classification is the strategy of treating the missing value of $*$ as $* = *$ in both strategies C and D; the repairing strategy has almost no effect: C and D differ very slightly with respect to this strategy.

9.5   Granular Rough Mereological Classifiers Using Residuals

Rough inclusions used in sects. 9.2 – 9.4 in order to build classifiers do take, to a certain degree, into account the distribution of values of attributes among objects, by means of the parameters ε and the catch radius r_{catch}.

The idea that metrics used in classifier construction should depend locally on the training set is, e.g., present in classifiers based on the idea of nearest neighbor, see, e.g., a survey in Polkowski [18]: for nominal values, the metric VDM (Value Difference Metric) in Stanfill and Waltz [22] takes into account conditional probabilities $P(d = v|a_i = v_i)$ of decision value given the attribute value, estimated over the training set Trn, and on this basis constructs in the value set V_i of the attribute a_i a metric $\rho_i(v_i, v_i') = \sum_{v \in V_d} |P(d = v|a_i = v_i) - P(d = v|a_i = v_i')|$. The global metric is obtained by combining metrics ρ_i for all attributes $a_i \in A$ according to one of many-dimensional metrics, e.g., Minkowski metrics, see Ch. 2.

This idea was also applied to numerical attributes in Wilson and Martinez [24] in metrics $IVDM$ (Interpolated VDM) and $WVDM$ (Windowed VDM). A modification of the $WVDM$ metric based again on the idea of using probability densities in determining the window size was proposed as $DBVDM$ metric.

In order to construct a measure of similarity based on distribution of attribute values among objects, we resort to residual implications, see Ch.4, sect. 3. As shown in Polkowski [14], $\Rightarrow_T$ does induce a rough inclusion on the interval $[0, 1]$

$$\mu_{\rightarrow T}(u, v, r) \text{ if and only if } x \Rightarrow_T y \geq r \qquad (9.18)$$

This rough inclusion can be transferred to the universe U of an information system; to this end, first, for given objects u, v, and $\varepsilon \in [0, 1]$, factors

$$dis_\varepsilon(u, v) = \frac{|\{a \in A : |a(u) - a(v)| \geq \varepsilon\}|}{|A|} \qquad (9.19)$$

and

$$ind_\varepsilon(u, v) = \frac{|\{a \in A : |a(u) - a(v)| < \varepsilon\}|}{|A|} \qquad (9.20)$$

are introduced.

The weak variant of rough inclusion $\mu_{\rightarrow T}$ is defined, see Polkowski [14], as

$$\mu_T^*(u, v, r) \text{ if and only if } dis_\varepsilon(u, v) \rightarrow_T ind_\varepsilon(u, v) \geq r \qquad (9.21)$$

Particular cases of this similarity measure induced by, respectively, t–norm min, t–norm $P(x, y)$, and t–norm L are, see Ch. 6, sect. 7

1. For $T = M(x, y) = min(x, y)$, $x \Rightarrow_{min} y$ is y in case $x > y$ and 1 otherwise, hence, $\mu_{min}^*(u, v, r)$ if and only if $dis_\varepsilon(u, v) > ind_\varepsilon(u, v) \geq r$ with $r < 1$ and 1 otherwise;
2. For $t = P$, where $P(x, y) = x \cdot y$, $x \Rightarrow_P y = \frac{y}{x}$ when $x \neq 0$ and 1 when $x = 0$, hence, $\mu_P^*(u, v, r)$ if and only if $\frac{ind_\varepsilon(u, v)}{dis_\varepsilon(u, v)} \geq r$ with $r < 1$ and 1 otherwise;
3. For $t = L$, $x \Rightarrow_L y = min\{1, 1 - x + y\}$, hence, $\mu_L^*(u, v, r)$ if and only if $1 - dis_\varepsilon(u, v) + ind_\varepsilon(u, v) \geq r$ with $r < 1$ and 1 otherwise.

These similarity measures will be applied in building granules and then in data classification. Tests are done with the Australian credit data set; the results are validated by means of the 5–fold cross validation (CV–5). For each of t–norms: M, P, L, three cases of granulation are considered, viz.,

1. *Granules of training objects* (GT);
2. *Granules of rules induced from the training set* (GRT);
3. *Granules of granular objects induced from the training set* (GGT).

In this approach, training objects are made into granules for a given ε. Objects in each granule g about a test object u, vote for decision value at u as follows: for each decision class c, the value

$$p(c) = \frac{\sum \text{training object v in g falling in c } w(u, v)}{\text{size of c in training set}} \qquad (9.22)$$

is computed where the weight $w(u, v)$ is computed for a given t–norm T as

$$w(u, v) = dis_\varepsilon(u, v) \to_T \ ind_\varepsilon(u, v) \tag{9.23}$$

The class c* assigned to u is the one with the largest value of p.

Weighted voting of rules in a given granule g for decision at test object u goes according to the formula $d(u) = argmaxp(c)$, where

$$p(c) = \frac{\sum \text{rule in g pointing to } c \ w(u, r) \cdot support(r)}{\text{size of c in training set}} \tag{9.24}$$

where weight $w(u, r)$ is computed as

$$dis_\varepsilon(u, r) \to_T \ ind_\varepsilon(u, r) \tag{9.25}$$

The optimal (best) results in terms of accuracy of classification are collected in the table of Fig. 9.9.

met	T	ε_{opt}	macc	mcov
GT	M	0.04	0.848	1.0
GT	P	0.06	0.848	1.0
GT	L	0.05	0.846	1.0
GRT	M	0.02	0.861	1.0
GRT	P	0.01	0.851	1.0
GGT	M	0.05	0.855	1.0
GRT	P	0.01	0.852	1.0

Fig. 9.9 5-fold CV; Australian; residual metrics. met=method of granulation, T=t–norm, ε_{opt}=optimal ε, macc=mean accuracy,mcov=mean coverage

9.6 Granular Rough Mereological Classifiers with Modified Voting Parameters

An interesting modification of voting schemes of above sections was proposed and tested in Artiemjew [4]. It consists in weighted voting with modified weight computing scheme, viz., the procedure is now as follows. It is considered in [4] in five cases, of which we include here two with best results, i. e., cases 4 and 5. For each attribute a, each training object v, and each test object u, we denote with the symbol $\rho_{trn}(u, v)$ the quotient

$$\frac{||a(u) - a(v)||}{|max_{training \ set}a - min_{training \ set}a|} \tag{9.26}$$

where $max_{training \ set}a$, $min_{training \ set}a$ are, respectively, the maximal and the minimal values of the attribute a over the training set; the symbol $||.||$ stands for the Euclidean distance in attribute value spaces. Augmented values of weights are computed in cases 4, 5 in two variants: (a) when $\rho_{trn}(u, v) \geq \varepsilon$ and (b) when $\rho_{trn}(u, v) \leq \varepsilon$.

We have in Case 4

$$
w(u,v) = \begin{cases} (a) \ w(u,v) + \rho_{trn}(u,v) \cdot \varepsilon + ||a(u) - a(v)|| \\ (b) \ w(u,v) + rho_{trn}(u,v) \cdot \varepsilon \end{cases} \tag{9.27}
$$

and in Case 5

$$
w(u,v) = \begin{cases} (a) \ w(u,v) + \rho_{trn}(u,v) \\ (b) \ w(u,v) + \rho_{trn(u,v)} \cdot \varepsilon \end{cases} \tag{9.28}
$$

Voting procedure consists in computing values of parameters

$$
p_1 = \frac{\sum v \text{ in positive class } w(u,v)}{\text{cardinality of positive class}} \tag{9.29}
$$

and respectively, p_2 by means of (9.29) with 'positive' replaced by 'negative'; when $p_1 < p_2$, the test object u is classified into the positive class, otherwise it is classified into the negative class. Optimal results are shown in the table of Fig. 9.10. As coverage is 1.0 in each case, we do not show it in the table. For comparison, we insert also results by RSES exhaustive algorithm and by RSES implemented k–NN method.

Case	ε_{opt}	maxacc	minacc
Case4	0.62	0.905	0.861
Case5	0.35 − 0.37	0.906	0.880
RSESexh	−	0.862	0.819
RSESk − NN	−	0.884	0.841

Fig. 9.10 Parameterized voting. CV–5. ε_{opt}=optimal ε for maxacc, maxacc=max max fold accuracy, minacc=min max fold accuracy

References

1. Artiemjew, P.: Classifiers from granulated data sets: Concept dependent and layered granulation. In: Proceedings RSKD 2007. Workshop at ECML/ PKDD 2007, pp. 1–9. Warsaw University Press, Warsaw (2007)
2. Artiemjew, P.: On classification of data by means of rough mereological granules of objects and rules. In: Wang, G., Li, T., Grzymala-Busse, J.W., Miao, D., Skowron, A., Yao, Y. (eds.) RSKT 2008. LNCS (LNAI), vol. 5009, pp. 221–228. Springer, Heidelberg (2008)
3. Artiemjew, P.: Rough mereological classifiers obtained from weak variants of rough inclusions. In: Wang, G., Li, T., Grzymala-Busse, J.W., Miao, D., Skowron, A., Yao, Y. (eds.) RSKT 2008. LNCS (LNAI), vol. 5009, pp. 229–236. Springer, Heidelberg (2008)
4. Artiemjew, P.: On Strategies of Knowledge Granulation with Applications to Decision Systems. L.Polkowski (supervisor). PhD Dissertation. Polish–Japanese Institute of Information Technology, Warszawa (2009)
5. Arkin, H., Colton, R.R.: Statistical Methods. Barnes and Noble, New York (1970)
6. Bazan, J.G.: A comparison of dynamic and non–dynamic rough set methods for extracting laws from decision tables. In: Polkowski, L., Skowron, A. (eds.) Rough Sets in Knowledge Discovery, vol. 1, pp. 321–365. Physica Verlag, Heidelberg (1998)
7. Bazan, J.G., Nguyen, H.S., Nguyen, S.H., Synak, P., Wróblewski, J.: Rough set algorithms in classification problems. In: Polkowski, L., Tsumoto, S., Lin, T.Y. (eds.) Rough Set Methods and Applications. New Developments in Knowledge Discovery in Information Systems, pp. 49–88. Physica Verlag, Heidelberg (2000)
8. Bruning, J.L., Kintz, B.L.: Computational Handbook of Statistics, 4th edn. Allyn and Bacon, Columbus (1997)
9. Grzymała-Busse, J.W., Hu, M.: A Comparison of Several Approaches to Missing Attribute Values in Data Mining. In: Ziarko, W.P., Yao, Y. (eds.) RSCTC 2000. LNCS (LNAI), vol. 2005, pp. 378–385. Springer, Heidelberg (2001)
10. Michalski, R.: Pattern recognition as rule–guided inductive inference. IEEE Transactions on Pattern Analysis and Machine Intelligence PAMI 2(4), 349–361 (1990)
11. Nguyen, S.H.: Regularity analysis and its applications in Data Mining. In: Polkowski, L., Tsumoto, S., Lin, T.Y. (eds.) Rough Set Methods and Applications. New Developments in Knowledge Discovery in Information Systems, pp. 289–378. Physica Verlag, Heidelberg (2000)

12. Polkowski, L.: Formal granular calculi based on rough inclusions (a feature talk). In: Proceedings of IEEE 2005 Conference on Granular Computing, GrC 2005, pp. 57–62. IEEE Press, Beijing (2005)

13. Polkowski, L.: A model of granular computing with applications (a feature talk). In: Proceedings of IEEE 2006 Conference on Granular Computing, GrC 2006, pp. 9–16. IEEE Press, Atlanta (2006)

14. Polkowski, L.: Granulation of knowledge in decision systems: The approach based on rough inclusions. The method and its applications. In: Kryszkiewicz, M., Peters, J.F., Rybiński, H., Skowron, A. (eds.) RSEISP 2007. LNCS (LNAI), vol. 4585, pp. 69–79. Springer, Heidelberg (2007)

15. Polkowski, L.: On the idea of using granular rough mereological structures in classification of data. In: Wang, G., Li, T., Grzymala-Busse, J.W., Miao, D., Skowron, A., Yao, Y. (eds.) RSKT 2008. LNCS (LNAI), vol. 5009, pp. 213–220. Springer, Heidelberg (2008)

16. Polkowski, L.: A Unified approach to granulation of knowledge and granular computing based on rough mereology: A Survey. In: Pedrycz, W., Skowron, A., Kreinovich, V. (eds.) Handbook of Granular Computing, ch. 16, John Wiley and Sons Ltd, Chichester (2008)

17. Polkowski, L.: Granulation of Knowledge: Similarity Based Approach in Information and Decision Systems. In: Meyers, R.A. (ed.) Encyclopedia of Complexity and System Sciences. Springer, Berlin (2009) Article 00 788

18. Polkowski, L.: Data-mining and Knowledge Discovery: Case Based Reasoning, Nearest Neighbor and Rough Sets. In: Meyers, R.A. (ed.) Encyclopedia of Complexity and System Sciences. Springer, Berlin (2009) Article 00 391

19. Polkowski, L., Artiemjew, P.: On granular rough computing: Factoring classifiers through granulated decision systems. In: Kryszkiewicz, M., Peters, J.F., Rybiński, H., Skowron, A. (eds.) RSEISP 2007. LNCS (LNAI), vol. 4585, pp. 280–289. Springer, Heidelberg (2007)

20. Polkowski, L., Artiemjew, P.: On granular rough computing with missing values. In: Kryszkiewicz, M., Peters, J.F., Rybiński, H., Skowron, A. (eds.) RSEISP 2007. LNCS (LNAI), vol. 4585, pp. 271–279. Springer, Heidelberg (2007)

21. RSES, http://mimuw.edu.pl/logic/~rses/ (last entered 01. 04. 2011)

22. Stanfill, C., Waltz, D.: Toward memory–based reasoning. Communications of the ACM 29, 1213–1228 (1986)

23. UCI (University of California at Irvine) Repository, http://archive.ics.uci.edu.ml/ (last entered 01. 04. 2011)

24. Wilson, D.R., Martinez, T.R.: Improved heterogeneous distance functions. Journal of Artificial Intelligence Research 6, 1–34 (1997)

25. Wróblewski, J.: Adaptive aspects of combining approximation spaces. In: Pal, S.K., Polkowski, L., Skowron, A. (eds.) Rough Neural Computing. Techniques for Computing with Words, pp. 139–156. Springer, Berlin (2004)

Author Index

Term Index